An Introduction to Organometallic Chemistry

An Introduction to Organometallic Chemistry

A. W. Parkins and R. C. Poller

King's College London

MACMILLAN

First published 1986

Published by
Higher and Further Education Division
MACMILLAN PUBLISHERS LTD
Houndmills, Basingstoke, Hampshire RG21 2XS
and London
Companies and representatives
throughout the world

Printed in Hong Kong

Distributed in the U.S.A. and its
Dependencies by
OXFORD UNIVERSITY PRESS, INC.
200 Madison Avenue
New York, N.Y. 10016
ISBN: 0-19-505061-4

ISBN 0-333-36433-3 Pbk

CONTENTS

PREFACE

The aim of this book is to provide a concise introduction to organometallic chemistry. The book is intended primarily for senior undergraduate and post-graduate students. Teachers and research workers may also find it useful to consult for specialised information.

We have used an integrated approach to discuss the organometallic chemistry of both transition-metal and main-group elements, making comparisons and contrasts where appropriate. We are aware that organic and inorganic chemists view the subject somewhat differently but have attempted to write a balanced non-partisan account. In writing a relatively short book which gives an overall impression of an immense field, we have had to make some difficult choices. This has led to some chapters being presented in a closely packed style, although we have written the chapters dealing with the principles of the subject in a more discursive explanatory manner. After some thought we decided to give a substantial number of references to key reviews and research papers, so that the book can serve as a guide to the literature of organometallic chemistry.

We are most grateful to Dr D. A. Armitage and Professor A. G. Davies for reading the complete manuscript and making helpful suggestions. We also thank Mrs E. A. Moor for her virtuoso typing of the text and diagrams.

It is unlikely, despite such excellent help, that the book is free from errors. The responsibility for these rests with us, and we welcome comments and criticisms from readers.

London, October, 1984   A. W. P.
                                                                                 R. C. P.

ABBREVIATIONS USED

Ac	acetyl
acac	acetylacetonate anion
ACMP	*o*-anisylcyclohexylmethylphosphine
AIBN	azo*bis*isobutyronitrile
BASF	Badische Anilin und Soda Fabrik
Bu	*n*-butyl
CHIRAPHOS	chiral 2,3-*bis*(diphenylphosphino)butane
COD	cycloocta-1,5-diene
COT	cyclooctatetraene
Cp	cyclopentadienyl
Cy	cyclohexyl
DIOP	chiral 2,3-*O*-isopropylidene-2,3-dihydroxy-1,4-*bis*(diphenylphosphino)butane
DMSO	dimethyl sulphoxide
dppe	1,2-*bis*(diphenylphosphino)ethane
e.s.r.	electron spin resonance
Et	ethyl
HMB	hexamethylbenzene
HMPT	hexamethylphosphoric triamide
HOMO	highest occupied molecular orbital
HSAB	hard and soft acids and bases
ICI	Imperial Chemical Industries
iPr	isopropyl
IUPAC	International Union of Pure and Applied Chemistry
L	unidentate ligand
LUMO	lowest unoccupied molecular orbital
Me	methyl
n.m.r.	nuclear magnetic resonance
Np	neopentyl
Ph	phenyl
phen	1,10-phenanthroline
Pr	*n*-propyl
py	pyridine
r.t.	room temperature
sBu	*s*-butyl
solv	solvent

tBu	*t*-butyl
TCNE	tetracyanoethene
TMED	N,N,N',N'-tetramethylethylenediamine
THF	tetrahydrofuran
Tol	*p*-tolyl
Ts	*p*-toluenesulphonyl

1

INTRODUCTION

1.1 SCOPE

We adopt the usual convention that an organometallic compound is a compound which contains one or more carbon–metal bonds. This is a less rigid definition than it appears, since we take a rather broad view of the term metal. While many elements can be classified very easily as metallic or non-metallic, others such as germanium or antimony are borderline and sometimes referred to as metalloids. We are including the chemistry of these elements and, when the compounds are sufficiently interesting, the net is extended even more widely to include, for example, such essentially non-metallic elements as boron and silicon. Despite their importance as starting materials, we are omitting any specific discussion of binary metal carbonyls since there is usually ample treatment of these compounds in standard textbooks of inorganic chemistry.

1.2 HISTORY

If one man could be said to have founded organometallic chemistry that man must surely be Edward Frankland (1825–1899). He began work in this field accidentally while pursuing other goals – a not uncommon phenomenon in chemistry. In the 19th century, as he himself expressed it (Frankland, 1902), 'The isolation of the alcohol (= alkyl) radicals was, at this time, the dream of many chemists. . . . I was also smitten with the fever and determined to try my hand at the solution of the problem'. His experimental approach was to treat organic halides with metals, as for example in equation (1.1).

$$2EtI + Zn \longrightarrow ZnI_2 + EtEt \tag{1.1}$$

What he isolated was, of course, the products of further reaction of the very unstable alkyl radicals, in this case butane. Using the same reactants, Frankland also isolated, on 12 July 1849, diethylzinc (equation 1.2) and went on to make other organic derivatives of zinc and also of tin, mercury and boron.

$$2EtI + 2Zn \longrightarrow Et_2Zn + ZnI_2 \tag{1.2}$$

He was aware of, and exploited, the value of organozinc compounds as reagents in synthesis. Their use became widespread and they remained pre-eminent until displaced, after 1900, by Grignard reagents. The term 'organometallic' for compounds containing direct carbon–metal bonds (as opposed to other metal-containing organic compounds, such as sodium ethoxide) was coined by Frankland. Experiments of this nature convinced Frankland of the importance of the concept of constant combining power, or valency, of the elements. This idea was accepted by Kekulé (1858) who, on the basis of a quadrivalent carbon atom, laid the foundations for the modern theory of structural chemistry.

General chemical knowledge before Frankland's time was insufficiently advanced to allow earlier experimenters to develop the subject. The credit for making the first organometallic compound goes to William Zeise who, in 1830, first in a paper to the University of Copenhagen and subsequently in journals (Zeise, 1831) reported the preparation and careful analysis of what is now known as Zeise's salt, $K[PtCl_3(C_2H_4)].H_2O$. It is true that cacodyl (tetramethyldiarsine, $Me_2AsAsMe_2$) had been obtained by Cadet in 1760, but it was not until many years later that this material was characterised and its composition established by Bunsen (1843).

As the subject developed, compounds containing carbon, σ-bonded to an increasing range of main group metals, were prepared. Frankland had noted that alkylzinc compounds reacted with transition metal halides, but was unable to isolate a product. We now know that σ-bonded organotransition metal compounds are considerably less stable than their main group metal analogues (p. 19). The first example appears to be trimethylchloroplatinum, Me_3PtCl, reported in 1907 by Pope and Peachey. After the isolation of Zeise's salt there were very few reports of the preparation of what are now termed π-bonded organometallic compounds. It was not until the first half of the 20th century that pioneering work in Germany, by Hein (aromatic complexes), Hieber (metal carbonyls) and Reppe (catalysis of carbon monoxide and alkyne reactions), laid the foundation for subsequent developments. However, the discovery of ferrocene (I) in 1951 is generally seen as the mark of a new dawn in transition metal chemistry. This was followed by a tremendous increase in research activity on organotransition metal compounds. The explanations of the sandwich structure of ferrocene and of bonding in alkene complexes in terms of the Dewar–Chatt model were succeeded by other preparative achievements such as that of dibenzenechromium (1955), cyclobutadiene complexes (1958), bis(cyclooctadiene)- nickel (1960), carbene complexes (1964) and carbyne complexes (1973).

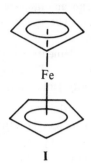

I

Important developments in catalysis, which is the major industrial application of organotransition metal chemistry, are hydroformylation (1938), Ziegler–Natta polymerisation of alkenes (1955), the Wacker process (1959), olefin metathesis (1964), and the rhodium catalysed carbonylation of methanol (1970).

This very brief survey has only mentioned some of the highlights of what is now a large subject. An elegant survey of our present understanding of the subject in its simplicity and complexity is given in Hoffmann's Nobel Lecture (Hoffmann, 1982).

To summarise, organometallic chemistry predates the birth of structural organic chemistry, but has played a part in its development. From its beginnings organometallic chemistry has challenged existing theories of structure and simultaneously provided, as reagents and catalysts, immensely powerful tools for synthesis. In addition, the preparation of compounds containing metal–carbon bonds demonstrated that there were, and are, no intrinsic differences between inorganic and organic chemistry despite the persistence of this ancient division.

1.3 NOMENCLATURE

According to the IUPAC nomenclature rules (Rigaudy and Klesney, 1979) for σ-bonded compounds containing only organic groups and hydrogen bonded to the metal, the names of the ligands are arranged alphabetically before the metal. Thus Bu_3GeEt is tributylethylgermanium and $MeBeH$ is hydridomethylberyllium. The oligomeric or polymeric nature of compounds can be ignored, as when $(BuLi)_n$ is referred to as butyllithium and $(Me_3Al)_2$ as trimethylaluminium. However, if it is necessary to emphasise the dimeric character of the latter, it is called di-μ-methyl-tetramethyldialuminium (p. 47), the Greek letter μ indicating the bridging groups.

When anionic groups are present, then three closely related nomenclature systems are approved. These are given below and applied to $BuSnClBr_2$.

(a) The names of the organic radicals are stated (in alphabetical order), followed by the metal and then the anions: butyltin dibromide chloride.

(b) All of the ligands, organic and anionic, are cited, alphabetically, before the metal: dibromo(butyl)chlorotin. (The brackets are optional, but useful to separate ligands.)

(c) For the Group IVb and Vb elements (the latter being in the trivalent state) the names can be based on those of the hydrides, typified by germane, GeH_4, and arsine, AsH_3: dibromo(butyl)chlorostannane.

Usage tends to favour system (c) for the less metallic elements. Where organometallic prefixes are required, these are based on the hydride names, for example $Me_3SiCH_2CH_2CH_2COOH$ is 4-trimethylsilylbutanoic acid.

For π-bonded compounds it is important to be able to indicate in the name the number of carbon atoms bound to the metal. The most widely used method of doing this is to quote the hapto number of the ligand (Cotton, 1975) which is now symbolised by the Greek letter, η (previously the letter h was used), followed by a superscript number which indicates the number of bound atoms. Thus the prefix η^3-allyl signifies that all three carbon atoms of the allyl group are bound to the metal. The term hapto derives from a Greek word, *haptein*, meaning to fasten. Examples are shown in formulae **II–IV**, where both full and

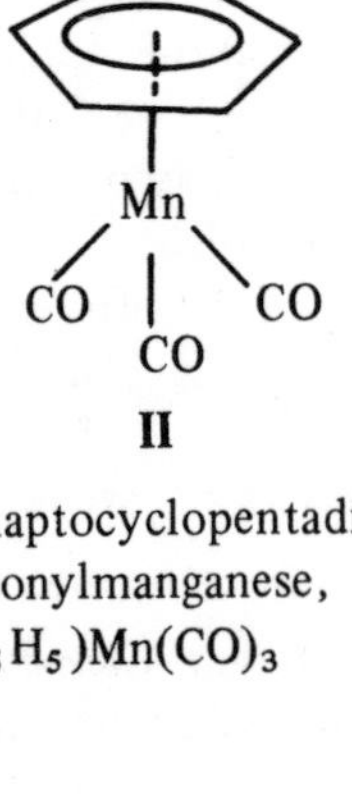

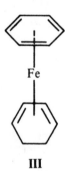

II

pentahaptocyclopentadienyl-
tricarbonylmanganese,
$(\eta^5\text{-}C_5H_5)Mn(CO)_3$

III

(hexahaptobenzene)(tetrahapto-
cyclohexa-1, 3-diene)iron,
$(\eta^6\text{-}C_6H_6)(\eta^4\text{-cyclohexa-1,3-diene})Fe$

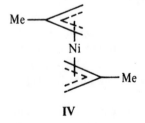

IV

bis(trihapto-2-methylallyl)nickel,
$(\eta^3\text{-}CH_2CHMeCH_2)_2Ni$

abbreviated names are shown. This system replaces an earlier, less precise, convention by which compound **II** would be termed (π-cyclopentadienyl)tricarbonylmanganese.

1.4 CONVENTIONS

It is not only in the field of nomenclature that more than one convention is used in chemistry. To avoid ambiguities we state here some conventions we have adopted.

1.4.1 The Eighteen Electron Rule

Most π-complexes of the transition metals conform to an empirical rule (known as the eighteen electron rule) which assumes that the metal atom accepts sufficient electrons from the ligands to enable it to attain the electronic configuration of the next noble gas. This means that the valence shell of the metal atom contains 18 electrons. Thus, the sum of the number of metal d-electrons plus the electrons conventionally regarded as being supplied by the ligands (see table 1.1) is 18. Application of the rule is illustrated with compounds **V–VII**. The rule can be used to indicate feasible structures, or for a given structure, to find the oxidation state of the metal. However, the empirical nature of this rule is stressed together with the fact that there are many exceptions to it. In particular, metals at the

Table 1.1 Electrons assumed to be supplied by ligands for the eighteen electron rule

Ligand	No. of electrons supplied
R^-, Ar^-, H^-, Cl^-, CN^-, $RCH=CH_2$, CH_2:, CO, PR_3, $P(OR)_3$, R_3N, RCN, R_2O	2
η^3-allyl$^-$, $RCH=CHCH=CHR^1$	4
η^5-$C_5H_5^-$, C_6H_6, η^7-$C_7H_7^+$(tropylium), $RCH=CHCH=CHCH=CHR^1$	6

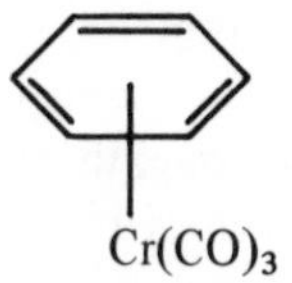

$$
\begin{aligned}
Cr(0), d^6 &= 6 \\
1 \times C_6H_6 &= 6 \\
3 \times CO &= \underline{6} \\
&\quad\; 18
\end{aligned}
$$

Cr(CO)₃

V

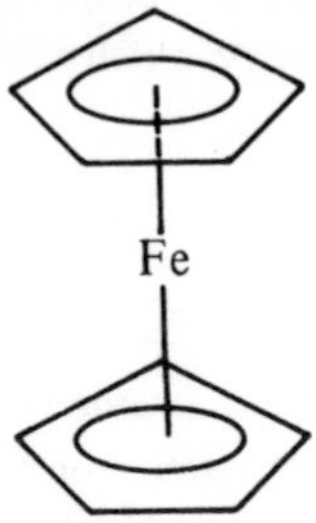

$$\begin{aligned}
\text{Fe(II)}, d^6 &= 6 \\
2 \times \eta^5\text{-}C_5H_5{}^- &= \underline{12} \\
&\ 18
\end{aligned}$$

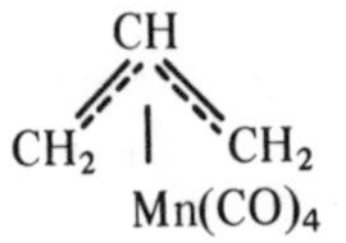

VI

$$\begin{aligned}
\text{Mn(I)}, d^6 &= 6 \\
\eta^3\text{-allyl}^- &= 4 \\
4 \times CO &= \underline{8} \\
&\ 18
\end{aligned}$$

VII

end of the transition series, such as rhodium, nickel, palladium and platinum, tend to form 16-electron compounds. Many catalyst systems are based on complexes of these metals which alternate between 16-electron and 18-electron compounds in the catalytic cycle.

1.4.2 Periodic table

We follow the widespread, but not universal, convention where, for groups III to VII of the periodic table the sub-groups containing the transition elements are labelled 'a', while for groups I and II the transition elements are labelled 'b'.

1.4.3 Additions to alkenes and Markownikoff's rule

Addition reactions, particularly of metal hydrides to alkenes, are discussed in various sections of this book and we make here some relevant general observations. In the reaction shown in equation (1.3), the species X—H can be said to have added to the alkene or, alternatively, the alkene can be regarded as having inserted into the X—H bond.

$$X–H + RCH=CH_2 \longrightarrow RCHXCH_3 \tag{1.3}$$

Thus the difference between addition and insertion is semantic rather than chemical and we shall employ both terms.

If, in equation (1.3), X = Cl and R = an alkyl group, then we have an illustration of Markownikoff's rule in which the hydrogen of the addendum molecule becomes attached to the alkene carbon carrying most hydrogens. The rule applies to ionic electrophilic additions to unsubstituted alkenes and should not be extended to other systems. It is rationalised in terms of the relative stability of the intermediate carbonium ion (scheme *i*). The terms Markownikoff addition or

$$RCH{=}CH_2 \quad H{-}X \quad \xrightarrow{\text{slow}} \quad R\overset{+}{C}HCH_3 \quad + \quad X^- \quad \xrightarrow{\text{rapid}} \quad RCHXCH_3$$

Scheme *i*

anti-Markownikoff addition are used to describe the regiospecificity in any reaction or sequence of reactions which result in the addition of some HX species to an alkene. Used in this sense the terms have no mechanistic implications. Thus, the overall procedure in scheme *ii* is referred to as anti-Markownikoff hydration.

$$MeCH = CH_2 \; + \; {>}B{-}H \; \longrightarrow \; MeCH_2CH_2B{<} \; \xrightarrow[\text{NaOH}]{H_2O_2} \; MeCH_2CH_2OH$$

Scheme *ii*

1.5 LITERATURE

Specific and general references are given at the end of each chapter but, in addition to this, some general discussion of the increasingly voluminous literature of organometallic chemistry is appropriate.

A guide to the literature of organotransition metal chemistry for the period 1950-1970 has been published and subsequently extended to include later material (Bruce, 1972, 1973, 1974). A similar survey is available for the main group organometallic literature covering the years 1950-1973 (Smith and Walton, 1975). It is to be hoped that these valuable guides will be brought up to date.

1.5.1 Textbooks

A major addition to the literature was made in 1982 by the publication, in nine volumes, of *Comprehensive Organometallic Chemistry* (Wilkinson *et al.*, 1982). Approximately one and a half volumes are devoted to organic compounds of the main group metals and some four and a half volumes to those of the transition metals. The applications of organometallic compounds to synthesis are covered in Volumes 7 and 8. The final volume consists of a series of indexes to the other 7800 pages of text and it includes an index of review articles and specialised texts. This book is a valuable reference work and contains many excellent articles.

An even larger work, with parts still appearing, is *Metallorganische Verbindungen*, which comprises Volume 13 of *Methoden der Organischen Chemie* (commonly known as Houben-Weyl) (Müller, 1970). This organometallic section is divided into 9 parts, each a substantial tome. For example, part 13/1 deals, in 904 pages, with organic compounds of Li, Na, K, Rb, Cs, Cu, Ag and Au. Part 13/3, which is devoted to boron, is further subdivided into 13/3a (1982), 13/3b (1983) and 13/3c (in preparation) and it is estimated that 13/3 will eventually contain more than 2500 pages.

Another German language publication, the monumental work on inorganic chemistry *Gmelin Handbuch der Anorganischen Chemie* (Gmelin, 1926 onwards), now includes over 40 volumes devoted to organometallic compounds. The treatment is very thorough and the work is revised at intervals by the publication of supplementary volumes which are now in English. While an up-to-date treatment of certain metals is still awaited, the organic chemistry of the following elements has been discussed in detail in a series of volumes published since 1970: Fe, Co, Ni, Sn, Sb, Ti, Ag, Bi, Cr, V, Zr, Hf, Ru, Ta, Nb. The well-known 'Chemistry of the Functional Group' series has recently been extended into organometallic chemistry with the publication of *The Chemistry of the Metal–Carbon Bond*, Vol. 1 (Hartley and Patai, 1982). This volume deals primarily with structure, preparation, thermochemistry and characterisation of organometallic compounds; other topics will be treated in later volumes.

A useful format is the monograph devoted to the organic chemistry of a particular metal or group of closely related metals. We note below some examples of books of this type (comprehensive lists of these books can be obtained from the sources given above):

Li	*The Chemistry of Organolithium Compounds*, by B. J. Wakefield, Pergamon Press, Oxford, 1974
Al	*Organoaluminium Compounds*, by T. Mole and E. A. Jeffery, Elsevier, Amsterdam, 1972
Si	*Organosilicon Compounds*, by C. Eaborn, Butterworth, London, 1960
Ge	*The Organic Compounds of Germanium*, by M. Lesbre, P. Mazerolles and J. Satgé, Wiley, New York, 1971
Sn	*The Chemistry of Organotin Compounds*, by R. C. Poller, Logos Press, London/Academic Press, New York, 1970; *The Organic Chemistry of Tin*, by W. P. Neumann, Wiley, London, 1970; *Organotin Compounds*, Vols 1–3, edited by A. K. Sawyer, Marcel Dekker, New York, 1971
Pb	*The Organic Compounds of Lead*, by H. Shapiro and F. W. Frey, Wiley, New York, 1968

As, Sb, Bi   *Organometallic Compounds of Arsenic, Antimony and Bismuth*, by G. O. Doak and L. D. Freedman, Wiley, New York, 1970

The following transition elements are covered in a series of monographs published by Academic Press (New York):

Ti, Zr, Hf   *The Organometallic Chemistry of Titanium, Zirconium and Hafnium*, by P. C. Wailes, R. S. P. Coutts and H. Weigold, 1974

Cr   *Organochromium Chemistry*, by R. P. A. Sneeden, 1975

Fe   *The Organic Chemistry of Iron*, edited by E. A. Koerner von Gustorf, F.-W. Grevels and I. Fischler, Vol. 1, 1978; Vol. 2, 1981

Ni   *The Organic Chemistry of Nickel*, by P. W. Jolly and G. Wilke, Vol. 1, 1974; Vol. 2, 1975

Pd   *The Organic Chemistry of Palladium*, by P. M. Maitlis, Vol. 1, 1971; Vol. 2, 1971.

1.5.2 Reviews and Journals

Authoritative review articles on various aspects of the subject appear in *Advances in Organometallic Chemistry* (Academic Press, New York) Vols 1 (1964) to 22 (1983). The *Journal of Organometallic Chemistry Library* is devoted to review articles and reports of conferences. A series which emphasised experimental aspects, and which appeared for seven years but then ceased to issue new volumes, is *Organometallic Reactions and Synthesis* (edited by E. J. Becker and M. Tsutsui), Vols 1 (1970) to 6 (1977). Other review publications, such as *Advances in Inorganic Chemistry and Radiochemistry* (Academic Press, New York) contain occasional articles on organometallic chemistry and most of the papers in *Reviews on Silicon, Germanium, Tin and Lead Compounds* (Freund, Tel-Aviv) deal with organic compounds of these elements.

There are two annual selective reviews of the general literature of organometallic chemistry. In the Royal Society of Chemistry's *Specialist Periodical Reports on Organometallic Chemistry* the material is classified partly by subgroup of the periodic table (main group metals) and partly by ligand type (transition metals). For example, Vol. 11, published in 1983, deals with the 1981 literature and contains over 4000 references. The reviews appearing in the *Journal of Organometallic Chemistry* generally deal separately with each metal or group of metals. Recently published articles in this series dealing with (a) B, (b) Sb, (c) Bi, (d) Mn, Tc, Re, (e) Co, Rh, Ir, (f) transition metals in organic synthesis, are to be found in Vol. 261, published in 1984, covering the 1982 literature. For reasons which are not clear, the corresponding surveys of the Group IVb elements only are now published, not in the main journal, but in *Journal of Organometallic Chemistry Library*.

Finally, although research papers on organometallic chemistry are widely dispersed throughout the literature of inorganic, organic, and physical chemistry, there are two journals which specialise in this field. The older is the *Journal of Organometallic Chemistry*, first published in 1963 by Elsevier, and the newer arrival, called *Organometallics*, was introduced in 1982 by the American Chemical Society.

REFERENCES

Bruce, M. I. (1972). *Adv. organomet. Chem.* **10**, 273

Bruce, M. I. (1973). *Adv. organomet. Chem.* **11**, 447

Bruce, M. I. (1974). *Adv. organomet. Chem.* **12**, 380

Bunsen, R. W. (1843). *Annalen* **46**, 1

Cotton, F. A. (1975). *J. organomet. Chem.* **100**, 29

Frankland, E. (1902). *Sketches from the Life of Edward Frankland: 1825-1899*. Spottiswoode, London

Gmelin's Handbuch der Anorganischen Chemie. Springer-Verlag, Berlin (1926 onwards)

Hartley, F. R., and Patai, S. (eds) (1982). *The Chemistry of the Metal–Carbon Bond*, Vol. 1. Wiley–Interscience, Chichester

Hoffmann, R. (1982). *Angew. Chem. int. Edn Engl.* **21**, 711

Kekulé, A. (1858). *Annalen* **106**, 129

Müller, E. (ed.) (1970). *Methoden der Organischen Chemie*, 4th edn, Vol. 13/1, Georg Thieme Verlag, Stuttgart

Pope, W. J., and Peachey, S. J. (1907). *Proc. chem. Soc.* **23**, 80

Rigaudy, J., and Klesney, S. P. (1979). *Nomenclature of Organic Chemistry*. Pergamon Press, Oxford, p. 360

Smith, J. D., and Walton, D. R. M. (1975). *Adv. organomet. Chem.* **13**, 453

Wilkinson, G., Stone, F. G. A., and Abel, E. W. (eds) (1982). *Comprehensive Organometallic Chemistry*, Pergamon Press, Oxford

Zeise, W. C. (1831). *Ann. Phys. (Poggendorff)* **21**, 497; *Z. Phys. Chem. (Schweigger)* **62**, 393; **63**, 121

GENERAL READING

Hoffmann, R. (1982). Building bridges between inorganic and organic chemistry, *Angew. Chem. int. Edn Engl.* **21**, 711

Thayer, J. S. (1975). Organometallic chemistry: an historical perspective, *Adv. organomet. Chem.* **13**, 1

2

PREPARATION OF ORGANOMETALLIC COMPOUNDS

To give a systematic account of methods for the synthesis of organometallic compounds, which gives some idea of their enormous range and diversity, yet remains sufficiently concise for an introductory text, is difficult. No system is entirely satisfactory, but the one adopted here is to divide the methods into five major categories with less important procedures grouped into a sixth, miscellaneous division.

It will become clear that many methods are applicable to both main group as well as transition metals. In these cases, examples from the main group metals are given first.

As most organometallic compounds react with water and oxygen, the preparations discussed below are generally carried out under anhydrous conditions with protection from the atmosphere.

2.1   THE DIRECT REACTION: STARTING WITH THE METAL

The oxidative addition of an organic halide to a metal is referred to as the direct reaction. This is a very important reaction which is the most widely used method of entry into organometallic chemistry.

When the metal is monovalent, an equivalent amount of metal halide is also produced (equation 2.1).

$$2M + RX \xrightarrow{\hspace{3cm}} RM + MX \qquad (2.1)$$
$$(M = Li, Na, K, Cs; X = Cl, Br, I)$$

Of the Group I metals, lithium gives the most useful organic compounds. On the laboratory scale, either a hydrocarbon or diethyl ether is used as reaction medium. Reaction is more rapid in ether, but the solvent is attacked by the reagent

$$RLi + Et_2O \xrightarrow{\hspace{1cm}} [EtOCHLiCH_3] + RH$$
$$\swarrow$$
$$EtOLi + CH_2 = CH_2$$

Scheme *i*

(scheme *i*). Another side reaction is coupling of the product with organic halide to give a hydrocarbon (equation 2.2).

$$RLi + RX \xrightarrow{\hspace{1cm}} RR + LiX \qquad (2.2)$$

Both of these undesired reactions are minimized by cooling to $0\,^{\circ}C$ or lower. Reaction (2.2) is favoured when $X = I$ (with the fortunate exception of methyl iodide) and, for this reason, chlorides or bromides are generally preferred. Only the relatively covalent organolithium compounds are soluble in organic solvents. The heavier alkali metals give ionic, insoluble, highly reactive organic compounds which are used in industry because of their lower cost.

In the case of a divalent metal, the organometallic compound is the sole product if we discount side reactions (equation 2.3).

$$M + RX \xrightarrow{\hspace{1cm}} RMX \qquad (2.3)$$

Illustrative examples of this reaction are shown in equations (2.4)–(2.7), and we should note that structures of the products are more complicated than is indicated by the simple RMX formulation (see, for example, the structure of the Grignard reagent in chapter 3).

$$\text{Mg} + \text{PhBr} \longrightarrow \text{PhMgBr} \tag{2.4}$$

$$\text{Zn} + \text{EtI} \longrightarrow \text{EtZnI} \tag{2.5}$$

$$\text{Hg} + \text{CH}_2\text{=CHCH}_2\text{I} \longrightarrow \text{CH}_2\text{=CHCH}_2\text{HgI} \tag{2.6}$$

$$\text{Ca} + \text{MeI} \longrightarrow \text{MeCaI} \tag{2.7}$$

The preparation of a Grignard reagent, such as shown in equation (2.4), is carried out in a donor solvent, usually diethyl ether, and it is often necessary to initiate the reaction by addition of small amounts of substances such as iodine or methyl iodide. The synthesis of organozinc halides (equation 2.5) also requires the presence of a donor solvent. Simple alkyl and aryl halides are unreactive towards mercury, but the labile allyl iodide rapidly gives allylmercury iodide in a reaction (equation 2.6) which was first described in 1855 (Zinin, 1855). At 300°C, mercury reacts with the atypical halogen compound, pentafluoroiodobenzene, to give dipentafluorophenylmercury, $(\text{C}_6\text{F}_5)_2\text{Hg}$, in 75 per cent yield, rather than the organomercury halide (Birchall et al., 1967). Donor solvents are also required for the preparation of organocalcium compounds and the reaction shown in equation (2.7) was carried out in THF at 0°C. The low temperature was necessary to minimise coupling reactions, as well as attack of solvent by the product (Kawabata et al., 1973).

In Group III, aluminium reacts with the lower alkyl chlorides to give the sesquihalides, the method being most successful with methyl and ethyl chlorides (equation (2.8); Adkins and Scanley, 1951). Bromo- and iodo-alkanes react slowly with indium to give mixtures of alkylindium halides but pentafluoroiodobenzene reacts abnormally (as it does with mercury) to give tris(pentafluorophenyl)indium.

The methylsilicon chlorides needed for the preparation of silicones are obtained by the direct reaction from methyl chloride and silicon at elevated temperatures in the presence of copper (equation (2.9); Rochow, 1971). The corresponding phenylsilicon chlorides can be prepared from chlorobenzene and silicon containing 10 per cent silver or, less expensively, from silicon containing copper and a second metal such as zinc to minimise side reactions.

$$3\text{MeCl} + 2\text{Al} \longrightarrow \text{Me}_3\text{Al}_2\text{Cl}_3 \tag{2.8}$$

$$2\text{MeCl} + \text{Si} \xrightarrow{\text{Cu, 280-300°C}} \text{Me}_2\text{SiCl}_2 \quad 90 \text{ per cent} \tag{2.9}$$

Similarly, dimethyltin dichloride is prepared on the commercial scale from methyl chloride and tin (equation 2.10).

$$2\text{MeCl} + \text{Sn} \xrightarrow[180-190°C]{\text{Ph}_3\text{MePBr,}} \text{Me}_2\text{SnCl}_2 \tag{2.10}$$

The method is unsuccessful with higher alkyl chlorides, but many iodides react in the presence of catalysts (Murphy and Poller, 1980).

2.1.1 Mechanistic Considerations

Mechanistic studies in this area have been sparse for a number of reasons, not least because the heterogeneous nature of the reactions makes kinetic measurements difficult.

It is most unlikely that all metals which give organometallic compounds with organic halides react by the same mechanism. Nevertheless, these reactions have many features in common. For example, in equations (2.1) and (2.3) the reactivity sequence is RI > RBr > RCl and this is compatible with carbon–halogen bond cleavage occurring in the rate-determining step. Alkyl compounds react more readily than aryl, and halogeno compounds which dissociate to give stable radicals (such as allyl and benzyl halides) are particularly reactive. There is considerable evidence for the involvement of radicals in the formation of organic derivatives of lithium, magnesium and tin. The essential steps seem to be (a) a one-electron transfer from metal to organic halide, and (b) a simultaneous, or subsequent, heterolysis of the C–X bond in a rate-determining step (scheme *ii*).

$$M \quad R\text{—}X \longrightarrow M^+ + R^{\boldsymbol{\cdot}} + X^- \qquad \text{(rate determining)}$$

or

$$M \quad R\text{—}X \longrightarrow M^+ + RX^{\boldsymbol{\cdot}-}$$

$$RX^{\boldsymbol{\cdot}-} \longrightarrow R^{\boldsymbol{\cdot}} + X^- \qquad \text{(rate determining)}$$

Subsequent steps depend on the valency of the metal:

Monovalent M $\left\{\begin{array}{l} M^+ + X^- \longrightarrow MX \\ M \quad R^{\boldsymbol{\cdot}} \longrightarrow RM \end{array}\right.$

Divalent M $\left\{\begin{array}{l} M^{\boldsymbol{\cdot}+} + X^- \longrightarrow MX^{\boldsymbol{\cdot}} \\ R^{\boldsymbol{\cdot}} + MX^{\boldsymbol{\cdot}} \longrightarrow RMX \end{array}\right.$

Tetravalent M $\quad [RMX] + RX \longrightarrow R_2MX_2$

Scheme *ii*

The ions and radicals produced then interact to give the organometallic product in a manner which depends upon the valency of the metal. A tetravalent metal such as tin gives, initially, an unstable tin(II) species, $[RSnX]$, which then undergoes a second oxidative addition to give the R_2SnX_2 product.

The direct reaction succeeds with metals of low ionisation potential. Efficient solvation of the ionic intermediates M^+ and X^- is necessary and the widespread use of Lewis base solvents and additives in these reactions suggests that solvation of the metal cation is particularly important.

The overall energy changes accompanying reactions (2.1) or (2.3) would be obtained by comparing the free energies of the products with those of the reactants but, unfortunately, these quantities are often not available. It is nevertheless clear that the energy inputs needed to break down the metal lattice, ionise the metal, and break the C–X bond must be set against the energy released on formation of the M–C and M–X bonds. Since metal-carbon bonds are generally weak, we might expect that the formation of the M–X bond would be a major influence in promoting the reaction. With this reasoning we would predict that chlorides would give best results since bond stabilities are in the order MCl > MBr > MI. As we have already seen, the actual reactivity order is RI > RBr > RCl, indicating that kinetic rather than thermodynamic factors are often dominant. Nevertheless, influences which are likely to increase the negative free energy change in the reaction must be considered favourable. Thus, the use of alloys with sodium or magnesium of the primary metal resulting in the formation of the relatively stable sodium or magnesium halides often gives improved yields (equations 2.11-2.13). Reaction (2.11) is the basis of one of the commercial methods for preparation of the anti-knock tetraethyllead. (Best results are obtained by using the 1 : 1 alloy and recycling unreacted lead.)

$$4EtCl + 4NaPb \longrightarrow Et_4Pb + 4NaCl + 3Pb \qquad (2.11)$$

$$4EtBr + Mg_2Sn \longrightarrow Et_4Sn + 2MgBr_2 \qquad (2.12)$$

$$6EtCl + Al_2Mg_3 + 2Et_2O \longrightarrow 2Et_3Al.OEt_2 + 3MgCl_2 \qquad (2.13)$$

Use of metals in the vapour state not only removes the thermodynamically unfavourable metal lattice energy term, but also makes the metal kinetically more reactive (Timms and Turney, 1977; Blackborow and Young, 1979). The technique consists of heating the metal, usually at temperatures in excess of $1600\,^\circ C$ in a high vacuum $(10^{-7}-10^{-8}$ atm). The metal vapour, and the other gaseous reactants, are condensed at a cooled surface where reaction occurs. Very successful results have been obtained by reaction of transition metal vapours with arenes and other unsaturated donor species, as in the examples shown in equations (2.14) (Timms and Turney, 1977), (2.15) (Skell and McGlinchey, 1975) and (2.16) (Simons et al., 1976) (see also p. 33). (In equations (2.14)-(2.16) an asterisk indicates that the condensed metal vapour is used.)

$$2C_6H_6 + Cr^* \longrightarrow Cr(C_6H_6)_2 \qquad (2.14)$$

$$3CH_2{=}CHCH{=}CH_2 + Mo^* \longrightarrow Mo(C_4H_6)_3 \qquad (2.15)$$

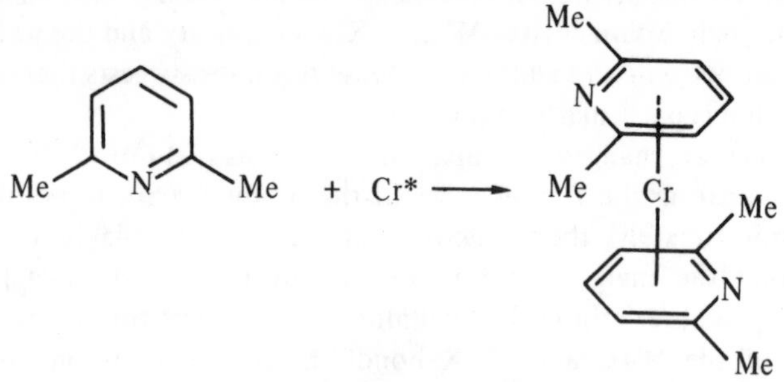

$$(2.16)$$

Reactions with organic halides have also been reported and many perfluoro-organic derivatives of transition metals have been prepared. Thus, in scheme *iii*,

$$Pd^* + CF_3Br \longrightarrow [CF_3PdBr] \xrightarrow{PEt_3} trans\text{-}CF_3Pd(PEt_3)_2Br$$

Scheme *iii*

palladium atoms react with bromotrifluoromethane to give, initially, a co-ordinatively unsaturated product which is converted to a stable adduct with triethylphosphine. In contrast, main group metal atoms produced by this technique are not particularly reactive and the method is rarely used for the preparation of organic compounds of these metals.

A simpler technique for making metals more reactive is to prepare them in a very finely divided state by reduction of a metal salt using potassium (Rieke, 1977). Thus magnesium, produced by reduction of anhydrous magnesium di-chloride with potassium in THF at reflux temperature is a very fine, black, pyrophoric powder. This material, unlike magnesium turnings, will react with chlorobenzene in diethyl ether to give phenylmagnesium chloride in high yield. Another striking result is the direct preparation of the otherwise inaccessible organomagnesium fluorides, as exemplified in equation (2.17), where the yield was determined by successive treatment of the product with carbon dioxide and dilute acid followed by isolation of the *p*-toluic acid (p-MeC$_6$H$_4$COOH) formed.

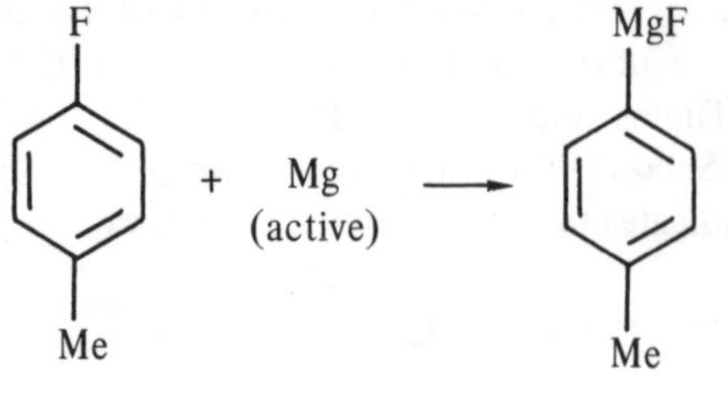

$$(2.17)$$

69 per cent

Reduction of diiodo*bis*(triethylphosphine)nickel(II) dissolved in THF gave a black slurry of finely divided metal which reacted rapidly with bromopentafluorobenzene to give bromopentafluorophenyl*bis*(triethylphosphine)nickel(II) (equation 2.18).

$$C_6F_5Br + \underset{(active)}{Ni} + 2Et_3P \longrightarrow C_6F_5-\overset{\displaystyle PEt_3}{\underset{\displaystyle PEt_3}{Ni}}-Br \qquad (2.18)$$

60 per cent

Despite these and other successes of this technique, it should be noted that it requires stoicheiometric quantities of very high purity (and very high cost!) potassium and by no means are all metals activated. Thus, tin powders produced by reduction of tin(II) chloride do not show enhanced reactivity to chloroalkanes.

It is convenient to consider here some preparations which involve oxidative additions of organic halides to compounds containing metals in lower valency states (equation 2.19).

$$RX + MY_n \longrightarrow RMY_nX \qquad (2.19)$$

There is evidence that, as indicated in scheme *ii*, the direct synthesis of dialkyltin dichlorides proceeds via oxidative addition of alkyl chloride to the initially formed tin(II) compound (equation 2.20).

$$RCl + [RSnCl] \longrightarrow R_2SnCl_2 \qquad (2.20)$$

Reaction (2.19) is not common in main group chemistry, but there is another example, also involving tin, which is the preparation of an alkyltin trihalide, by the catalysed addition of alkyl halide to tin(II) halide (equation (2.21); Bulten, 1975). More examples of oxidative addition are to be found amongst the transition metals and a number of four- and five-coordinate d^8 complexes give stable adducts with alkyl (and acyl) halides.

$$RX + SnX_2 \xrightarrow{R_3'Sb} RSnX_3 \qquad (2.21)$$

Thus the six-coordinate iridium(III) compound shown in equation (2.22) is obtained by addition of methyl bromide to an iridium(I) species (Collman and Roper, 1968).

$$(2.22)$$

2.2 STARTING WITH ORGANOMETALLIC REAGENTS

To form organic derivatives of metals which do not undergo the direct reaction it is often convenient to use organometallic reagents made from metals which do – chiefly magnesium and lithium. This section is subdivided by considering the nature of the compound which reacts with the organometallic reagent.

2.2.1 Reaction of Organometallic Reagents with Metal Halides

This is a widely used method which can be expressed in general terms by equation (2.23).

$$R-M + M'-X \rightleftharpoons R-M' + M-X \tag{2.23}$$

The most important factor in determining the position of equilibrium is the readiness with which metals give up their electrons to form ions. For reactions in solution, this is measured by the standard reduction potential in volts (E_0). Metals with high negative values of E_0, such as sodium, lose their electrons very readily. These metals are more likely to exist as salts than metals with lower negative or positive values of E_0 which prefer to exist as organometallic compounds where the electrons are retained. Thus, if we rank metals in order of standard reduction potentials, beginning with the most negative and ending with the most positive values, the equilibrium in equation (2.23) will be to the right-hand side if M is higher than M' in this ranking. Some illustrative values are given in table 2.1, and these, of course, refer to unit activities in aqueous solution (Ball and Norbury, 1974). The examples of preparations by this method given in equations (2.24)–(2.27) are in general accord with the data in table 2.1, even

Table 2.1   Standard reduction potentials

Electrode		E_0 (volts)
$2H^+ + 2e^-$	$\rightleftharpoons H_2$ (g)	0 (standard)
$Li^+ + e^-$	$\rightleftharpoons Li$ (s)	-3.03
$Na^+ + e^-$	$\rightleftharpoons Na$ (s)	-2.71
$Mg^{2+} + 2e^-$	$\rightleftharpoons Mg$ (s)	-2.37
$Al^{3+} + 3e^-$	$\rightleftharpoons Al$ (s)	-1.66
$Zr^{4+} + 4e^-$	$\rightleftharpoons Zr$ (s)	-1.53
$Zn^{2+} + 2e^-$	$\rightleftharpoons Zn$ (s)	-0.76
$Sn^{4+} + 4e^-$	$\rightleftharpoons Sn$ (s)	0
$Cu^+ + e^-$	$\rightleftharpoons Cu$ (s)	$+0.52$
$Hg^{2+} + 2e^-$	$\rightleftharpoons Hg$ (l)	$+0.79$

though these reactions are not carried out in water. It is assumed that the ranking order of reduction potentials would be the same in other solvents.

$$2C_6F_5Li + HgCl_2 \longrightarrow (C_6F_5)_2Hg + 2LiCl \tag{2.24}$$

$$3BuMgCl + AlCl_3 \xrightarrow{Et_2O} Bu_3 Al.OEt_2 + 3MgCl_2 \qquad (2.25)$$

$$4PhCH_2MgCl + ZrCl_4 \longrightarrow (PhCH_2)_4Zr + 4MgCl_2 \qquad (2.26)$$

$$4Bu_3Al + 3SnCl_4 + 4R_3N \longrightarrow 3Bu_4Sn + 4AlCl_3.R_3N \qquad (2.27)$$

The low-cost trialkylaluminium reagents are specially favoured for large-scale work, but alkylation of tin(IV) chloride stops short of completion because of complex formation between partially alkylated tin compounds and aluminium trichloride (such as $Bu_3SnCl.AlCl_3$). Addition of a more powerful donor than the alkyltin chlorides allows reaction to go to completion (equation 2.27). Even in cases where the position of equilibrium is unfavourable it may be possible to force the reaction to proceed in the required direction. For example, continuous removal of a minor product will ensure that it eventually becomes a major product.

It is pertinent to mention here a closely related reaction in which a mixture of a metal halide and an organic halide is treated with sodium (equation 2.28).

$$2Na + R-X + M-X \longrightarrow R-M + 2NaX \qquad (2.28)$$

The reaction is loosely referred to as 'Wurtz-type' because of its similarity to the true Wurtz reaction in which alkyl residues are coupled by treatment of the alkyl iodides with sodium (equation 2.29).

$$2RI + 2Na \longrightarrow R-R + 2NaI \qquad (2.29)$$

An example of reaction (2.28) is the preparation of tetrabutyltin from dibutyl-tin dichloride (equation 2.30).

$$Bu_2SnCl_2 + 2BuCl + 4Na \longrightarrow Bu_4Sn + 4NaCl \qquad (2.30)$$

It is generally not practicable to start with tin(IV) chloride since it is reduced to tin(II) and tin(0) by the sodium, though it has been reported (Rulewicz et al., 1973) that this problem can be overcome by careful control over the size of the sodium particles. Another way of avoiding inadvertent reduction of the starting material is to react chlorobutane with sodium to give butylsodium before introduction of the tin(IV) chloride. This procedure gives excellent yields of tetrabutyltin (scheme iv; Owen and Poller, 1983).

$$BuCl + 2Na \longrightarrow BuNa + NaCl$$

$$4BuNa + SnCl_4 \longrightarrow Bu_4Sn \text{ (91 per cent)} + 4NaCl$$

Scheme iv

The reaction of organometallic compounds with metal halides (equation 2.23) is an important route to organic derivatives of the transition metals. Before considering some examples, we will digress to note that alkyl transition metal compounds are often unstable. They are particularly susceptible to decomposition by the β-elimination pathway in which the alkyl group is lost as an alkene, a

β-hydrogen atom being transferred from carbon to metal (equation (2.31); Davidson *et al.*, 1976).

$$M-\overset{|}{\underset{|}{C}}-\overset{|}{\underset{\underset{H}{|}}{C}}- \quad \longrightarrow \quad M-H \;+\; \overset{\diagdown}{\diagup}C\!=\!C\overset{\diagup}{\diagdown} \tag{2.31}$$

More stable compounds result from alkyl groups which lack β-hydrogen atoms, such as methyl or trimethylsilylmethyl (Me_3SiCH_2). The bulk of the latter group may confer additional stability by shielding the metal from attack. Enhanced stability also results when Lewis bases are coordinated to the metal. This is because the rate-determining step in reaction (2.31) involves a transition state in which both carbon and hydrogen occupy coordination sites at the metal (scheme v). This reaction will, therefore, be less likely if the metal has all coordination

$$M-\overset{|}{\underset{|}{C}}-\overset{|}{\underset{\underset{H}{|}}{C}}- \quad \rightleftharpoons \quad \left[\begin{array}{c} M\text{---}\overset{|}{C}- \\ \vdots \quad \overset{|}{} \\ H\text{---}\overset{|}{C}- \end{array}\right]^{\ddagger} \quad \rightleftharpoons \quad M\overset{\diagup C\diagdown}{\underset{H\diagup}{\underset{}{\overset{|}{\parallel}}}}\underset{\diagdown}{C}$$

$$\rightleftharpoons \quad M-H \;+\; \overset{\diagdown}{\diagup}C\!=\!C\overset{\diagup}{\diagdown}$$

Scheme v

sites filled (Wilkinson, 1974). In reactions (2.32) and (2.33) organolithium reagents are used to make alkylpalladium and alkyltitanium compounds; the former are stabilized by the presence of triethylphosphine groups and the latter by π-bonded cyclopentadienyl residues.

$$2MeLi + (Et_3P)_2PdBr_2 \longrightarrow (Et_3P)_2PdMe_2 + 2LiBr \tag{2.32}$$

$$2Me_3SiCH_2Li + (\eta^5\text{-}C_5H_5)_2TiCl_2 \longrightarrow (\eta^5\text{-}C_5H_5)_2Ti(CH_2SiMe_3)_2$$
$$+ \; 2LiCl \tag{2.33}$$

Even when an alkyl group does not contain β-hydrogen atoms, complications may arise from reactions of the α-hydrogen atoms. This subject is discussed on p. 147.

Another problem which can arise when transition metal halides react with organometallic reagents is that they may be reduced, often to the metal. Thus reaction between iron(III) chloride and phenylmagnesium bromide gave biphenyl and iron metal (equation 2.34).

$$6PhMgBr + 2FeCl_3 \longrightarrow 3PhPh + 2Fe + 3MgCl_2 + 3MgBr_2 \qquad (2.34)$$

In fact this is a method for the preparation of biaryls and it was when it was used in an attempt to prepare dihydrofulvalene (I) that ferrocene was discovered (equation (2.35), M = Fe; Kealy and Pauson, 1951).

I

$$2C_5H_5MgBr + MCl_3 \xrightarrow[(M = Fe, Cr, V)]{} (\eta^5\text{-}C_5H_5)_2M \qquad (2.35)$$

Preparations of the analogous chromium and vanadium compounds, Cp_2Cr and Cp_2V, can be carried out similarly; in all three cases the metal is reduced from the tri- to the di-valent state (equation 2.35). In a closely related procedure, η^3-allyl compounds are prepared by reaction between an allyl Grignard reagent and a metal halide, as exemplified by the formation of diallylnickel (equation (2.36); Wilke et al., 1966).

$$2CH_2{=}CHCH_2MgBr + NiBr_2 \longrightarrow (\eta^3\text{-}C_3H_5)_2Ni + 2MgBr_2 \qquad (2.36)$$

2.2.2 Reaction of Organometallic Reagents with Metal Compounds or Free Metals

More rarely, groups other than halogen are replaced by an organic residue as in the preparation of pentaphenylantimony from triphenylantimony diacetate (equation (2.37); Wittig and Hellwinkel, 1964).

$$2PhLi + Ph_3Sb(OAc)_2 \longrightarrow Ph_5Sb \text{ (80 per cent)} + 2AcOLi \qquad (2.37)$$

Another example is the alkylation of methyl dialkylborinates, R_2BOMe, in preference to the chloro compound R_2BCl, since use of the latter often caused isomerisation of the incoming alkyl group (equation (2.38); Kramer and Brown, 1974).

$$R_2BOMe + R'Li \longrightarrow R_2BR' + MeOLi \qquad (2.38)$$

These reactions are conveniently carried out in pentane and the soluble product removed by decantation from the insoluble lithium methoxide.

Occasionally, a rather inaccessible compound may be prepared by interaction between two different organometallic species as in the general case shown in equation (2.39).

$$M-R + M'-R' \rightleftharpoons M-R' + M'-R \qquad (2.39)$$

This method is useful for making certain organolithium compounds from derivatives of less electropositive metals. Thus vinyl-, allyl- and many substituted methyl-lithium species are prepared from the corresponding tin, mercury, boron or silicon derivatives (Wakefield, 1974; Krief, 1980). An example is given in equation (2.40), where ready removal of the insoluble tetraphenyltin facilitates product isolation.

$$(CH_2=CH)_4 Sn + 4PhLi \longrightarrow 4CH_2=CHLi + Ph_4 Sn \qquad (2.40)$$

Another method of limited application is reaction of an organometallic compound with a metal (equation 2.41).

$$R-M + M' \rightleftharpoons R-M' + M \qquad (2.41)$$

Such a reaction should be favoured if the stability of $R-M'$ exceeds that of $R-M$. In practice, most work has been done using organomercury derivatives, which have been shown to give organometallic compounds with a wide range of metals, including lithium, sodium, beryllium, magnesium, zinc, cadmium, aluminium, gallium, indium and tin. Despite problems in handling the toxic organomercury compounds, this is often the method of choice for the preparation of pure dialkyl- and diaryl-magnesium compounds free from metal halides and donor solvents (equation 2.42).

$$sBu_2 Hg + Mg \longrightarrow sBu_2 Mg + Hg \qquad (2.42)$$

2.2.3 Reaction of Organometallic Reagents with Organic Halides

Although a general equation (equation (2.43), X = halogen) is given, this method is mainly used for the preparation of organolithium compounds from organic halides which do not readily react with the metal.

$$R-M + R'-X \rightleftharpoons R'-M + R-X \qquad (2.43)$$

Usually, best yields are obtained with bromides in donor solvents. The equilibrium favours the metal derivative of the most acidic hydrocarbon and we can expect reaction (2.43) to proceed from left to right if the acidities are $R'H > RH$ (Wakefield, 1974). This method is often successful for the preparation of aryl-lithium compounds from aryl bromides and butyllithium, but coupling (to give in this case ArBu) and metallation (see section 2.3) may cause lowered yields. These, and other side reactions, can be minimised by using brief reaction times and low temperatures. An example is the preparation of 3-lithiopyridine carried out at $-35\,^{\circ}C$ (equation 2.44) to avoid the competing addition of butyllithium to the $C=N$ bond which predominates at room temperature (compare equation (4.17) and scheme *viii* in section 4.3).

$$\text{BuLi} + \underset{N}{\overset{Br}{\bigcirc}} \xrightarrow{-35\,^{\circ}\text{C}} \underset{N}{\overset{Li}{\bigcirc}} + \text{BuBr} \qquad (2.44)$$

In general aryllithium compounds carrying a wide range of functional substituents can be prepared in good yields by carrying out the reaction at $-100\,^{\circ}\text{C}$ (Parham and Bradsher, 1982). However, in other examples, better yields are obtained by heating and readers are advised to consult the review by Jones and Gilman (1951), which contains a wealth of experimental detail.

A related synthesis, used mainly to attach an organic residue to a metal which already carries organic groups, is to use a metallate anion, $R_n M^-$, to effect nucleophilic substitution of halogen from an organic halide. An example is shown in equation (2.45).

$$\text{Ph}_3\text{GeLi} + \text{PhCH}_2\text{CH}_2\text{Br} \longrightarrow \text{Ph}_3\text{GeCH}_2\text{CH}_2\text{Ph} + \text{LiBr} \qquad (2.45)$$

The method has also been used for making boron, silicon, tin and lead compounds; other metals such as sodium or magnesium can be used for making σ-bonded organotransition metal compounds (King, 1964). An example is the preparation of benzylmanganese pentacarbonyl (equation 2.46).

$$\text{PhCH}_2\text{Br} + \text{NaMn(CO)}_5 \longrightarrow \text{PhCH}_2\text{Mn(CO)}_5 + \text{NaBr} \qquad (2.46)$$

We note that the same product can be obtained by carrying out the reaction in the reverse sense using nucleophilic displacement of bromine from manganese by the benzyl anion (equation 2.47).

$$\text{PhCH}_2\text{MgBr} + \text{BrMn(CO)}_5 \longrightarrow \text{PhCH}_2\text{Mn(CO)}_5 + \text{MgBr}_2 \qquad (2.47)$$

2.2.4 Reaction of Organometallic Reagents with Unsaturated Compounds

The more reactive organometallic compounds such as lithium alkyls add to alkenes (equation 2.48), but these reactions are of most interest in the context of polymerisation and are, accordingly, dealt with in chapter 9.

$$\text{RLi} + \underset{/}{\overset{\backslash}{C}}{=}\underset{\backslash}{\overset{/}{C}} \longrightarrow R-\overset{|}{\underset{|}{C}}-\overset{|}{\underset{|}{C}}-\text{Li} \qquad (2.48)$$

However, addition to a carbonyl group which is bonded to a transition metal is an important route to carbene complexes (scheme vi; Fischer, 1976).

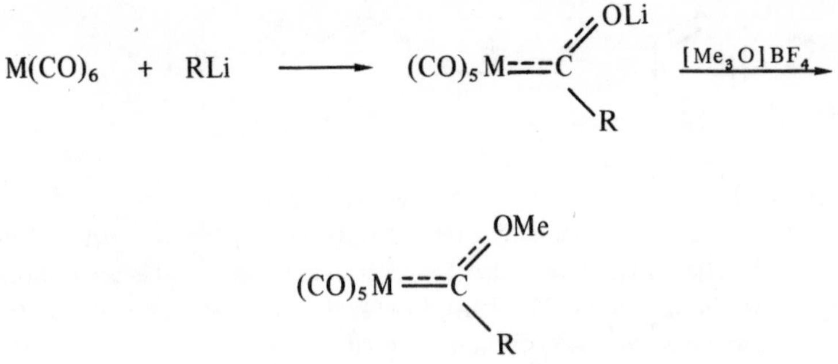

$$M(CO)_6 \;+\; RLi \;\longrightarrow\; (CO)_5 M\!=\!\!=\!\!=\!\!C\!\!\begin{array}{l}\nearrow OLi \\[2pt] \searrow R\end{array} \quad \underrightarrow{[Me_3O]BF_4}$$

$$(CO)_5 M\!=\!\!=\!\!=\!\!C\!\!\begin{array}{l}\nearrow OMe \\[2pt] \searrow R\end{array}$$

Scheme *vi* (R = alkyl or aryl; M = Group·VI transition metal)

The reaction is generally successful for carbonyls of the Groups VI and VII metals and recently a general procedure for iron compounds has been reported (Semmelhack and Tamura, 1983). Aminocarbene complexes can be made from reaction of chromium hexacarbonyl, iron pentacarbonyl or nickel tetracarbonyl with lithium diethylamide (scheme *vii*; Fischer *et al.*, 1972).

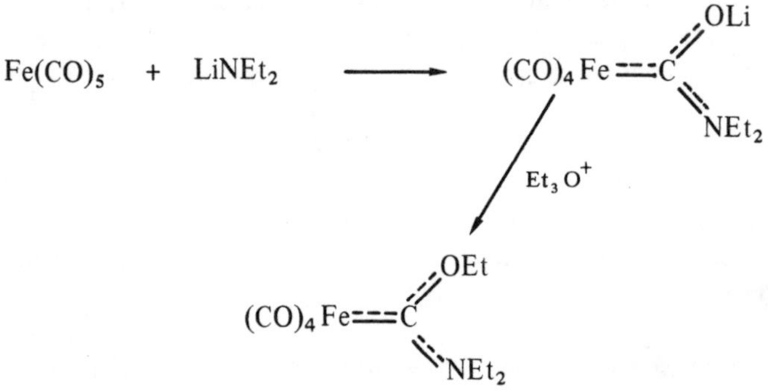

$$Fe(CO)_5 \;+\; LiNEt_2 \;\longrightarrow\; (CO)_4 Fe\!=\!\!=\!\!C\!\!\begin{array}{l}\nearrow OLi \\[2pt] \searrow NEt_2\end{array}$$

$$\Big/ Et_3 O^+$$

$$(CO)_4 Fe\!=\!\!=\!\!C\!\!\begin{array}{l}\nearrow OEt \\[2pt] \searrow NEt_2\end{array}$$

Scheme *vii*

If the carbene complexes are then treated with a boron halide they are transformed into carbyne complexes (equation (2.49); Fischer, 1976).

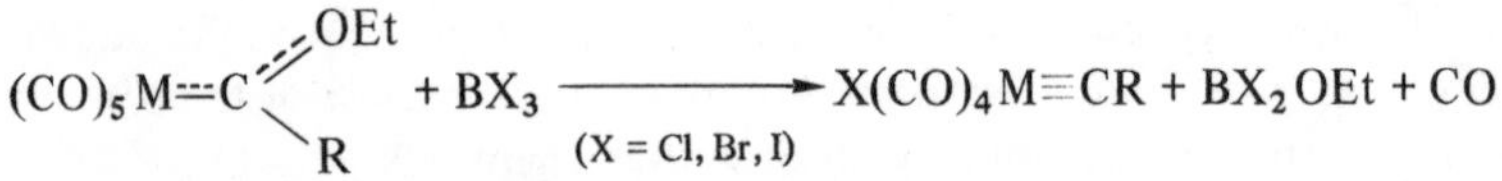

$$(CO)_5 M\!=\!\!C\!\!\begin{array}{l}\nearrow OEt \\[2pt] \searrow R\end{array} \;+\; BX_3 \;\underrightarrow{\;(X=Cl,\,Br,\,I)\;}\; X(CO)_4 M\!\equiv\!CR + BX_2 OEt + CO \quad (2.49)$$

The tungsten carbyne bromide can then undergo replacement of bromine by the cyclopentadiene anion and this is accompanied by the loss of two carbonyl groups (equation (2.50); Fischer *et al.*, 1977).

$$Br(CO)_4 W\!\equiv\!CR + NaC_5 H_5 \;\longrightarrow\; (\eta^5\text{-}C_5 H_5)(CO)_2 W\!\equiv\!CR + 2CO + NaBr \quad (2.50)$$

2.3 METALLATION: DIRECT REPLACEMENT OF HYDROGEN BY A METAL

This reaction involves direct replacement of hydrogen by a metal (equation 2.51).

$$R–H + M–R' \rightleftharpoons R–M + R'–H \tag{2.51}$$

Since an organometallic reagent is used, the reaction could have been discussed in section 2.2; however, since the method not only has a specific name but, more importantly, provides a route to organometallic compounds which does not require the presence of halogen or other reactive group, it is treated separately. For reaction (2.51) to proceed from left to right the acidity of RH should exceed that of R'H. This method is used mainly (but not exclusively) for the preparation of lithium compounds and, since the alkanes have very low acidities, lithium alkyls are normally used. The arenes with their higher acidities are appropriate substrates and the method is particularly valuable for the preparation of aryllithium compounds (equation 2.52).

R = MeO, Me$_2$N, CONMe$_2$, SO$_2$Me, etc.   (2.52)

It will be noted that the lithium atom enters *ortho* to both electron withdrawing and electron donating substituents. This has been explained either in terms of the inductive effect of the methoxyl group rendering the *ortho* hydrogens more acidic or by coordination of oxygen to lithium. Studies of lithiation of p-substituted anisoles, in which the directing effects of various groups were in competition with those of methoxyl, showed that the ability of groups to direct lithium into the *ortho* position is in the order OMe > NMe$_2$ > F. If coordination effects were dominant, the order would be NMe$_2$ > OMe > F, whereas control by combined inductive and resonance effects would result in the order F > OMe > NMe$_2$. It was therefore concluded that both coordinative and electronic effects are important (Slocum and Jennings, 1976). A tertiary amide group in an aromatic system takes precedence over other substituents in directing a lithium atom into the *ortho* position and this fact has been exploited in a series of syntheses of polysubstituted aromatic compounds (Beak and Snieckus, 1982). The reaction is by no means confined to replacement of aromatic hydrogen atoms, acidic aliphatic hydrogens can also be replaced, as exemplified in equation (2.53).

$$MeCN + BuLi \longrightarrow LiCH_2CN + BuH \tag{2.53}$$

The reactivity of the butyllithium reagent is increased by the presence of donor solvents and this effect is particularly striking if a chelate, such as tetramethylethylenediamine (TMED) is added. Thus, although benzene itself is not attacked by butyllithium, the BuLi.TMED complex gives phenyllithium in 100 per cent yield in 34 min at $60\,^{\circ}$C (Langer, 1974). n-Alkyllithium compounds usually exist as hexamers in hydrocarbon solvents and one function of the donor species may be to effect some depolymerisation though, in concentrated solutions, TMED complexes of $(BuLi)_4$ exist. Another activating influence of the donor is to increase the polarity of the Li–C bond, as indicated in the resonance structures **IIa** and **IIb**. Finally, in some BuLi.TMED preparations used for the polymerisation of alkenes (p. 232) the active species was shown to be lithiated TMED (Halasa *et al.*, 1980).

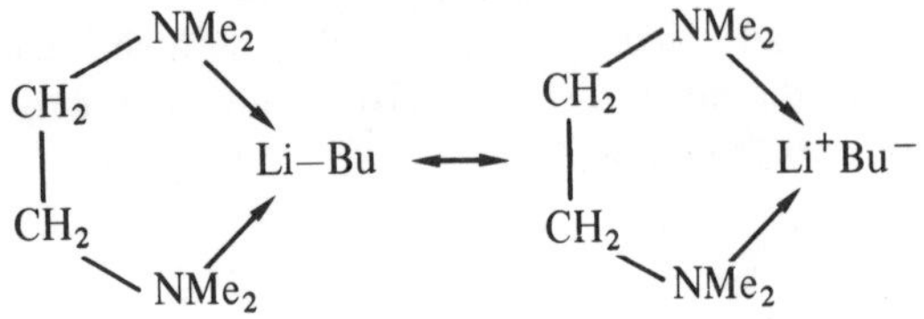

IIa   **IIb**

Organopotassium compounds can be made by treating the hydrocarbon with a mixture of n-butyllithium and potassium t-butoxide. For example, this reagent converts toluene into benzylpotassium in high yield: sodium and rubidium derivatives can be made similarly.

Another variation of this method is to replace the organolithium reagent by a lithium amide as in equation (2.54), where β-picoline is rapidly lithiated at the methyl group by lithium diisopropylamide in the presence of hexamethylphosphoric triamide.

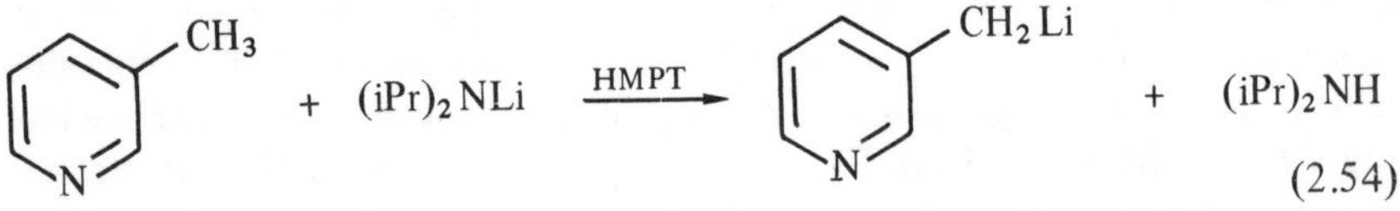

$$+ \; (iPr)_2NLi \; \xrightarrow{HMPT} \qquad + \; (iPr)_2NH \qquad (2.54)$$

More limited preparative use is made of Grignard reagents for metallation of relatively acidic hydrocarbons such as acetylenes (equation 2.55) and cyclopentadiene (equation 2.56).

$$EtMgBr + RC{\equiv}CH \longrightarrow RC{\equiv}CMgBr + EtH \qquad (2.55)$$

$$(2.56)$$

Note that the stable form of cyclopentadiene is the dimer, dicyclopentadiene, formed from a spontaneous Diels-Alder reaction. The monomer is generated as required by heating dicyclopentadiene at its boiling point (equation 2.57).

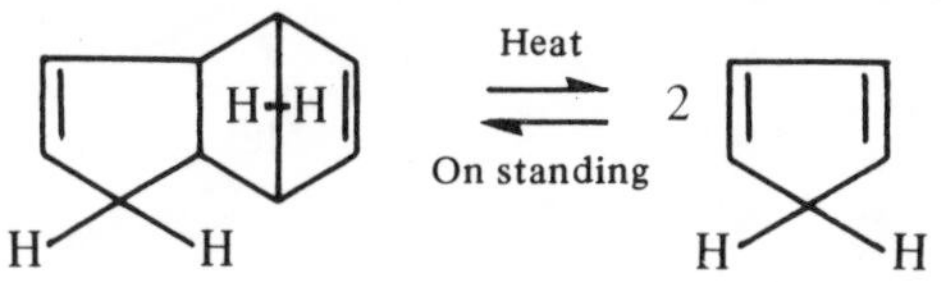

Heat
On standing
2
H
H
H
H
H
H

(2.57)

As well as the Grignard reagent, the sodium derivative, obtained from cyclopentadiene and sodium in THF, is widely used for preparing σ- and π-cyclopentadienyl derivatives of other metals.

Mercuration of reactive aromatic systems has some importance in synthetic organic chemistry. It differs from lithiation in that the reagent is an inorganic mercury compound (equation 2.58).

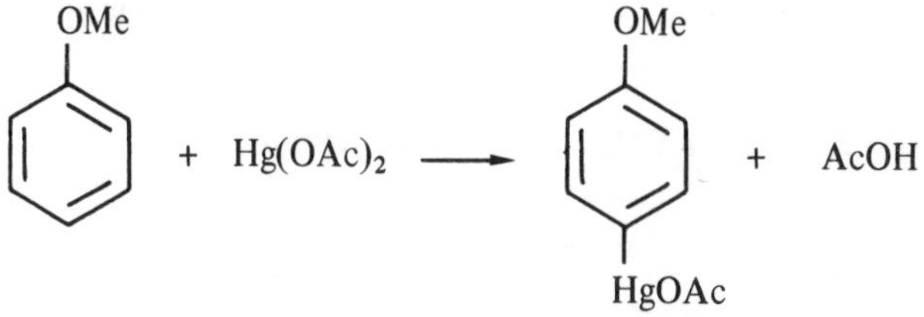

OMe
+ Hg(OAc)2
OMe
HgOAc
+ AcOH

(2.58)

Many of these reactions show a low degree of regioselectivity, due partly to reversibility, so that the first formed products are isomerised. Regioselectivity is, therefore, favoured by short reaction times (Wardell, 1983). Somewhat similar is the formation of arylthallium compounds by metallation using thallium *tris*-(trifluoroacetate) (equation (2.59); Taylor and McKillop, 1970).

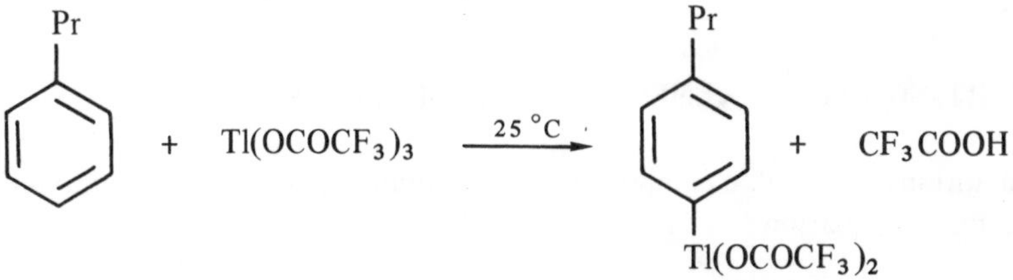

Pr
+ Tl(OCOCF3)3
25 °C
Pr
Tl(OCOCF3)2
+ CF3COOH

(2.59)

When metallation occurs intramolecularly at a ligand coordinated to the metal atom, the method is termed cyclometallation. This is an important route to many σ-bonded derivatives of transition metals (Bruce, 1977). Typically, formation of a complex with *N*-substituted benzylamines promotes attack by the metal at an *ortho* carbon leading to simultaneous metallation and five-membered ring formation (equation 2.60).

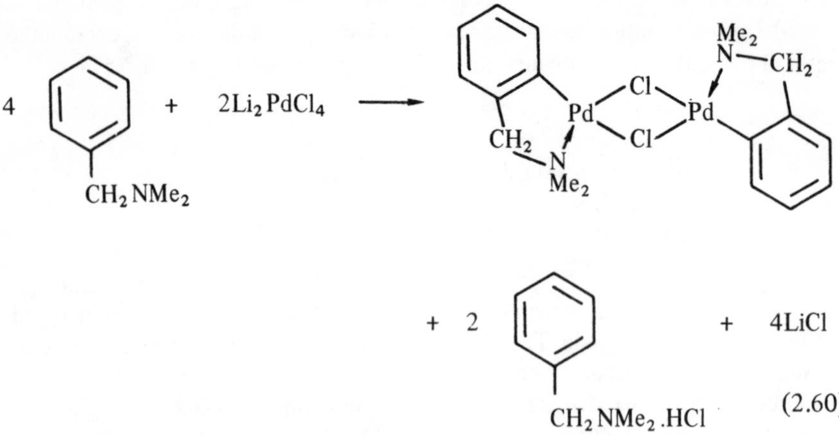

$$4 \left[\text{benzene ring with } CH_2NMe_2\right] + 2Li_2PdCl_4 \longrightarrow \left[\text{dimeric Pd complex}\right]$$

$$+ \ 2 \ \left[\text{benzene ring}\right] \quad + \quad 4LiCl$$

$$CH_2NMe_2 \cdot HCl \qquad (2.60)$$

More rarely, metallation of aliphatic side chains of donor ligands has been demonstrated, as in the example given in equation (2.61), where a neopentyl group of a tertiary phosphine is metallated by platinum giving, again, a five-membered ring system (Mason *et al.*, 1976).

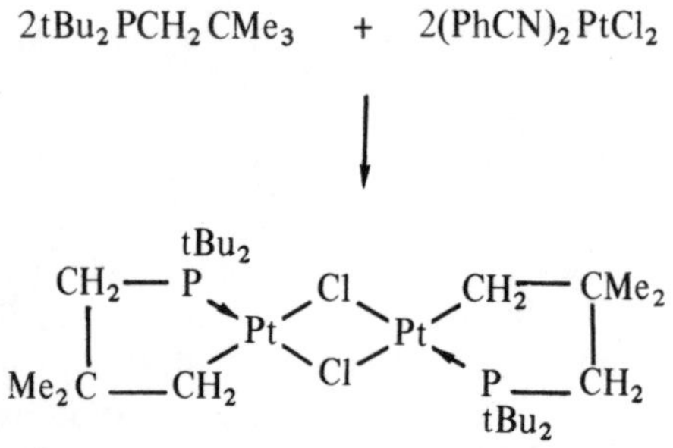

$$2tBu_2PCH_2CMe_3 \ + \ 2(PhCN)_2PtCl_2$$

$$\downarrow$$

$$(2.61)$$

$$\left[\text{dimeric Pt complex}\right]$$

2.4 STARTING WITH ALKENES, ALKYNES OR ARENES

Simple additions of metallic compounds to unsaturated systems are expressed in general terms by equation (2.62).

$$\underset{\diagup}{\overset{\diagdown}{}}C{=}C\underset{\diagdown}{\overset{\diagup}{}} \ + M{-}Y \longrightarrow M{-}\overset{|}{\underset{|}{C}}{-}\overset{|}{\underset{|}{C}}{-}Y \qquad (2.62)$$

This reaction is also called an insertion and is most widely used for main group metal compounds. In transition metal chemistry the unsaturated species usually displaces one or more ligands from coordination sites at the metal to form a π-complex (equation 2.63).

$$\diagdown C = C \diagdown \quad + M-Y \longrightarrow M-\underset{\underset{/ \diagdown}{\overset{\diagup \diagdown}{\underset{\parallel}{C}}}}{\overset{\diagdown \diagup}{\overset{C}{}}} + Y \qquad (2.63)$$

There are a few examples of π-complex formation, such as that occurring between silver nitrate and an alkene, as in equation (2.64) (Quinn and Tsai, 1969) in which the incoming alkene does not, apparently, displace a ligand.

$$\boxed{H\text{-}H} \quad + \quad 2AgNO_3 \quad \longrightarrow \quad NO_3Ag \text{---} \boxed{H\text{-}H} \text{---} AgNO_3 \qquad (2.64)$$

However, the interaction between the Ag^+ and NO_3^- ions in the starting material (Lindley and Woodward, 1966) is substantially modified by the introduction of the alkene group and this modification can be regarded as a displacement. Indeed, where there is greater interaction between the silver cation and the anion, as in the chloride and sulphate, alkene complexes cannot be isolated.

2.4.1 Simple Additions

By far the largest group of reactions in this category is the addition of hydrides to alkenes (equation (2.62), Y = H). A significant example is hydroboration, that is the preparation of organoboranes from diborane and an alkene (equation 2.65).

$$B_2H_6 + 6MeCH_2CH=CH_2 \longrightarrow 2Bu_3B \qquad (2.65)$$

These reactions often show considerable regioselectivity, with boron going on to the least substituted carbon atom. Although hydrogen is more electronegative than boron so that the bond polarity is $B^{\delta+}-H^{\delta-}$, Markownikoff's rule is not applicable since no simple carbonium ion intermediate is involved. Also, bulky substituents on the alkene may influence regioselectivity, but the reaction is stereospecific and gives the *cis* adduct. Although some disagreement over details of the mechanism persists, it seems likely that prior π-complex formation occurs, as suggested in scheme *viii*. The organoboranes made by this procedure are rarely

Scheme *viii*

isolated, but are used as reactive intermediates in laboratory transformations of alkenes (section 8.1.4). The Zr—H bond in $(\eta^5\text{-}C_5H_5)_2ZrHCl$ adds to alkenes in a similar manner to give σ-bonded alkylzirconium compounds which are also useful synthetically (p. 199).

Many industrial-scale reactions of alkenes utilise organoaluminium intermediates made in an analogous manner, though the aluminium hydride is produced *in situ* from the elements (equation (2.66); Ziegler, 1968).

$$2Al + 3H_2 + 6RCH{=}CH_2 \longrightarrow 2Al(CH_2CH_2R)_3 \qquad (2.66)$$

If R = H in reaction (2.66), then oligomerisation occurs by insertion of further ethene molecules into the Al–C bonds, though it is possible to make triethylaluminium by first making diethylaluminium hydride and adding this to ethene under mild conditions.

It is of interest that organomagnesium compounds can be made by addition of magnesium hydride to alkenes, but the reaction is of little preparative value as the hydride is itself produced by reduction of diethylmagnesium (Ashby and Ainslie, 1983), and a better method is by reaction between magnesium and an organomercury compound (p. 22).

Addition of hydrides of metals of Group IVb are important synthetic procedures. Hydrosilation of alkenes occurs readily in the presence of very small amounts $(10^{-5}$–10^{-8} mol per mol of reactant) of chloroplatinic acid (equation 2.67).

$$Cl_3SiH + RCH{=}CH_2 \xrightarrow{\;H_2PtCl_6\;} Cl_3SiCH_2CH_2R \qquad (2.67)$$

Non-terminal alkenes give the adduct from the terminal isomer and a mechanism has been advanced in which the active catalyst is considered to be a platinum(II) complex of the unsaturated substrate (scheme *ix*; Harrod and Chalk, 1977).

Scheme *ix*

Chlorostannanes will only add to reactive double bonds as in equation (2.68), where the reagent is obtained by reacting an ethereal suspension of tin with gaseous hydrogen chloride.

$$`H_2SnCl_2` + 2CH_2{=}CHCOOEt \longrightarrow Cl_2Sn(CH_2CH_2COOEt)_2 \qquad (2.68)$$

In reactions (2.69) and (2.70), organostannanes are used to prepare compounds containing functional groups which could not be obtained by the use of Grignard or organolithium reagents (because of interaction between the M–C bond and the functional group).

$$Ph_3SnH + CH_2{=}CHOCOMe \longrightarrow Ph_3SnCH_2CH_2OCOMe \qquad (2.69)$$

$$Et_2SnH_2 + 2CH_2{=}CHCN \longrightarrow Et_2Sn(CH_2CH_2CN)_2 \qquad (2.70)$$

Addition of an alkene to the complex rhodium hydride shown in equation (2.71) is an important step in a catalytic hydrogenation cycle of the type discussed in section 9.1.

$$\underset{}{>}C{=}C\underset{}{<} + (Ph_3P)_2RhClH_2 \longrightarrow (Ph_3P)_2RhClHC{-}CH \qquad (2.71)$$

Finally, to illustrate the point that simple additions to alkenes are not confined to hydrides, an example of solvomercuration is given in equation (2.72).

$$RCH{=}CH_2 + Hg(OAc)_2 \xrightarrow[\text{THF}]{\text{H}_2\text{O}} RCH(OH)CH_2HgOAc + AcOH \qquad (2.72)$$

2.4.2 Preparation of π-Bonded Compounds

The classical example is the preparation of Zeise's salt in which a chloride ion is displaced from platinum by ethene (equation (2.73); Chock *et al.*, 1973).

$$CH_2{=}CH_2 + PtCl_4^{2-} \xrightarrow{\text{SnCl}_2} C_2H_4PtCl_3^- + Cl^- \qquad (2.73)$$

This reaction has been extended to dienes such as cycloocta-1,5-diene (equation (2.74); Chatt *et al.*, 1957).

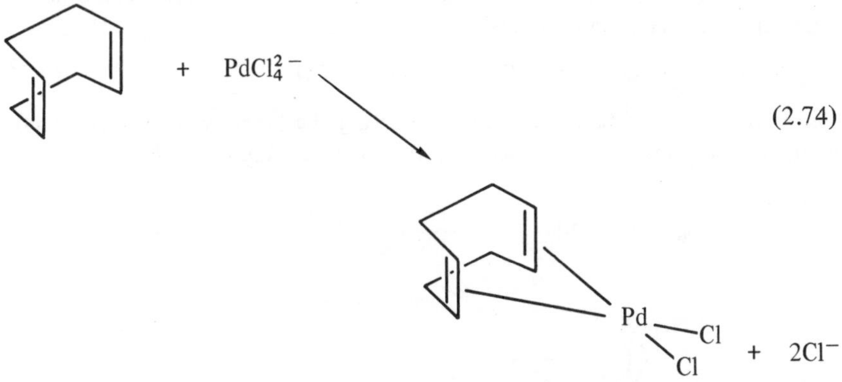

$$(2.74)$$

Deprotonation of alkene–palladium adducts formed in this manner gives π-allyl complexes (scheme x; Ketley and Braatz, 1968). This type of complex is an intermediate in the palladium-catalysed alkylation at the allylic positions of double

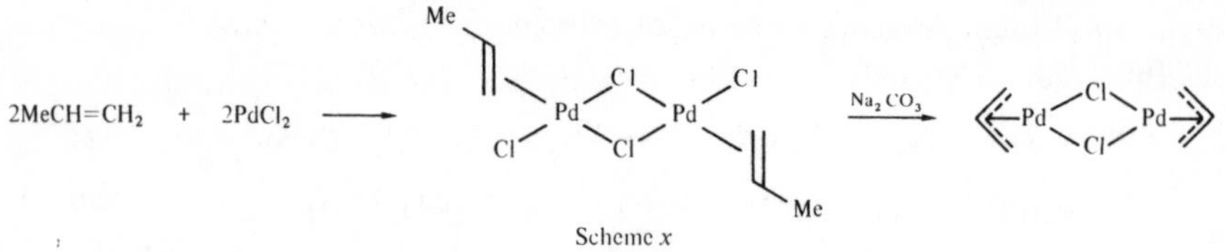

$$2MeCH{=}CH_2 \;+\; 2PdCl_2 \;\longrightarrow\; \text{[chloro-bridged Pd complex]} \xrightarrow{Na_2CO_3} \text{[Pd chloro-bridged dimer]}$$

Scheme x

Scheme x

bonds in complex molecules. These reactions allow considerable control over regio- and stereo-selectivity (Trost, 1980). Iron pentacarbonyl reacts with butadiene in two stages, firstly to give butadieneiron tricarbonyl, and then *bis*(butadiene)iron monocarbonyl (scheme *xi*; Koerner von Gustorf *et al.*, 1971).

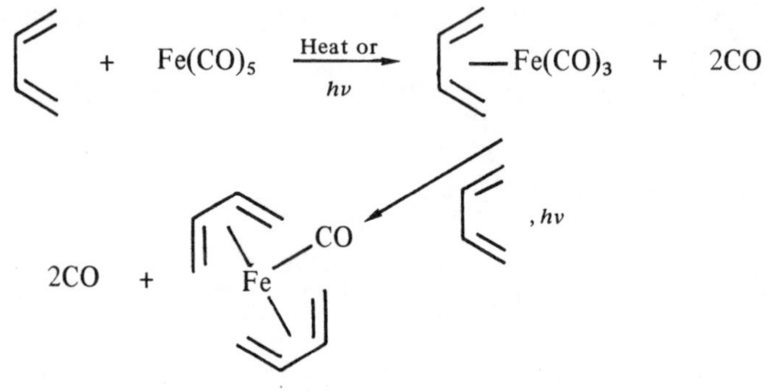

$$\text{[butadiene]} \;+\; Fe(CO)_5 \;\xrightarrow[h\nu]{\text{Heat or}}\; \text{[butadiene]}{-}Fe(CO)_3 \;+\; 2CO$$

$$2CO \;+\; \text{[bis(butadiene)Fe(CO)]} \;\xleftarrow{\,,h\nu}\; \text{[butadiene-Fe(CO)_3]}$$

Scheme *xi*

Arenes react directly with chromium hexacarbonyl (equation 2.75) but the method has the disadvantage that high temperatures and long reaction times are necessary to get reasonable yields.

$$C_6H_6 + Cr(CO)_6 \longrightarrow C_6H_6Cr(CO)_3 + 3CO \qquad (2.75)$$

Better results are obtained by first converting the carbonyl compound to triamminetricarbonylchromium (scheme *xii*; Moser and Rausch, 1974).

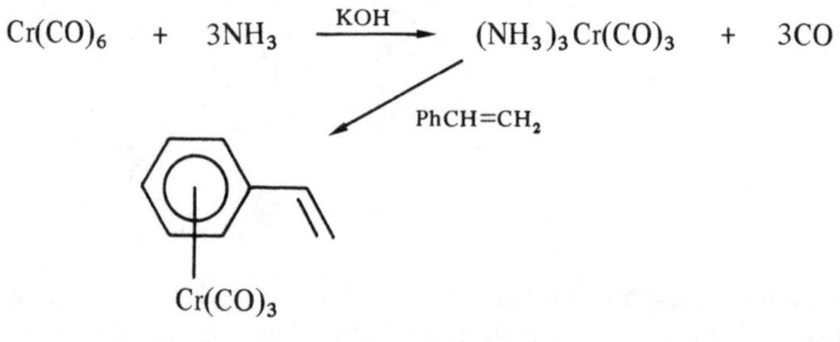

$$Cr(CO)_6 \;+\; 3NH_3 \;\xrightarrow{KOH}\; (NH_3)_3Cr(CO)_3 \;+\; 3CO$$

$$\Big\downarrow \text{PhCH=CH}_2$$

[styrene–Cr(CO)$_3$ complex]

$Cr(CO)_3$

Scheme *xii*

Many of these additions are accompanied by reduction of the metal as in the conversion of rhodium(III) chloride to di-μ-chlorobis(η^4-1,5-cyclooctadiene)-dirhodium(I) (equation 2.76) in which ethanol is the reducing agent (Giordano and Crabtree, 1979).

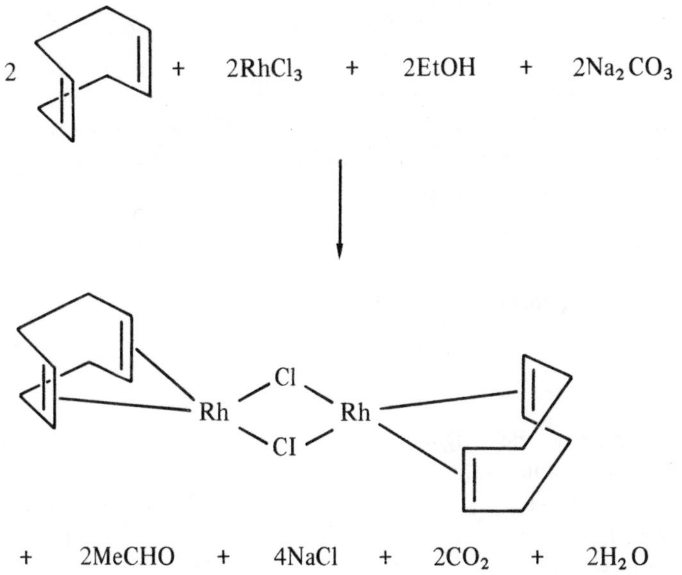

$$2 \quad + \quad 2RhCl_3 \quad + \quad 2EtOH \quad + \quad 2Na_2CO_3 \qquad\qquad (2.76)$$

$$+ \quad 2MeCHO \quad + \quad 4NaCl \quad + \quad 2CO_2 \quad + \quad 2H_2O$$

In the preparation of the dibenzenechromium cation, aluminium is used to reduce trivalent chromium to the monovalent state (equation 2.77).

$$3CrCl_3 + 2Al + AlCl_3 + 6C_6H_6 \longrightarrow 3[Cr(C_6H_6)_2][AlCl_4] \qquad (2.77)$$

Further reduction of this cation with dithionite gives dibenzenechromium(0) (equation (2.78); Fischer, 1960).

$$2[Cr(C_6H_6)_2]^+ + S_2O_4^{2-} + 4OH^- \longrightarrow 2Cr(C_6H_6)_2 + 2SO_3^{2-} + 2H_2O \quad (2.78)$$

Dibenzenechromium(0) can also be made by condensing together benzene and chromium vapour using the technique referred to on p. 15 (Timms and Turney, 1977). The metal vapour technique is the only successful procedure for preparing dinaphthalenechromium (Elschenbroich and Möckel, 1977).

Bis(cyclooctadiene)nickel, Ni(COD)$_2$, is a useful reagent in preparative organic chemistry; for example it is used in the coupling of aryl halides to give biaryls. It is prepared by the action of the diene on nickel-diacetylacetonate, the metal being reduced from Ni(II) to Ni(0) by triethylaluminium (equation (2.79); Bogdanovič *et al.*, 1966).

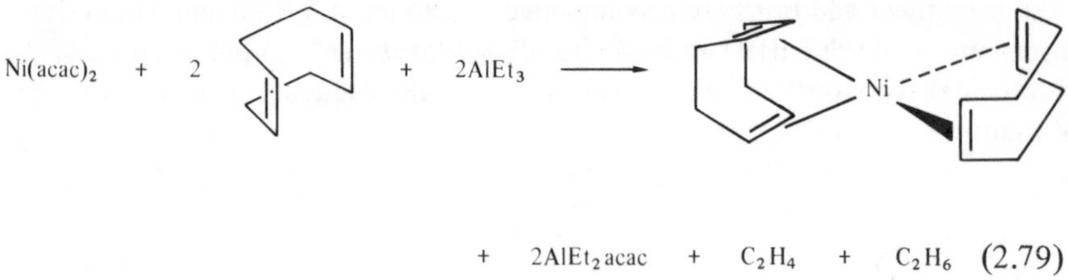

$$+ \quad 2AlEt_2acac \quad + \quad C_2H_4 \quad + \quad C_2H_6 \quad (2.79)$$

A general method of reducing the transition metal is to use an isopropyl Grignard reagent. Thus, irradiation with ultraviolet light of a mixture of isopropylmagnesium bromide, iron(III) chloride and cyclohexadiene gave the cyclohexadienebenzeneiron(0) complex (equation 2.80).

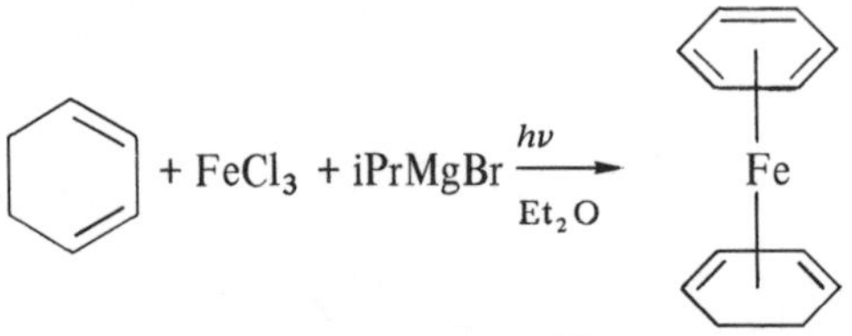

$$(2.80)$$

It is believed that, initially, the Grignard reagent alkylates the iron and this is followed by homolytic fission of the Fe–C bonds. This frees the coordination positions, allowing them to be occupied by the diene. The isopropyl radicals, presumably, abstract hydrogen atoms from the diene, leading to benzene formation (Fischer and Müller, 1962).

In a related reaction, ruthenium trichloride was treated with zinc dust and cycloocta-1,5-diene giving cycloocta-1,5-dienecycloocta-1,3,5-trieneruthenium(0) (equation 2.81).

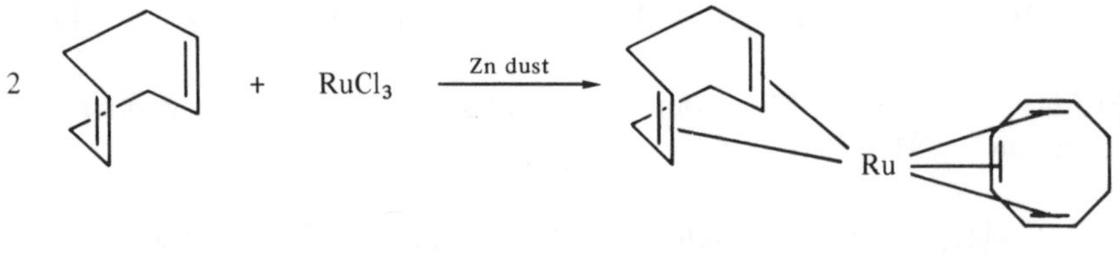

$$(2.81)$$

In this case cyclooctene was a by-product, indicating that the diene partly disproportionates to the mono- and tri-ene (Pertici *et al.*, 1980). Electrochemical reduction of the metal followed by addition to the unsaturated compound has also been used (section 2.5.1).

Examples of the use of transition metal compounds as catalysts in the hydrogenation of alkenes to alkanes are numerous but, recently, the reverse process has been demonstrated. An iridium compound, in the presence of an alkene as hydrogen acceptor, catalysed the dehydrogenation of cyclopentane to cyclopentadiene to give a product in which the cyclopentadienyl anion was complexed to the metal (equation (2.82); Crabtree *et al.*, 1982).

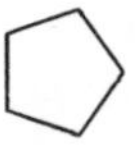

$$\text{(cyclopentane)} + [H_2Ir(PPh_3)_2S_2]^+ BF_4^- + 3tBuCH{=}CH_2 \xrightarrow{-2S} [(\eta^5\text{-}C_5H_5)IrH(PPh_3)_2]^+ BF_4^-$$

$$S = H_2O \text{ or } Me_2CO$$

$$+ 3tBuCH_2CH_3$$

$$(2.82)$$

Transition metal carbonyl compounds add to acetylenes, but these reactions are sometimes complicated by migration of CO groups from the metal to the organic residue. Dicobalt octacarbonyl reacts with diphenylacetylene to give a binuclear complex bridged by the acetylene (equation (2.83); Dickson and Fraser, 1974).

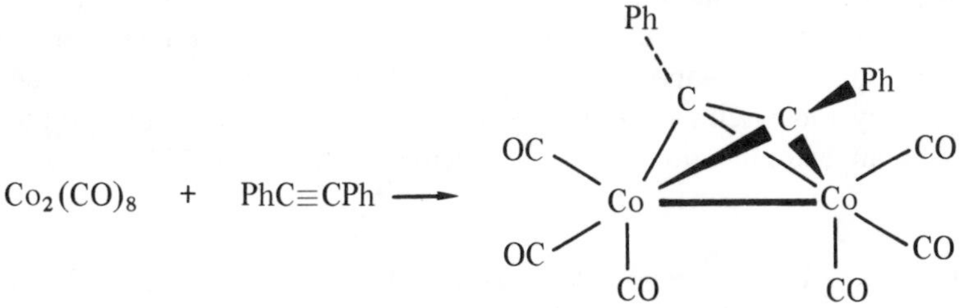

$$Co_2(CO)_8 \quad + \quad PhC{\equiv}CPh \longrightarrow$$

$$(2.83)$$

Carbonyl group migration is more common with iron compounds; for example, iron pentacarbonyl and diphenylacetylene react to give one product arising from ligand dimerisation and one from dimerisation and carbonylation (equation 2.84).

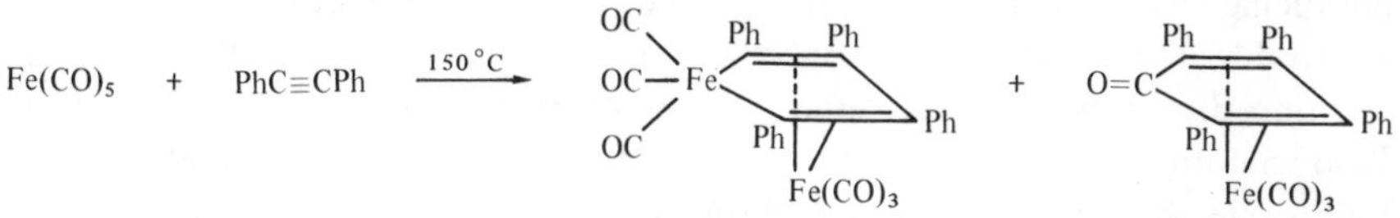

$$Fe(CO)_5 \quad + \quad PhC{\equiv}CPh \xrightarrow{150\,^\circ C}$$

$$(2.84)$$

2.5 ELECTROCHEMICAL METHODS

While a small number of organometallic compounds are prepared on an industrial scale using electrochemical processes, these methods are currrently of minor importance though they have considerable potential (Lehmkuhl, 1973).

2.5.1 Cathodic Processes

Electrolyses of solutions of organic halides are a special case of the direct reaction in which loss of electrons from the metal is aided by making it the cathode in a cell (equation 2.85).

$$nRX + M + ne^- \longrightarrow R_nM + nX^- \qquad (2.85)$$

This method has been used to prepare alkyl derivatives of lead, thallium and mercury (Tedoradze, 1975). Although the commercial process for the preparation of tetramethyl- and tetraethyl-lead is anodic (see below) there has, nevertheless, been considerable interest in the use of cathodic processes for the preparation of these compounds (Mengoli, 1979). Yields of tetraethyllead approaching 100 per cent have been obtained using a propylene carbonate solution of ethyl bromide containing tetraethylammonium salts as electrolyte and a lead cathode. There are a few examples of the preparation of mercury and lead alkyls by electrolysis of aqueous acid solutions of aldehydes and ketones, but the yields are poor. A remarkable preparation is that of *tetrakis*(2-cyanoethyl)tin in 60 per cent yield by electrolysis of acrylonitrile in aqueous alkali at a tin cathode. Attempts to extend this preparation using other nitriles and other metals have met with only limited success. As exemplified in equation (2.86), electrochemical reduction of certain transition metal compounds in an electrolyte containing unsaturated ligands such as cyclooctatetraene (COT) or cycloocta-1,5-diene, etc., leads to the formation of complexes in moderate to good yields (Lehmkuhl, 1973).

$$ZrCl_4 + 2e^- + COT \xrightarrow{\text{THF}} Cl_2 ZrCOT.THF + 2Cl^- \qquad (2.86)$$

2.5.2 Anodic Processes

The Nalco Chemical Company in the United States has, for many years, been producing tetramethyl- and tetraethyl-lead electrochemically using sacrificial lead electrodes. The electrolyte consists of a solution of methyl (or ethyl) magnesium chloride dissolved in dibutyl ether (or some other relatively involatile ether). Product formation occurs by loss of an electron from the alkyl carbanion at the lead anode, the radical thus generated then attacks the metal. Magnesium deposited at the cathode reacts with the alkyl chloride present in the electrolyte to reform the Grignard reagent. Numerous variants of the basic process have been proposed, these include replacement of the Grignard reagent by other carbanion sources, such as sodium tetraethylaluminate, organozinc compounds, or organoboron complexes.

As with the cathodic reactions, examples of electrochemical synthesis of organic derivatives of transition metals are rare, but we note that ferrocene has been produced by electrolysis of a solution of cyclopentadienylthallium in dimethylformamide using an iron anode (equation (2.87); Valcher and Aluni, 1968).

$$Fe + 2TlC_5H_5 \longrightarrow (\eta^5\text{-}C_5H_5)_2Fe + 2Tl^+ + 2e^- \qquad (2.87)$$

2.6 MISCELLANEOUS METHODS

σ-Aryl derivatives of certain metals and metalloids can be made from arenediazonium salts (Wulfman, 1978). Thus, treatment of an arenediazonium chloride with mercury(II) chloride gives the diazonium trichloromercurate which decomposes in the presence of copper powder to give the arylmercury chloride. If excess copper in the presence of aqueous ammonia is used the product is the diarylmercury (scheme *xiii*; Larock, 1976). Where Ar is a substituted phenyl

$$ArN_2^+Cl^- \; + \; HgCl_2 \longrightarrow ArN_2^+HgCl_3^-$$

$$\overset{2Cu}{\diagup} \qquad \overset{3Cu,NH_3}{\diagdown}$$

$$ArHgCl \; + \; 2CuCl \; + \; N_2 \qquad\qquad \tfrac{1}{2}Ar_2Hg \; + \; \tfrac{1}{2}Hg \; + \; 3CuCl \; + \; N_2$$

Scheme xiii

group, this method has the advantage over the mercuration reaction (p. 27) in that a single regioisomer is produced.

Perhaps the widest use of this method is in Group V, since it has been known for many years that addition of a diazonium salt solution to aqueous alkaline sodium arsenite gives the arylarsonic acid (equation 2.88).

$$ArN_2Cl + Na_3AsO_3 \xrightarrow[\text{2.H}^+]{\text{1.NaOH}} ArAsO_3H_2 \qquad (2.88)$$

Where aqueous conditions are undesirable, it is possible to carry out the diazotisation in absolute ethanol using amyl nitrite and dry hydrogen chloride. The complex salt, ArN_2AsCl_4, obtained by addition of arsenic trichloride is subsequently decomposed by the addition of copper powder. Either version of this reaction can be modified to yield a diarylarsinic acid by decomposing the diazonium salt in the presence of arylarsenic(III) species (equation 2.89).

$$ArN_2Cl + Ar'AsO_2Na_2 \xrightarrow[\text{2.H}^+]{\text{1.NaOH}} ArAr'AsO_2H \qquad (2.89)$$

The corresponding aryl-antimony and -bismuth compounds can be made similarly.

In this method of preparing Group V compounds, the metal becomes oxidised from the tri- to the penta-valent state. It may, therefore, seem reasonable to extend it to Group IV by treating, say, benzenediazonium chloride with tin(II) chloride until one recalls that this causes reduction to phenylhydrazine. However, some success in preparing diaryltin dichlorides has been obtained by decomposing diazonium hexachlorostannates with zinc (or tin) powder (scheme *xiv*).

$$2ArN_2Cl + SnCl_4 \longrightarrow [ArN_2]_2^+ [SnCl_6]^{2-}$$
$$\Big\downarrow 2Zn$$
$$Ar_2SnCl_2 + 2ZnCl_2 + 2N_2$$

Scheme *xiv*

An interesting transition metal reaction which suggests that this method, with suitable modifications, may have wider applications, is the formation of chloro-*bis*(triethylphosphine)phenylplatinum(II). The benzenediazonium salt was treated with a platinum hydride when addition occurred to give, after deprotonation, a phenylazoplatinum compound which decomposed in the presence of alumina, giving the phenylplatinum species and nitrogen (scheme *xv*; Parshall, 1965).

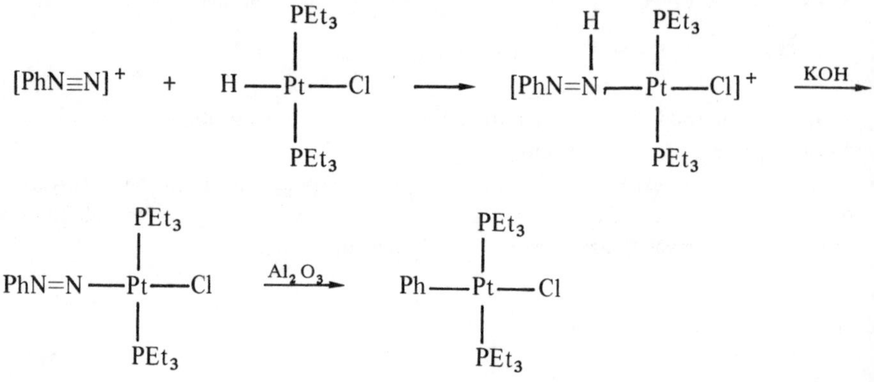

$$[PhN{\equiv}N]^+ \quad + \quad H-\underset{\underset{PEt_3}{|}}{\overset{\overset{PEt_3}{|}}{Pt}}-Cl \quad \longrightarrow \quad [PhN{=}N{-}\underset{\underset{PEt_3}{|}}{\overset{\overset{H \quad PEt_3}{|}}{Pt}}-Cl]^+ \quad \xrightarrow{KOH}$$

$$PhN{=}N-\underset{\underset{PEt_3}{|}}{\overset{\overset{PEt_3}{|}}{Pt}}-Cl \quad \xrightarrow{Al_2O_3} \quad Ph-\underset{\underset{PEt_3}{|}}{\overset{\overset{PEt_3}{|}}{Pt}}-Cl$$

Scheme *xv*

Finally, two methods which can be described as insertion and extrusion reactions are useful in certain circumstances. The insertion of carbenes into metal–halogen bonds gives α-haloalkylmetal compounds; two examples are shown in equations (2.90) (Barluenga *et al.*, 1979) and (2.91) (Yakubovich *et al.*, 1952).

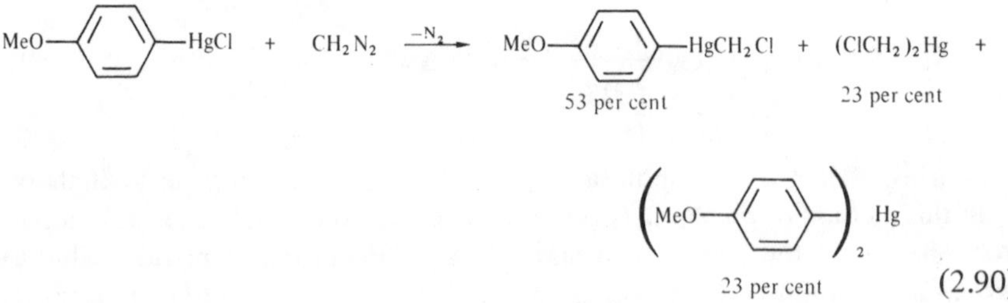

$$MeO{-}\langle\!\!\!\bigcirc\!\!\!\rangle{-}HgCl \;+\; CH_2N_2 \;\xrightarrow{-N_2}\; MeO{-}\langle\!\!\!\bigcirc\!\!\!\rangle{-}HgCH_2Cl \;+\; (ClCH_2)_2Hg \;+$$

53 per cent   23 per cent

$$\left(MeO{-}\langle\!\!\!\bigcirc\!\!\!\rangle{-} \right)_2 Hg$$

23 per cent   (2.90)

$$SnBr_4 + 3CH_2N_2 \longrightarrow BrSn(CH_2Br)_3 + 3N_2 \qquad (2.91)$$

(mainly)

The general form of the extrusion reaction is shown in equation (2.92), where X is usually carbon dioxide.

$$M{-}X{-}C \longrightarrow M{-}C + X \tag{2.92}$$

Thus certain metallic salts of carboxylic acids, particularly those containing electron withdrawing substituents, give carbon–metal bonds on heating, as exemplified in equations (2.93) and (2.94) (Deacon, 1970).

$$MeHgOCOC_6F_5 \xrightarrow[\text{pyridine}]{\text{Heat}} MeHgC_6F_5 + CO_2 \tag{2.93}$$

$$Bu_3SnOCOC{\equiv}CPh \longrightarrow Bu_3SnC{\equiv}CPh + CO_2 \tag{2.94}$$

The last example (equation 2.95) shows the preparation of an allyltin compound by extrusion of a ketone from the alkyltin derivate of an allyldialkylcarbinol (Gambaro et al., 1981).

$$Bu_3SnOCR_2CH_2CH{=}CH_2 \longrightarrow Bu_3SnCH_2CH{=}CH_2 + R_2CO \tag{2.95}$$

REFERENCES

Adkins, H., and Scanley, C. (1951). J. Am. chem. Soc. 73, 2854

Ashby, E. C., and Ainslie, R. D. (1983). J. organomet. Chem. 250, 1

Ball, M. C., and Norbury, A. H. (1974). Physical Data for Inorganic Chemists. Longman, London, p. 118

Barluenga, J., Campos, P. J., Garcia-Martin, J. C., Roy, M. A., and Asensio, G. (1979). Synthesis 893

Beak, P., and Snieckus, V. (1982). Acc. chem. Res. 15, 306

Birchall, J. M., Hazard, R., Haszeldine, R. N., and Wakalski, W. W. (1967). J. chem. Soc. (C) 47

Blackborow, J. R., and Young, D. (1979). Metal Vapour Synthesis in Organometallic Chemistry. Springer-Verlag, Berlin

Bogdanovič, B., Kröner, M., and Wilke, G. (1966). Annalen 699, 1

Bruce, M. I. (1977). Angew Chem. int. Edn Engl. 16, 73

Bulten, E. J. (1975). J. organomet. Chem. 97, 167

Chatt, J., Vallarino, L. M., and Venanzi, L. M. (1957). J. chem. Soc. 3413

Chock, P. B., Halpern, J., and Paulik, F. E. (1973). Inorg. Synth. 14, 90

Collman, J. P., and Roper, W. R. (1968). Adv. organomet. Chem. 7, 54

Crabtree, R. H., Mellea, M. F., Mihelcic, J. M., and Quirk, J. M. (1982). J. Am. chem. Soc. 104, 107

Davidson, P. J., Lappert, M. F., and Pearce, R. (1976). Chem. Rev. 76, 219

Deacon, J. B. (1970). Organomet. Chem. Revs A 5, 355

Dickson, R. S., and Fraser, P. J. (1974). Adv. organomet. Chem. 12, 323

Eisch, J. J. (1981). Organometallic Synthesis, Vol. 2. Academic Press, New York

Elschenbroich, C., and Möckel, R. (1977). Angew. Chem. int. Edn Engl. 16, 870

Fischer, E. O. (1960). *Inorg. Synth.* **6**, 132

Fischer, E. O. (1976). *Adv. organomet. Chem.* **14**, 1

Fischer, E. O., and Müller, J. (1962). *Z. Naturforsch. B* **17**, 776

Fischer, E. O., Kreissl, F. R., Winkler, E., and Kreiter, C. G. (1972). *Chem. Ber.* **105**, 588

Fischer, E. O., Lindner, T. L., Huttner, G., Friedrich, P., Kreissl, F. R., and Besenhard, J. O. (1977). *Chem. Ber.* **110**, 3397

Gambaro, A., Marton, D., Peruzzo, V., and Tagliavini, G. (1981). *J. organomet. Chem.* **204**, 191

Giordano, G., and Crabtree, R. H. (1979). *Inorg. Synth.* **19**, 218

Halasa, A. F., Schulz, D. N., Tate, D. P. and Mochel, V. D. (1980). *Adv. organomet. Chem.* **18**, 55

Harrod, J. F., and Chalk, A. J. (1977). In *Organic Synthesis via Metal Carbonyls*, Vol. 2 (I. Wender and P. Pino, eds). Wiley, New York, p. 673

Jones, R. G., and Gilman, H. (1951). In *Organic Reactions*, Vol. VI (R. Adams, ed.). Wiley, New York, p. 339

Kawabata, N., Matsumura, A., and Yamashita, S. (1973). *Tetrahedron* **29**, 1069

Kealy, T. J., and Pauson, P. L. (1951). *Nature (Lond.)* **168**, 1039

Ketley, A. D., and Braatz, J. (1968). *Chem. Commun.* 169

King, R. B. (1964). *Adv. organomet. Chem.* **2**, 157

Koerner von Gustorf, E., Pfajfer, Z., and Grevels, F.-W. (1971). *Z. Naturforsch. B* **26**, 66

Kramer, G. W., and Brown, H. C. (1974). *J. organomet. Chem.* **73**, 1

Krief, A. (1980). *Tetrahedron* **36**, 2596

Langer, A. W. (ed.) (1974). *Polyamine-chelated Alkali Metal Compounds*, Advances in Chemistry Series, no. 130. American Chemical Society, Washington, D.C.

Larock, R. C. (1976). *J. organomet. Chem. Library* **1**, 257

Lehmkuhl, H. (1973). *Synthesis*, 377

Lindley, P. F., and Woodward, P. (1966). *J. chem. Soc. (A)* 123

Luijten, J. G. A., and van der Kerk, G. J. M. (1964). *Rec. Trav. Chim. Pays-Bas* **83**, 295

Mason, R., Textor, M., Al-Salem, N., and Shaw, B. L. (1976). *J. chem. Soc. chem. Commun.* 292

Mengoli, G. (1979). *Rev. Silicon, Germanium, Tin Lead Compounds* **4**, 59

Moser, G. A., and Rausch, M. D. (1974). *Synth. React. inorg. Metal-org. Chem.* **4**, 37

Murphy, J. and Poller, R. C. (1980). *J. organomet. Chem. Library* **9**, 189

Negishi, E. I. (1980). *Organometallics in Organic Synthesis*, Vol. 1. Wiley, New York

Owen, D. W., and Poller, R. C. (1983). *J. organomet. Chem.* **255**, 173

Parham, W. E., and Bradsher, C. K. (1982). *Acc. chem. Res.* **15**, 300

Parshall, G. W. (1965). *J. Am. chem. Soc.* **87**, 2133

Pertici, P., Vitulli, G., Paci, M., and Porri, L. (1980). *J. chem. Soc. Dalton Trans.* 1961

Quinn, H. W., and Tsai, J. H. (1969). *Adv. inorg. Chem. Radiochem.* **12**, 217

Rieke, R. D. (1977). *Acc. chem. Res.* **10**, 301

Rochow, E. G. (1971). *Adv. organomet. Chem.* **9**, 1

Rulewicz, G., Trautner, K., and Thust, U. (1973). *Chem. Tech. (Leipzig)* **25**, 284

Semmelhack, M. F., and Tamura, R. (1983). *J. Am. chem. Soc.* **105**, 4099

Simons, L. H., Riley, P. E., Davis, R. E., and Lagowski, J. J. (1976). *J. Am. chem. Soc.* **98**, 1044

Skell, P. S., and McGlinchey, M. J. (1975). *Angew. Chem. int. Edn Engl.* **4**, 195

Slocum, D. W., and Jennings, C. A. (1976). *J. org. Chem.* **41**, 3653

Taylor, E. C., and McKillop, A. (1970). *Acc. chem. Res.* **3**, 338

Tedoradze, G. A. (1975). *J. organomet. Chem.* **88**, 1

Timms, P. L., and Turney, T. W. (1977). *Adv. organomet. Chem.* **15**, 53

Trost, B. M. (1980). *Acc. chem. Res.* **13**, 385

Valcher, S., and Aluni, E. (1968). *Ric. Sci.* **38**, 527

Wakefield, B. J. (1974). *The Chemistry of Organolithium Compounds*. Pergamon Press, Oxford

Wardell, J. E. (1983). In *Comprehensive Organometallic Chemistry*, Vol. 2 (G. Wilkinson, ed.). Pergamon Press, Oxford, p. 874

Wilke, G., Bogdanovič, B., Hardt, P., Heimbach, P., Keim, W., Kröner, M., Oberkirch, W., Tanaka, K., Steinrücke, E., Walter, D., and Zimmermann, H. (1966). *Angew. Chem. int. Edn Engl.* **5**, 151

Wilkinson, G. (1974). *Science* **185**, 109

Wittig, G., and Hellwinkel, D. (1964). *Chem. Ber.* **97**, 789

Wulfman, D. S. (1978). In *The Chemistry of Diazonium and Diazo Groups*, Part 1 (S. Patai, ed.). Wiley, Chichester, p. 296

Yakubovich, A. Ya., Makarov, S. P., and Gavrilov, G. I. (1952). *Zn. obshch. Khim.* **22**, 1788

Ziegler, K. (1968). *Adv. organomet. Chem.* **6**, 7

Zinin, N. (1855). *Annalen* **96**, 361

GENERAL READING

Blackborow, J. R., and Young, D. (1979). *Metal Vapour Synthesis in Organometallic Chemistry*. Springer-Verlag, Berlin

Eisch, J. J. (1981). *Organometallic Syntheses*, Vol. 2, *Nontransition-metal Compounds*. Academic Press, New York

Fischer, E. O., and Werner, H. (1966). *Metal π-Complexes*, Vol. 1. Elsevier, Amsterdam

King, R. B. (1965). *Organometallic Syntheses*, Vol. 1, *Transition-metal compounds*. Academic Press, New York

3

STRUCTURE AND BONDING
IN ORGANOMETALLIC
COMPOUNDS

3.1 σ-BONDED COMPOUNDS

The structure of alkyl derivatives of metals is influenced by similar considerations as apply to other σ-bonded systems (such as hydrides). The electronic configuration of the central atom and the presence or absence of lone pairs are the most important. In addition to these simple considerations, which apply to all compounds, two additional factors need to be considered. These are the possible occurrence of bridging by the organic groups between metal atoms, and metal–metal bonding. The bridges are important from a theoretical point of view as sometimes the carbon atoms in the bridge are bonded to five other atoms. The metal–metal bonded compounds are particularly numerous in the transition metal block and often have multiple bonds between the metal atoms.

Surveying the periodic table, we can expect σ-compounds of Group IVb to be the most straightforward. This is because the usual valency is four, and four covalent bonds give an octet of electrons, whereas Group IIIb compounds will be electron-deficient if monomeric. In Group IIa the problem of electron deficiency is even more severe if we think of trying to attain the electronic configuration of the next noble gas. It is in this group that the solvent frequently plays a vital role in stabilising the organometallic compounds. We shall now describe the structure and bonding of σ-bonded organic derivatives of each group in the periodic table.

3.1.1 Group I Derivatives

The most important compounds in this group are the organic derivatives of lithium. The compounds are generally employed in solution and therefore are not often encountered in a pure state. The Li–C bond is a strongly polarised covalent bond, so that the compounds serve as sources of anionic carbon (see chapter 4). The compounds differ considerably in structure, depending on the nature of the R group. Butyllithium is a liquid at room temperature, but methyl- and ethyl-lithium are solids. The X-ray structures of the latter two compounds have been carried out and both show tetrameric units $(LiCH_3)_4$ and $(LiCH_2CH_3)_4$.

In methyllithium the structure is made up of two interpenetrating tetrahedral arrays, so forming a distorted C_4Li_4 cube, I (Weiss and Lucken, 1964). Each

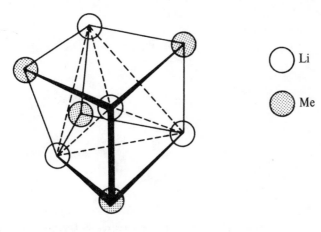

$\bigcirc$ Li

$\bullet$ Me

I

carbon atom is bonded to three lithium atoms on a triangular face of the tetrahedron. The Li—C distance is 2.31 Å and the Li—C—Li angle has the low value of 68°.† The Li—Li distance is 2.56 Å ($2 \times r_{cov}$ = 2.46 Å) and the C—C distance is 3.68 Å. Ethyllithium also contains tetrahedra of lithium atoms, but both structures are complicated by the relatively close Li···C distances between adjacent $(LiR)_4$ units (Dietrich, 1963). Ab initio molecular orbital calculations (Guest et al., 1972) suggest some direct Li—Li bonding within the tetrahedra, but the distance between the metal atoms and detailed studies of the vibrational spectra of $(tBuLi)_4$ indicate that this bonding is weak (Scovell et al., 1971).

The relatively complicated solid state structure is also found in solution. n-Alkyllithium compounds usually exist as hexamers in hydrocarbon solvents. In donor solvents the degree of association is reduced from six to four, and CH_3Li is tetrameric in THF. When complexing agents are added, the reagents often become more reactive. An x-ray structure of the adduct of PhLi and TMED showed a dimeric unit with bridging phenyl groups (Thoennes and Weiss, (1978), in contrast to the tetrameric structure of PhLi which crystallises from diethyl ether (Hope and Power, 1983). It seems quite probable that the increase in reactivity brought about by complexing agents is due to partial depolymerisation, but there are other effects that should be considered (Wakefield, 1974).

The alkyl derivatives of sodium, potassium, rubidium and caesium are much more ionic in character than their lithium analogues. X-ray studies (Weiss and Köster, 1977) show that KCH_3, $RbCH_3$ and $CsCH_3$ have the NiAs structure, although tetrameric units are present in $NaCH_3$.

There is some similarity between the alkyl derivatives of Cu(I) and those of Li. However, metal—metal bonding is more significant with copper. In planar $[CuCH_2SiMe_3]_4$, II, the Cu—Cu distance is only 2.42 Å, which is shorter than the Li—Li distance in $(MeLi)_4$ despite the larger atomic radius of copper (Jarvis et al., 1973).

II

---

†Throughout this text bond distances have been given in ångström units, these being commonly accepted in structural chemistry although not SI units; $1\,\text{Å} = 10^{-10}$ m.

In $[(dppe)_2Cu]$ $[(mesityl)_2Cu]$ the diarylcopper unit is linear (Leoni *et al.*, 1983). Gold alkyls adopt different geometries depending on the oxidation state. Au(I) alkyls, which are often stabilised by phosphines, are linear as in **III** (Gavens *et al.*, 1971). However, Au(III) alkyls, which again require an additional ligand for stabilisation are square planar, as in **IV** (Baker and Pauling, 1969). There are

$$CH_3 —Au \longleftarrow PPh_3$$

$$\begin{array}{c} F_5C_6 \diagdown \qquad \diagup C_6F_5 \\ Au \\ Cl \diagup \qquad \diagdown PPh_3 \end{array}$$

III   **IV**

also many polynuclear compounds containing Au–C bonds, and some of these also contain direct Au–Au bonds.

3.1.2 Group II Derivatives

Grignard reagents formally represented by RMgX are the most important organic derivatives of Group IIa. They have been extensively studied (Ashby *et al.*, 1974) and the structures are complicated. In 1929 Schlenk and Schlenk showed that if dioxan was added to a solution of a Grignard reagent in diethyl ether, most of the halide was precipitated as the dioxan complex of MgX_2. A species such as R_2Mg or RMgX has only four electrons in its valence shell and we can expect both association and/or solvation to occur. The increase in coordination number from two to four reduces the residual charge on the magnesium atom, which carries a partial positive charge, because of polarisation of the Mg–C bond.

A large volume of research work has demonstrated that, ignoring ionic forms, most common Grignard reagents can be represented by scheme *i*. The mono-

Scheme *i*

meric structure is favoured when X = Br or I and R = branched alkyl (for example tBu) and by low concentrations in powerfully donating solvents. X-ray structure determinations have been carried out on a few solvated or complexed Grignard reagents. The configuration at magnesium in $EtMgBr.2Et_2O$ is tetrahedral

(Guggenberger and Rundle, 1968). Similar equilibria occur in solutions of organozinc halides.

Since bridging can occur through carbon as well as halogen we will expect some association in dialkylmagnesium compounds. In fact, both Me_2Mg and Me_2Be have chain structures with methyl groups bridging the metal atoms, **V** (Weiss, 1964).

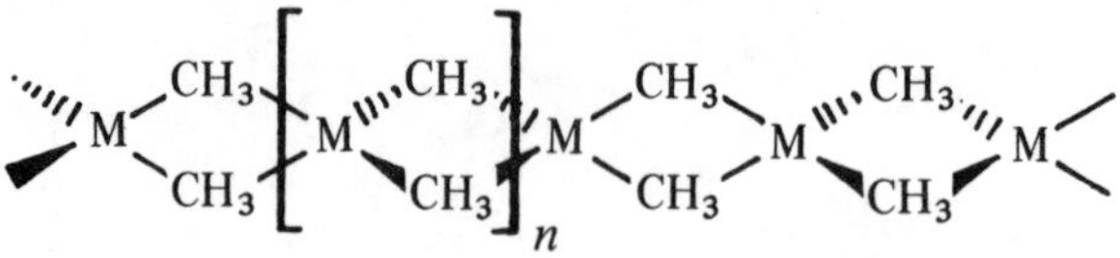

V, M = Be, Mg

The metal atoms are tetrahedral and bonded to the carbon atoms through three-centre bonds. Chelating ligands break up the chain of Me_2Mg and $Me_2MgTMED$ is monomeric, again with tetrahedral coordination at magnesium (Greiser et al., 1980).

In contrast to Me_2Be and Me_2Mg, the dimethyl derivatives of zinc, cadmium and mercury are liquids at room temperature. Studies on the vapour phase indicate discrete linear molecules, and an electron diffraction study of Me_2Hg gave an Hg–C distance of 2.083 Å.

3.1.3 Group III Derivatives

The monomeric trialkyls of this group, MR_3, have only six electrons in their valency shell. The electron deficiency confers Lewis acidity on the compounds, and this has been thoroughly investigated (Jensen, 1978). The tendency for the metals to attain an octet is so strong that some of the trialkyls exist as dimers, such as $Al_2(CH_3)_6$ with bridging alkyl groups. Diborane, B_2H_6 (**VI**), also has a dimeric structure and the molecular orbital description of its structure in terms of three-centre orbitals has served as a model for dimeric Group III alkyls.

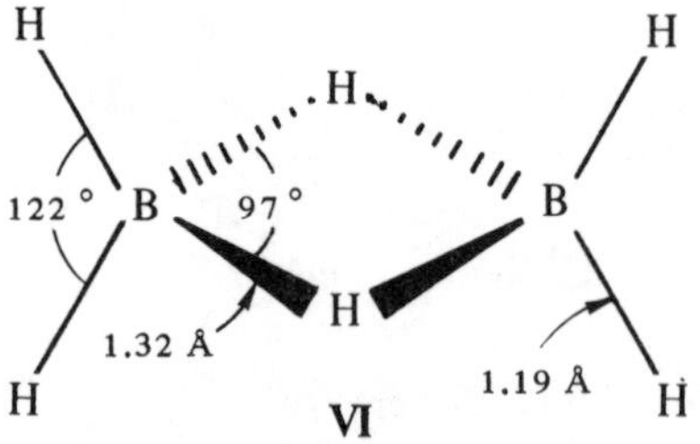

Although B_2H_6 and $Al_2(CH_3)_6$ are dimeric with roughly tetrahedral geometry about the Group III atom, trimethylboron is a planar monomer. The reason for the difference between boron and aluminium is often said to be related to the steric requirements of the four-membered ring and the larger size of aluminium.

Such arguments do not bear close examination, not least because the heavier elements gallium and indium both form monomeric trimethyl derivatives; hence it appears that a number of factors are involved (Muller and Otermat, 1965).

It is interesting to note the ligand found in the bridging position in the following mixed species, **VII–IX**; clearly methyl bridging does not occur when other

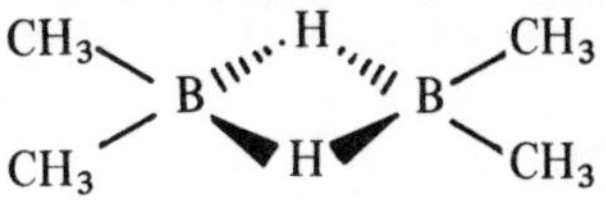

VII

VIII

IX

bridging groups are present. The x-ray structure of Al_2Me_6 has been determined at low temperature and this allowed the hydrogen atoms on the bridging methyl groups to be located (**X**; Huffman and Streib, 1971). The study revealed that these hydrogen atoms are not involved in direct interaction with the aluminium

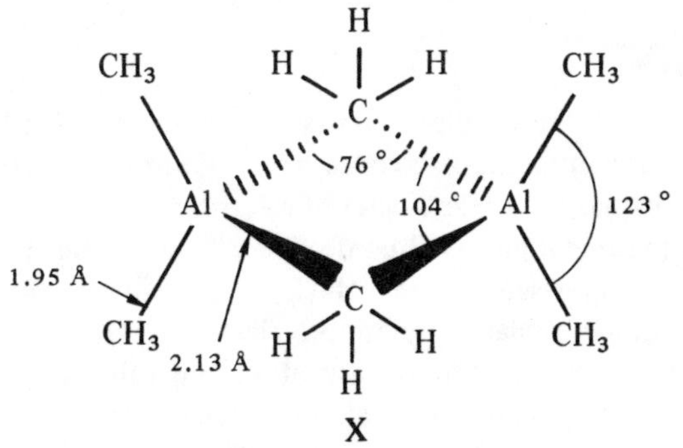

X

atoms. Although the bonding in B_2H_6 and Al_2Me_6 is similar in principle, there are differences. Molecular orbital calculations suggest that the d-orbitals contribute to some degree in the metal–metal bonding of the aluminium compounds. If the bridging group is a halogen rather than alkyl, the additional electrons provide electron density in an orbital which is antibonding with respect to the metal–metal bond. This explains the large difference in the Al–Al distance in Al_2Me_6 (2.61 Å) and Al_2Cl_6 (3.40 Å) (Mason and Mingos, 1973).

Although hydrogen and methyl can act as bridging groups in dimeric molecules, the association is reduced with higher normal or branched chain alkyl groups. Both triisopropylaluminium and tribenzylaluminium are monomeric in benzene solutions. However, unsaturation at the organic group increases the tendency towards association. With vinyl groups and acetylenic groups the tendency to association also extends to gallium (Oliver, 1977). Triphenylaluminium is dimeric in benzene and in the solid state. The X-ray structure of the dimethylphenylaluminium showed a dimer, **XI**, with the bridging phenyl groups, participating in spiro-type ring junctions.

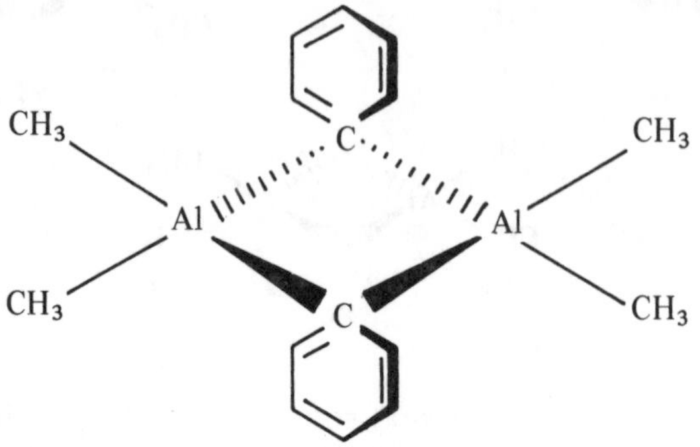

XI

3.1.4 Group IV Derivatives

The Group IV alkyls are the simplest case, so far as the relationship between structure and electronic configuration is concerned. A Group IVb element bonded to four organic groups has a complete octet of electrons and so will have a tetrahedral geometry at the metal atom. Thus Me_4Si molecules exist as tetrahedra in solid, liquid, solution and gaseous states. Me_4Sn is also a tetrahedral molecule, but if one methyl group is replaced by a more electronegative chlorine atom, a small positive charge is generated at the tin atom and this, coupled with the tendency of chlorine to bridge, gives an intermolecular interaction and so develops a chain structure in the solid with penta-coordination at tin, **XII** (Lefferts et al., 1982). The chain structure is relatively easy to break up, and in

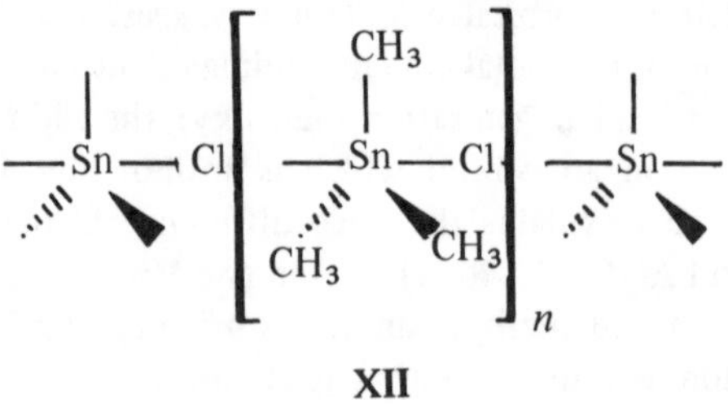

XII

dilute solutions in non-polar solvents free molecules are obtained. The association is stronger in the involatile Me_3SnF and weaker in Me_3SnBr and Me_3SnI, as expected from electronegativity considerations (Zubeita and Zuckerman, 1978). The Group IVb elements form compounds with metal–metal bonds. In $Ph_3Sn-SnPh_3$ the Sn–Sn distance is 2.77 Å (Preut et al., 1973). A special group of compounds of great theoretical interest are the dinuclear compounds $R_2Sn{\cdot}{\cdot}{\cdot}SnR_2$. Although superficially analogous to ethene derivatives the substituents are not coplanar and the Sn–Sn bond length shows no shortening compared to $Ph_3Sn-SnPh_3$ (Davidson et al., 1976).

Recently compounds containing a Si=C double bond have been prepared and characterised. The Si=C distance of 1.76 Å is 0.13 Å shorter than a Si–C single bond (Brook et al., 1982).

The tetraalkyl derivatives of titanium, zirconium and hafnium also possess tetrahedral geometry. The x-ray structure of tetrabenzylzirconium shows slight distortions from the expected tetrahedral angle at zirconium, but the angle at the α-carbon is only 91° instead of the expected value of 109°28' (Davies et al., 1971). There may be some π-bonding between the phenyl ring and the zirconium. In $Sn(benzyl)_4$ the corresponding angle is 111°, and in $Ti(benzyl)_4$ it is 103°.

3.1.5 Group V Derivatives

With this group we find alkyl compounds in which the oxidation state is less than the group number of 5. Antimony and bismuth form organic derivatives of composition MR_3 and MR_5. The trialkyls are pyramidal (Poller, 1979). The MR_5 species might be expected to take up a trigonal bipyramidal structure, and this is found in $Sb(p\text{-tolyl})_5$, **XIII**, but $Sb(Ph)_5$, **XIV**, is square pyramidal in the solid state, indicating that the energy difference between these geometries is very small (Brock and Ibers, 1975). The Group V transition metals form alkyls in the

XIII
trigonal bipyramid

XIV
square pyramid

+5 and lower oxidation states. $TaMe_5$ is a solid which sometimes explodes and so has not been examined by x-ray diffraction, but Me_3TaCl_2(bipy) is a capped trigonal prism (Drew and Wilkins, 1973).

A compound of vanadium with 2,6-dimethoxyphenyl ligands, **XV**, is particularly interesting, in that it contains what is apparently square planar carbon (Cotton and Millar, 1977). The compound also has a metal–metal bond, a

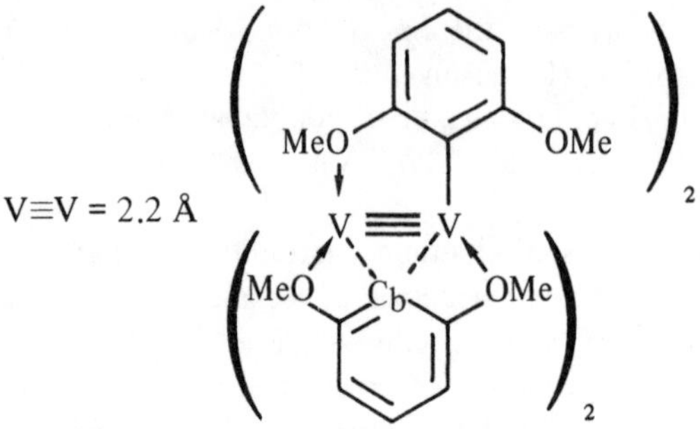

$$V \equiv V = 2.2 \text{ Å}$$

XV

feature which becomes increasingly common in Groups VI and VII. The bridging carbon atom C_b is apparently bonded to each V atom and to two carbon atoms in the phenyl ring. These five atoms are roughly planar.

3.1.6 Group VI Derivatives

Alkyls are known in several oxidation states. *Tetrakis*(2-methyl-2-phenylpropyl)-chromium, **XVI**, has been examined by x-ray crystallography. The ligands are arranged in a distorted tetrahedron around the chromium atom (Gramlich and

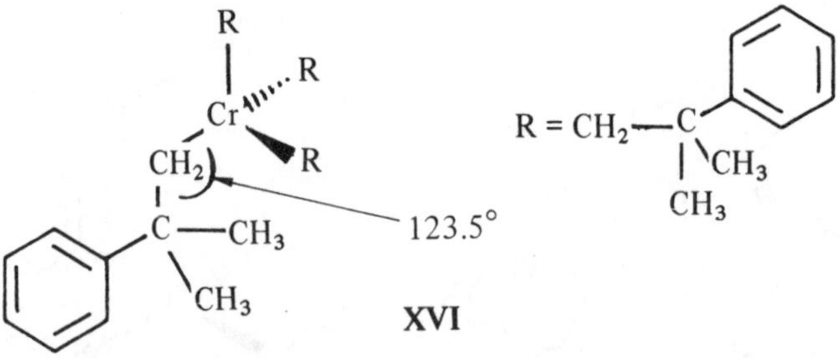

$$R = CH_2 - C \begin{matrix} \\ | \\ CH_3 \end{matrix} \begin{matrix} CH_3 \\ \\ \end{matrix}$$

XVI

Pfefferkorn, 1973). The angle at the methylene group is 123.5°, which is much larger than the tetrahedral angle, and distorted in the reverse sense compared with $Zr(benzyl)_4$. Me_6W is thought to be octahedral. Group VI has many examples of metal–metal bonding, and the dimethoxyphenyl derivative of chromium, **XVII**, has the lowest formal shortness ratio (formal shortness ratio = observed bond length ÷ sum of covalent radii) known (Cotton, 1978).

$$Cr \equiv Cr = 1.847 \text{ Å}$$

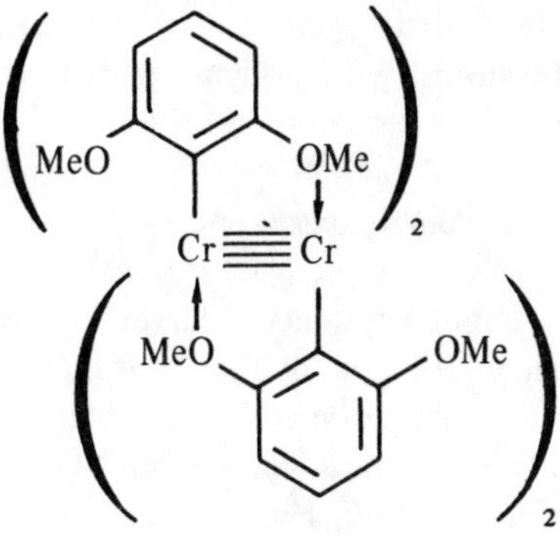

XVII

3.1.7 Group VII Derivatives

Rhenium σ-bonded derivatives have been studied by x-ray diffraction, which has revealed structures with one, two and three rhenium atoms.

In $(Et_2PhP)_2 RePh_3$, **XVIII**, the phosphine ligands take up axial and the phenyl

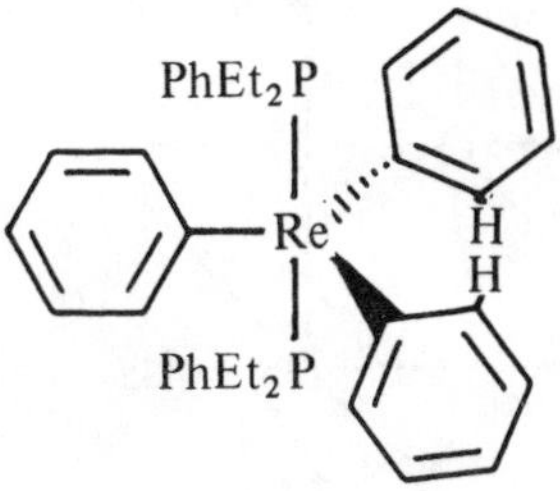

XVIII

groups equatorial positions in a trigonal bipyramid. The three phenyl groups are almost co-planar, complete planarity being prevented by steric interactions of the *ortho* hydrogen atoms (Carrol and Bau, 1978). The $Me_8 Re_2^{2-}$ ion, **XIX**, has the same geometry as the well-known $Re_2Cl_8^{2-}$ ion, but in the methyl compound the quadruple $Re \equiv Re$ bond has a length of only 2.178 Å, compared with 2.24 Å

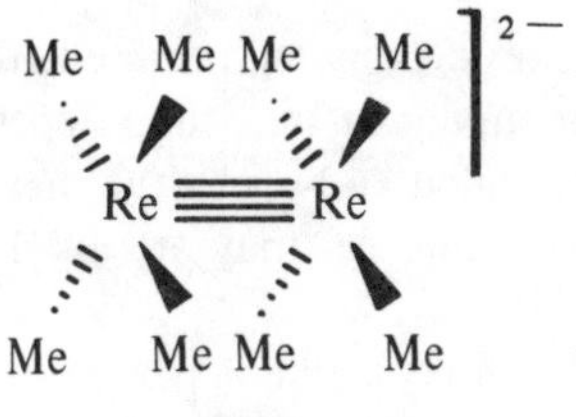

XIX

in the chloride (Cotton *et al.*, 1976). The compound $(Et_2PhP)_2Re_3Me_9$ has a triangle of Re atoms, with bridging methyl groups in the Re_3 plane (Edwards *et al.*, 1980).

3.1.8 Group VIII Derivatives

We shall not attempt a systematic survey of this group of nine elements. There is a large number of compounds that contain σ-bonded organic groups, but these are generally in combination with other ligands, especially phosphines, carbonyls or cyclopentadienyl groups. Earlier reports of the structure of $[Me_4Pt]_4$ have been shown (Cowan *et al.*, 1968) to be erroneous and the material investigated was the hydroxide, **XXa**, which has a structure similar to that of $[Me_3PtI]_4$, **XXb**.

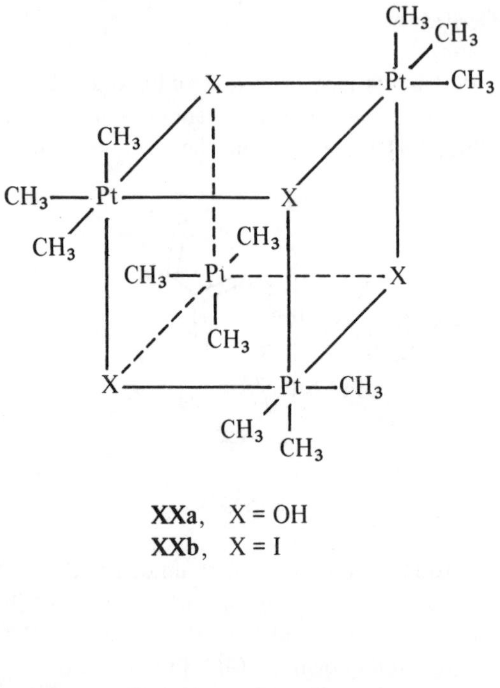

XXa, X = OH
XXb, X = I

3.2 CARBENE AND CARBYNE COMPLEXES

Carbene and carbyne complexes occupy an intermediate position between σ- and π-bonded compounds in our survey of structural types. They are σ in the sense that the carbon chains are bonded end-on to the metal atom, but the bonding does involve some M—C π-bonding, i.e. they are alkyl complexes with multiple M—C bonds.

The structure of phenyl(methoxy)carbene pentacarbonylchromium, **XXI**, was investigated by Mills and Redhouse (1968). The geometry at chromium is roughly

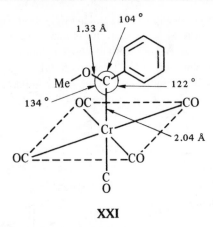

XXI

octahedral and the atoms attached to the carbene carbon atom are nearly co-planar with angles as shown. This plane bisects the OC–Cr–CO angle in the equatorial plane of the octahedron. The Cr–C carbene distance is 2.04 Å compared with a calculated value of 2.21 Å for a Cr(0)–C single bond and 1.91 Å for the double Cr=C bond in $Cr(CO)_6$. The $C_{carbene}$–O bond distance is 1.33 Å, midway between the values found in diethyl ether (1.43 Å) and acetone (1.23 Å). These bond distances, which show partial double bond character in both C–Cr and C–O, are often represented by writing the structure as **XXII**.

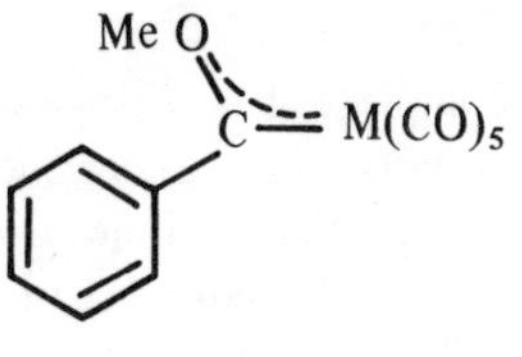

Me O
C $=$ $M(CO)_5$

XXII

$Ph_2CW(CO)_5$, in which the methoxy group is replaced by phenyl, has a $W–C_{carbene}$ distance of 2.15 Å, compared with a W=C bond length of 2.07 Å in $W(CO)_6$ (Casey *et al.*, 1977).

In the Pt(II) carbenes, **XXIII** (Badley *et al.*, 1969) and **XXIV** (Cardin *et al.*, 1971), the $Pt–C_{carbene}$ distances of 1.98 and 2.00 Å are intermediate between

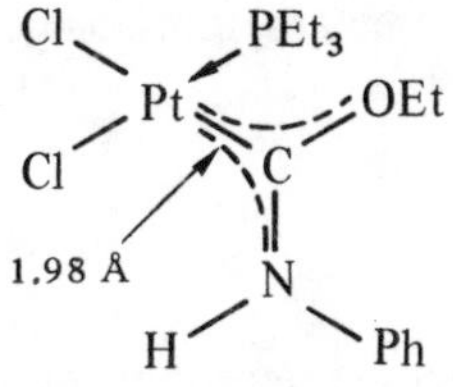

Cl   PEt₃
  Pt   OEt
Cl   C
1.98 Å   N
H   Ph

Et₃P   Cl   R
  Pt   N
Cl   C
2.00 Å   N
R

XXIII 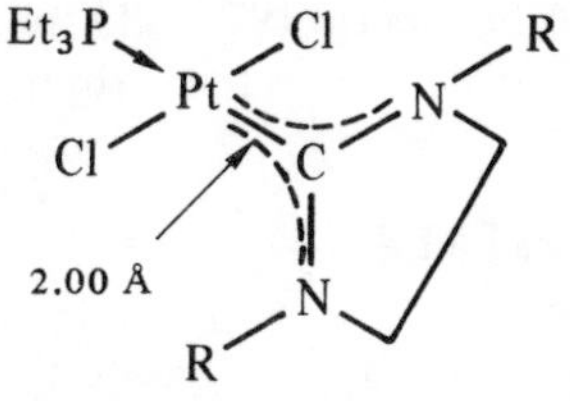   **XXIV**

Pt=C in Pt(II) carbonyls (typically 1.83 Å) and in Pt(II) alkyls (typically 2.07 Å). In carbyne complexes the M—C bond order is increased to 3. Thus, in **XXV**, the W≡C distance is only 1.90 Å, substantially shorter than in $W(CO)_6$, although slightly longer than the sum of triple bond radii taken from acetylene and tungsten metal–metal bonded complexes (Fischer and Schubert, 1975). In **XXVI**, the bond lengths are W—C, 2.26 Å; W=C, 1.94 Å; W≡C, 1.79 Å (Churchill and Youngs, 1979).

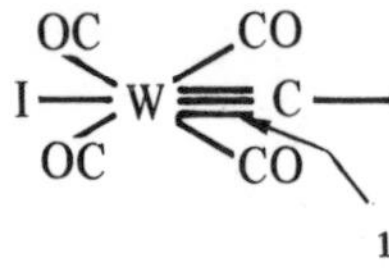

Bonding

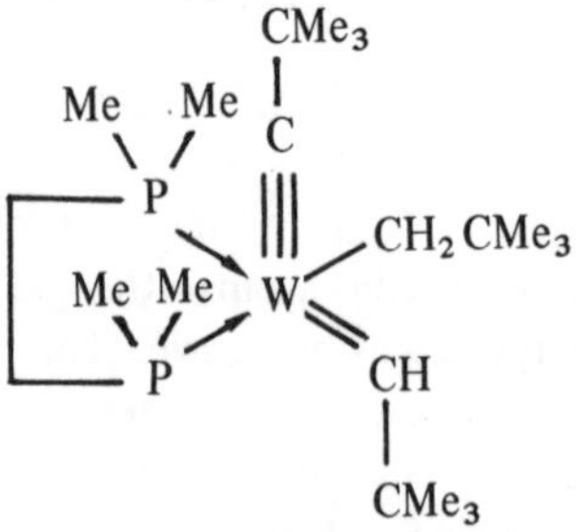

The bonding in metal carbene complexes has analogies with metal carbonyls and (free) alkenes. The trigonal framework of the uncomplexed carbene carbon comprises two σ-bonds to the substituents and a lone pair. There is also an empty *p*-orbital on the carbene carbon atom (see figure 3.1 opposite).

The bonding follows the familiar pattern of forward donation from the lone pair on the ligand, supplemented by back donation from the metal to an 'empty' orbital on the ligand. In this case the orbital on the ligand accepts electrons both from the metal and from the lone pairs on any heteroatom. This explains the bond length data given above. A molecular orbital scheme for chromium carbene complexes shows good correlation with the photoelectron spectra (Block and Fenske, 1977).

In carbyne complexes the triple bond comprises a σ- and two π-components (Fischer and Schubert, 1975). There is generally no heteroatom competing to donate electron density to the carbyne carbon.

3.3 π-COMPLEXES

Metal π-complexes adopt a considerable range of structures depending principally on the type of unsaturated hydrocarbon involved. The best known structure is the 'sandwich' or 'double-cone' structure of ferrocene. Most π-complexes con-

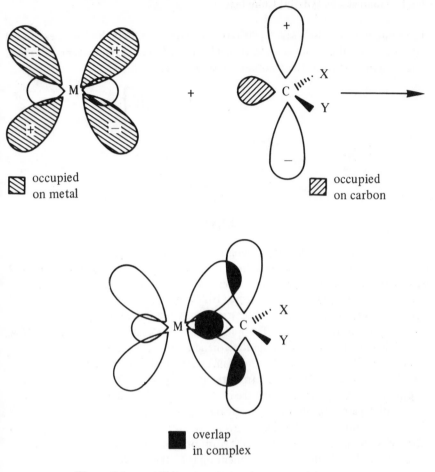

occupied
on metal

occupied
on carbon

overlap
in complex

Figure 3.1 Orbitals involved in metal–carbene bonding.

form with the eighteen electron rule, which requires the metal atom in the complex to have the electronic configuration of the next noble gas. The most important exceptions to this are complexes of second and third row transition metals, which often have 16 rather than 18 electrons.

Various approaches to the theoretical treatment of bonding in π-complexes have been made. We shall not describe the methods in detail, and have chosen molecular-orbital schemes which fit the experimental facts as well as possible. The photoelectron spectra of many organometallic compounds have been recorded, and these allow us to assess the results of theoretical calculations. However, it seems that in many π-complexes the ordering of the orbitals changes on ionisation (that is, Koopmans' theorem breaks down) so that exact correlation is impossible.

3.3.1 Mono-alkene (Olefin) Complexes

The structure of Zeise's salt $K[PtCl_3(C_2H_4)].H_2O$, **XXVII**, has been examined by both x-ray and neutron diffraction (Love *et al.*, 1975). The ethene is bonded side on with respect to the platinum–chlorine plane.

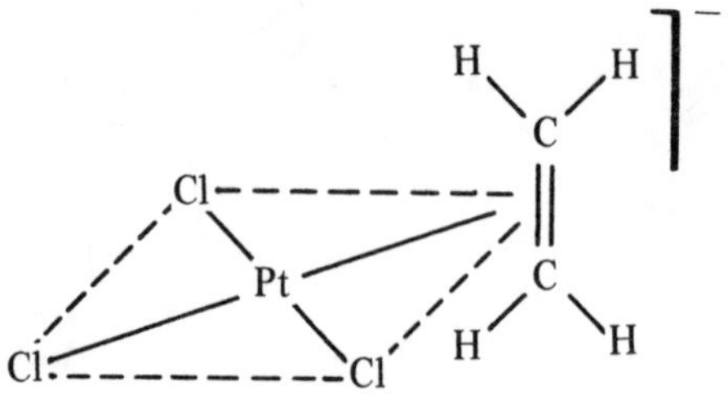

XXVII

An examination of the bond lengths shows some interesting features. The C–C distance of 1.37 Å is larger than in free ethene (1.34 Å). This agrees well with a reduced bond order exemplified by the fall in the C=C stretching frequency in Raman spectrum of ethene (1623 cm^{-1}) to the infrared frequency of Zeise's salt (1520 cm^{-1}). (The C=C stretching vibration is not active in the infrared spectrum of free ethene.) The mid-point of the coordinated ethene is 0.2 Å out of the coordination plane and the ethene C–C axis is tilted by 6° away from the vertical. However, both Pt–C distances are equal at 2.13 Å. The four hydrogen atoms are bent 0.16 Å away from the platinum atom. The Pt–Cl bond *trans* to ethene (2.34 Å) is longer than the *cis* Pt–Cl bonds (2.30 Å). This lengthening in the solid state (*trans*-influence) is related to the relative ease of substitution of the *trans* chlorine by other ligands (*trans*-effect). Examination of a number of structures of alkene complexes of transition metals shows the same general geometry, although obviously dependent on the other ligands.

In the case of $CH_2=CHCNFe(CO)_4$, **XXVIIIa**, in the solid state and $C_2H_4Fe(CO)_4$, **XXVIIIb**, in the vapour phase, the alkene occupies an equatorial

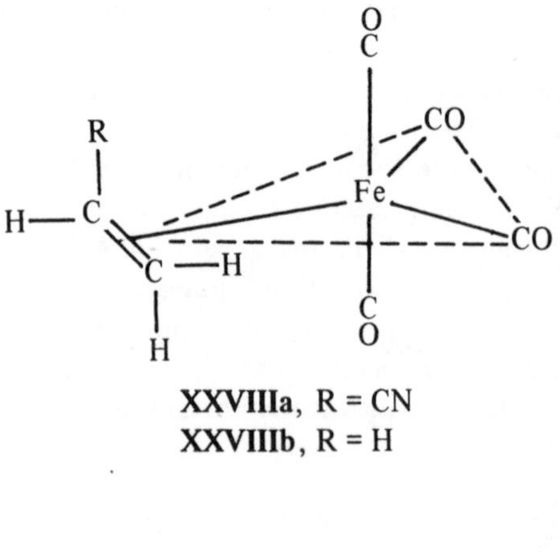

XXVIIIa, R = CN
XXVIIIb, R = H

position, and both carbon atoms of the alkene are in the equatorial plane of the trigonal bipyramid.

There is a considerable variation in C–C bond lengths with values over 1.40 Å quite common, and even exceeding 1.50 Å in the case of some tetracyanoethene complexes (Ittel and Ibers, 1976). The substituents are generally bent away from the metal.

The bonding of alkenes to transition metal ions is explained in terms of the Dewar–Chatt model (see figure 3.2). According to this model the bonding con-

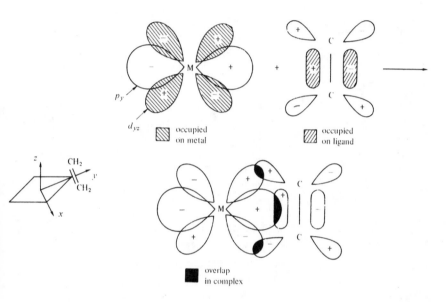

Figure 3.2 Orbitals involved in metal-alkene bonding (the paper is in the yz plane)

sists of two components: (a) 'forward donation' from the π-bonding orbital of the alkene into an empty orbital on the metal; (b) 'back donation' from a filled d-orbital on the metal into the π^*-orbital on the alkene, which is antibonding with respect to C–C bond.

The theory explains very elegantly why the C–C bond length increases on co-ordination. The very long C–C bonds in the tetracyanoethene complexes are caused because the electron withdrawing CN groups increase the amount of 'back donation' from the metal. In common with metal carbonyls for which a similar bonding scheme applies, there is synergic reinforcement of the forward and back donation. The interactions can be illustrated by a molecular orbital diagram (figure 3.3; Albright et al., 1979).

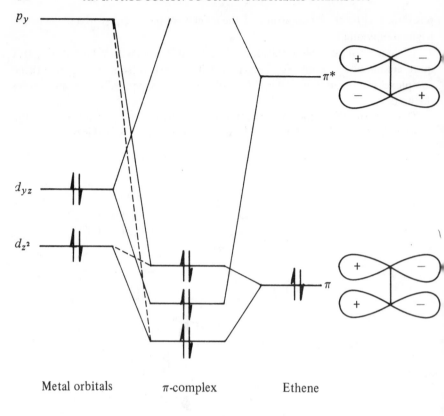

Metal orbitals π-complex   Ethene

Figure 3.3   A partial molecular orbital diagram of Zeise's salt.

The reader will probably realise that there is usually a second d-orbital perpendicular to the first (in this case d_{xy} perpendicular to d_{yz}) which can participate in the back-bonding. Thus, electronically a structure in which the C—C axis is rotated through 90° will be energetically reasonably favourable. For this reason the alkene can rotate in a propeller fashion in many olefin complexes. This effect can be observed by broadening of the n.m.r. signals. The activation energy for the rotation is about 30–80 kJ mol^{-1} (see p. 73).

3.3.2 Alkyne Complexes

The bonding in most alkyne complexes appears to be analogous to that of the alkene compounds. However, there is a possibility that in some complexes the bonding might be better represented as a metallacyclopropene, **XXIX**, rather than a π-complex, **XXX**.

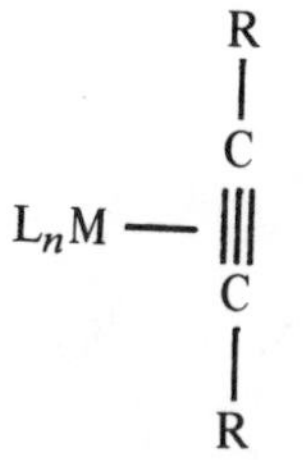

$$L_nM \longrightarrow \begin{matrix} R \\ | \\ C \\ ||| \\ C \\ | \\ R \end{matrix}$$

XXX

$$L_nM \begin{matrix} & R \\ & \diagup \\ & C \\ & \| \, (\theta \\ & C \\ & \diagdown \\ & R \end{matrix}$$

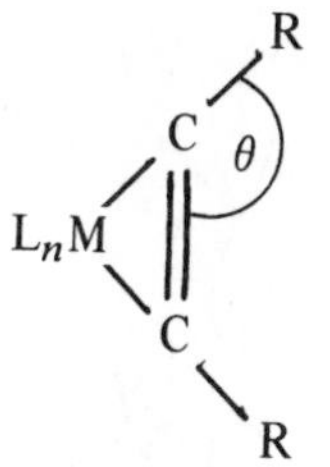

XXIX

The x-ray structure of the alkyne diol complex K$^+$[MeC(OH)EtC≡CEt(OH)-CMePtCl$_3$]$^-$ is shown in **XXXI**. It is formally analogous to Zeise's salt (Dubey, 1976). The C−C axis is perpendicular to the PtCl$_3$ plane and the C−C distance is

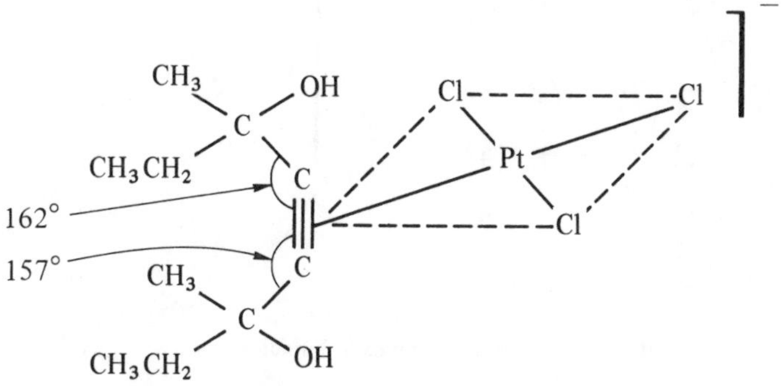

XXXI

1.23 Å. The average Pt−C distance is 2.14 Å. The substituents are bent away from the platinum, producing C−C−C angles of 162° and 157°. The OH groups are not coordinated to the platinum. A survey of a range of acetylene complexes of a variety of metals shows that C−C bond distances usually lie between 1.22 and 1.32 Å, and the bend-back angle θ is in the range 130-170° (Robinson, 1981).

3.3.3 Complexes of Conjugated Dienes

The structure and bonding of diene complexes is no different in principle from that in mono-olefin complexes. However, an examination of the C−C bond lengths in conjugated diene fragments bonded to transition metals has provided a particularly vivid picture.

Butadiene has two π-bonding and two π*-antibonding orbitals (figure 3.4). The energies of the orbitals increase in the order 1 → 4, and only 1 and 2 are occupied in uncomplexed butadiene.

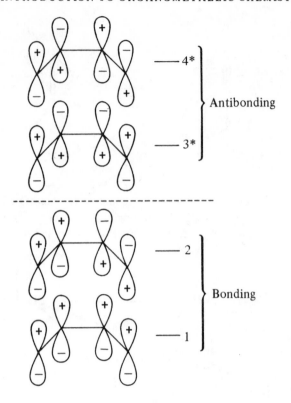

Figure 3.4 Molecular orbitals of butadiene.

When the butadiene is complexed to a transition metal, the bonding occurs through the familiar combination of forward and back donation. The antibonding orbital labelled 3 receives electron density by back donation from the metal. This orbital has lobes which correspond to a double bond between the two central carbon atoms. When electron density is donated into this orbital the central C—C bond should shorten, and this is found experimentally. In butadiene tricarbonyliron, **XXXII**, the C—C bond lengths are roughly equal and in the

$$\text{CH} \overset{}{-\!\!-} \text{CH}$$

$$\text{CH}_2 \diagup \qquad \overset{\|}{\text{CH}_2}$$

$$\text{Fe(CO)}_3$$

XXXII

range 1.45–1.46 Å. This compares with distances of 1.48 and 1.34 Å for the central and terminal bonds of free butadiene (Mills and Robinson, 1963). The iron atom is 2.06 and 2.14 Å from the internal and terminal carbon atoms respectively. When an electron withdrawing group is attached to the diene fragment, as in (all-*trans*-retinal)tricarbonyliron, **XXXIII**, there is increased back

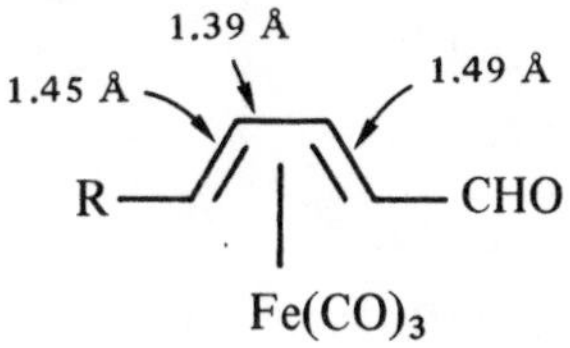

$$R = C_{16}H_{23}$$

XXXIII

donation into orbital 3 and the central C–C bond drops to 1.39 Å (Birch *et al.*, 1966). Replacement of the $Fe(CO)_3$ group by CpCo, which is isoelectronic, but a poorer electron acceptor, also leads to shortening of the central C–C bond (Churchill and Mason, 1967).

Cyclobutadiene complexes represent a special class of diene complexes. Their existence was predicted on theoretical grounds, before it was accomplished in the laboratory. The cyclobutadiene rings of transition metal complexes are square, although substituents on the ring inevitably lead to slight distortions from this geometry (Efraty, 1977).

3.3.4 η^3-Allyl Complexes (π-Allyl Complexes)

We shall discuss two structures in detail.

η^3-*Allylpalladium Chloride Dimer*

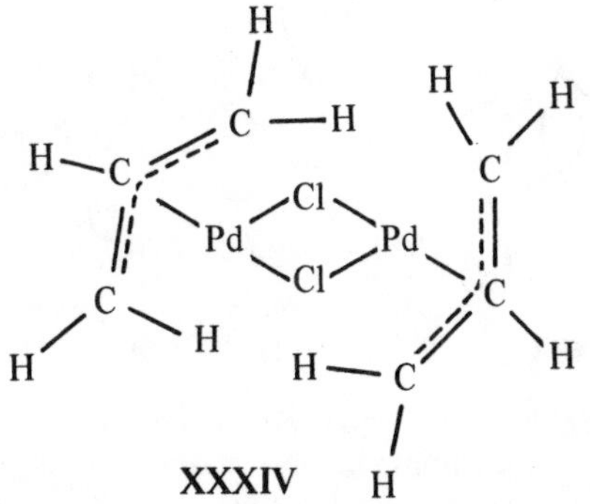

XXXIV

The η^3-allyl systems in this molecule, **XXXIV** (Smith, 1965), are related by a centre of symmetry in the molecule. The C–C–C plane makes an angle of 111° with the Pd$\overset{Cl}{\underset{Cl}{<>}}$Pd plane. Both C–C distances are equal at 1.38 Å and the

C–C–C angle is 120°. Each palladium atom is equidistant from all three carbon atoms of its respective allyl group at 2.11 Å.

Bis(methallyl)nickel

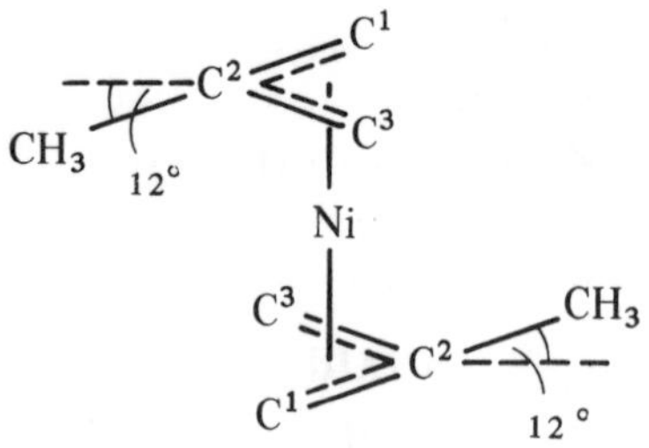

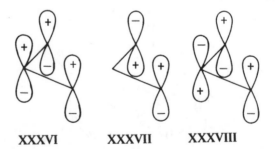

XXXV

In this molecule, **XXXV** (Uttech and Dietrich, 1965), the allyl groups are again related by a centre of symmetry, but this time are bonded to the same metal atom. The C–C distances within the π-bonded system are both 1.41 Å and the C–C–C angle within this system is 119°. The central carbon atom, C^2, is slightly closer (1.98 Å) to the nickel atom than the terminal atoms C^1 and C^3 (2.02 Å). The methyl groups are tilted towards the nickel atom by 12°. The π-bonded framework is somewhat like $(C_6H_6)_2Cr$ with half of each benzene ring removed.

Bonding

The π-orbitals of the allyl group are shown in **XXXVI–XXXVIII**.

XXXVI   XXXVII   XXXVIII

Unlike the very symmetrical sandwich molecules, such as ferrocene and dibenzenechromium, it is not immediately clear where the metal atom should be situated in order to maximise overlap with the allyl orbitals. Thus, both the extreme orientations, with the metal in the C–C–C plane, **XXXIX**, or perpendicular to it, **XL**, give good overlap for at least one orbital. The structure adopted by a complex will be a compromise between these two extremes and

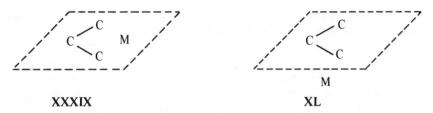

XXXIX   XL

will also be dependent on the other ligands on the metal atom (Kettle and Mason, 1966). In some cases there is good agreement with molecular orbital calculations and photoelectron spectra (for example, allylMn(CO)$_4$), but the case of (allyl)$_2$Ni is complicated by the fact that Koopmans' theorem does not apply.

The allyl orbital, **XXXVIII**, which is antibonding with respect to the carbon framework, does not overlap favourably with any metal orbital. This explains why the C–C distances in allyl groups are rarely above 1.45 Å The C–C distances are usually (Kaduk et al., 1977) in the range 1.36–1.42 Å. Some allyl groups are unsymmetrical, with the two limbs of the allyl fragment having unequal bond lengths.

3.3.5 η^5-Cyclopentadienyl Complexes

Recent x-ray (Seiler and Dunitz, 1979) and neutron diffraction studies on ferrocene have established, contrary to earlier work, that the C$_5$H$_5$ rings in the sandwich structure of ferrocene have an eclipsed configuration, **XLI**. The symmetry of the molecule is D_{5h}, and this persists in the vapour phase.

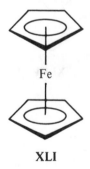

XLI

The average C–C bond length within the two planar parallel rings is 1.389 Å, which is very close to the value of 1.395 Å found in (free) benzene. The Fe–C distances are all equal at 2.03 Å and the neutron diffraction study showed that the hydrogen atoms are slightly tilted towards the iron atom. Cobaltocene has a staggered (D_{5d}) configuration in the solid state (Bünder and Weiss, 1975a). The C–C distance is increased to 1.41 Å and the Co–C distance is 2.096 Å.

Nickelocene is also staggered and the Ni–C distance is 2.185 Å (Seiler and Dunitz, 1980). These three metallocenes are considered to be covalently bonded with substantial overlap of the π-orbitals on the cyclopentadienyl ring with d-orbitals on the metal. $(Cp)_2 Mg$ has a sandwich structure with a staggered configuration (Bünder and Weiss, 1975b). The magnesium atom has a smaller ionic radius, but no appropriate d-orbitals, and the Mg–C distance is 2.304 Å, and this is thought to indicate that the bonding is ionic in character.

In addition to the compounds discussed above, which all have sandwich structures with parallel rings, a number of other structures should be considered. The gas-phase structures of $Cp_2 Sn$, **XLII**, and $Cp_2 Pb$, **XLIII**, have cyclopentadienyl rings which are side-on bonded, but not parallel (Connolly and Hoff, 1981). The Pb–C distance in **XLIII** is 2.762 Å, but in the solid phase the compound is associated. The structure of $Cp_2 Be$, **XLIV**, is rather enigmatic, because although the beryllium atom is between two parallel cyclopentadienyl rings, it appears that the centre of one of the rings is displaced from the C_5 axis of the other ring.

One final type of sandwich metallocene is the triple decker sandwich ion $Cp_3 Ni_2^+$, **XLV**.

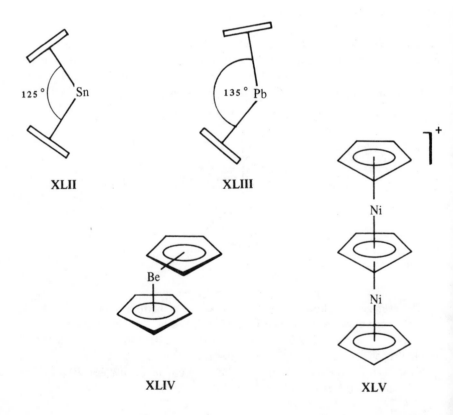

125°   Sn

135°   Pb

XLII   XLIII

Ni

Be

Ni

XLIV   XLV

The x-ray structure study (Dubler *et al.*, 1974) did not establish the relative configuration of the rings, but did give the distances $Ni-C_{terminal} = 2.085$ Å; and $Ni-C_{bridge} = 2.145$ Å. These values are close to those of nickelocene itself and suggest covalent bonding.

In addition to the sandwich structures, two other compounds should be mentioned. Manganocene, Cp_2Mn, is isomorphous with ferrocene above $159\,^{\circ}C$, but below this it has a chain structure which confers antiferromagnetic properties (Bünder and Weiss, 1978).

The structure of titanocene has been a puzzle for many years. Two structures, **XLVI** and **XLVII**, have now been reported for substances which were previously claimed to be titanocene (Pez and Armor, 1981). Neither compound is Cp_2Ti, although **XLVI** has the same empirical formula.

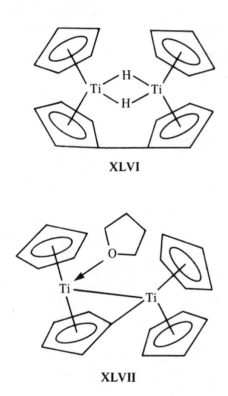

XLVI

XLVII

Bonding

There have been a large number of molecular orbital calculations on metallocenes. Ferrocene has been the most extensively studied and the diagram we give below (figure 3.8) agrees most closely with experimental observations both on ferrocene itself, and other sandwich metallocenes which are expected to be qualitatively analogous. Although the original calculations by Rohmer and Veillard (1975) were carried out in the D_{5d} point group, we present the results here for the recently established eclipsed configuration of the D_{5h} point group (Haaland, 1979).

The $C_5H_5^-$ anion has five molecular orbitals, three of which are occupied by two electrons, giving six π-electrons (figure 3.5). When two $C_5H_5^-$ rings are

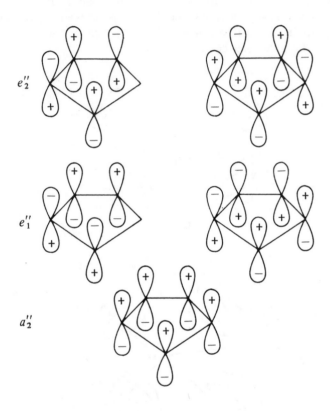

e_2''

e_1''

a_2''

Figure 3.5 Molecular orbitals of the $C_5H_5^-$ ion.

placed parallel, as in Cp_2Fe, they can interact in two ways to produce two arrangements with differing energies and symmetries. We show this for the two combinations of a_2'' molecular orbitals (figure 3.6).

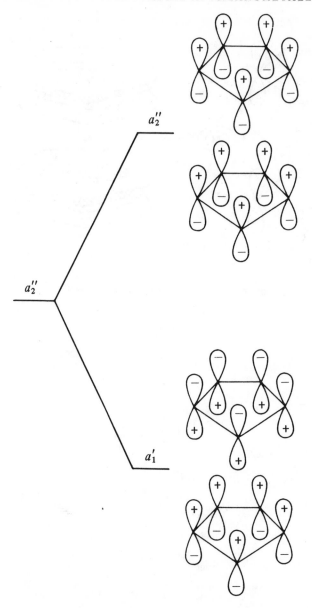

Figure 3.6 Interaction of two a_2'' orbitals.

The $(C_5H_5^-)_2$ orbitals interact with metal orbitals of the same symmetry. We show the overlap of the d_{z^2} orbital which has a a_1' symmetry in D_{5h} with the a_1' combination (figure 3.7).

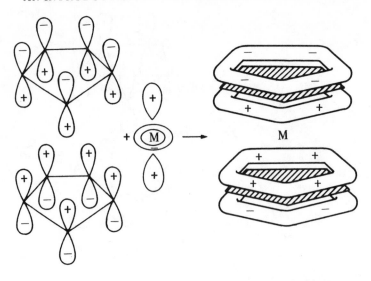

Figure 3.7   Formation of the a_1' bonding molecular orbital of ferrocene.

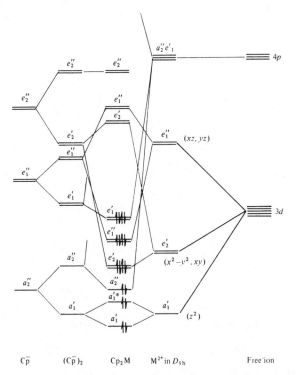

Figure 3.8   Molecular orbital diagram of a metallocene Cp_2M, with orbital occupancy shown for ferrocene. Only π-interactions are shown.

This a_1' orbital is the lowest in energy of the metal–ligand bonding. The complete diagram (figure 3.8) was obtained by splitting the $3d$ orbitals as required by D_{5h} symmetry before performing the calculations. The diagram is given with the 18-electron occupancy of ferrocene; it follows that the highest occupied molecular orbital is e_1', which agrees with the photoelectron spectrum of ferrocene and also the e.s.r. spectrum of the 17-electron ferrocinium ion. However, there are still some problems in assigning the photoelectron spectrum completely.

Figure 3.8 readily explains the paramagnetism of nickelocene. The orbital next higher in energy is the doubly degenerate e_2' orbital. Thus, in nickelocene, which has $20e^-$, the two electrons above the ferrocene complement go into this e_2' orbital with parallel spin (Hund's rule). These two unpaired electrons give rise to the magnetic moment of 2.86 Bohr magnetons.

3.3.6 η^6-Arene Complexes

The structure of dibenzene chromium, **XLVIII**, was the subject of controversy for about a decade following its discovery in 1955. During this period it was not certain whether all the C–C bond distances were equal. In 1966 a low temperature x-ray study established that the C–C distances are equal and that the molecule has D_{6h} symmetry (Keulen and Jellinek, 1966).

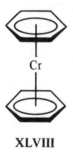

XLVIII

The structure is analogous to that of ferrocene, with a metal atom sandwiched between two parallel planar aromatic rings. The C–C bond distances are all 1.42 Å, compared with 1.395 Å in free benzene. The Cr–C distance is 2.15 Å.

Bonding

The bonding in $(C_6H_6)_2Cr$ is based on the overlap of molecular orbitals on the benzene ring with orbitals of appropriate symmetry on the metal atom. There are six π-orbitals, three bonding and three antibonding with respect to the carbon skeleton. The ordering of the orbitals given in figure 3.9 agrees with the majority of experimental observations, although an *ab initio* calculation (Guest *et al.*, 1975) reverses the order of the two highest filled orbitals. The photo-

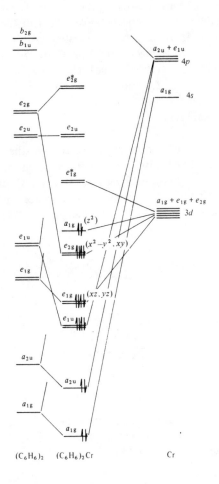

Figure 3.9 Molecular orbital diagram of $(C_6H_6)_2Cr$ with an 18-electron complement.

electron spectrum of $(C_6H_6)_2Cr$ has its lowest band at 5.45 eV, corresponding to electron loss from the a_{1g} orbital, and the e.s.r. spectrum of $(C_6H_6)_2Cr^+$ corresponds to a $^2A_{1g}$ state.

3.3.7 Cyclooctatetraene Complexes

Cyclooctatetraene forms π-complexes with transition metals in which it may be regarded as either a neutral molecule, C_8H_8, or a dianion, $C_8H_8^{2-}$. In the latter case there are 10 π-electrons, and the Hückel rule for aromaticity is satisfied.

Cyclooctatetraene forms both a mono- and a di-adduct with iron carbonyl, **XLIX** and **L** respectively, as well as several other more complicated compounds.

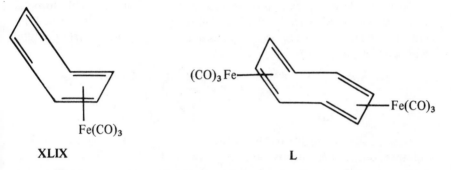

$Fe(CO)_3$

XLIX

$(CO)_3Fe$ — — $Fe(CO)_3$

L

In both these compounds, the cyclooctatetraene is a neutral ligand. In **XLIX** the ligand provides four electrons, and in **L** it provides four electrons to each iron atom, making eight in all. **XLIX** has the solid state structure shown (Dickens and Lipscomb, 1962) with all the C–C bond lengths in the bonded diene fragment approximately equal at 1.42 Å. However, the ^{1}H-n.m.r. spectrum shows only one resonance at room temperature.

The lanthanide and actinide elements form complexes with cyclooctatetraene in which the ligand acts as a dianion, $C_8H_8^{2-}$. The best known compound of this class is uranocene, **LI**.

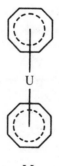

U

LI

Uranocene is a sandwich molecule with two planar C_8H_8 rings. The C—C bond lengths are all equal at 1.39 Å, and the two rings are eclipsed (Baker *et al.*, 1976). The bonding is analogous to that of ferrocene, but involves the *f*-orbitals on the uranium ion, which is formally in the IV oxidation state.

3.4 DYNAMIC BEHAVIOUR OF ORGANOMETALLIC COMPOUNDS IN SOLUTION

The structures we have discussed so far in this chapter have mainly been determined by diffraction methods in the solid state. Studies of organometallic compounds in solution by n.m.r. spectroscopy indicate that some unsaturated molecules undergo dynamic processes. In a few cases it is possible to obtain chemical evidence of dynamic behaviour. Thus the Grignard reagent obtained from crotyl bromide behaves (equation 3.1) as a mixture of **LII** and **LIII** on reaction with ketones (see p. 180).

$$CH_3-CH=CH-CH_2-MgBr \rightleftharpoons CH_3-\underset{\underset{MgBr}{|}}{CH}-CH=CH_2 \qquad (3.1)$$

$$\textbf{LII} \qquad\qquad\qquad\qquad \textbf{LIII}$$

It is unusual to be able to obtain chemical evidence of dynamic behaviour in solution, but it is frequently possible to study the process by n.m.r. spectroscopy. The n.m.r. spectra of the fluxional molecules, as they have come to be called, are often temperature dependent. It is possible to evaluate the energy barrier involved in the rearrangement by studying the spectra over a range of temperatures.

3.4.1 η^1-Cyclopentadienyl Compounds

Historically the η^1-cyclopentadienyl compounds were the first group of organometallic compounds for which fluxional behaviour was observed (Piper and Wilkinson, 1956). At room temperature the solution n.m.r. spectrum of $\eta^5 CpFe(CO)_2\eta^1 Cp$, **LIV**, shows two singlet peaks. On cooling, the resonance

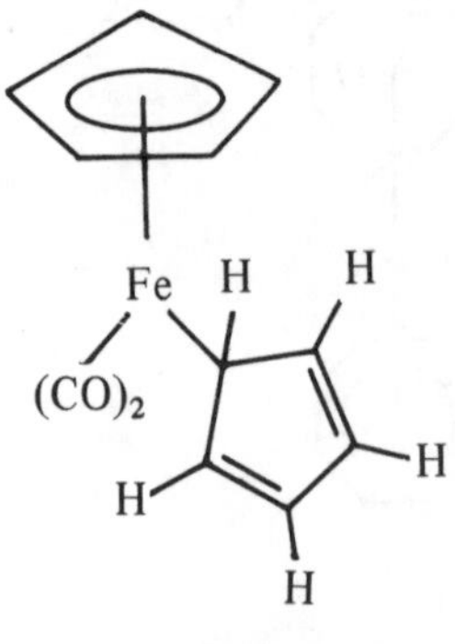

LIV

due to the σ-bonded cyclopentadienyl group splits into three multiplets, the peak of the η^5-group remaining unaffected. The site exchange process occurring in this system is a 1,2-shift of the iron atom around the σ-bonded ring (Cotton, 1968). Fluxional η^1-cyclopentadienyl systems are not restricted to the transition metals and Group IVb derivatives have been particularly widely studied. Again, the mechanism is a 1,2-shift of the metal around the ring (Bonny et al., 1982).

3.4.2 Mono-alkene Complexes

Temperature-dependent spectra of mono-alkene complexes were first reported in 1964. The molecular motion responsible for the temperature dependence has been shown to be a propellor-like rotation of the alkene about the coordination bond as axis, and not a rotation about the C—C axis. This has been established for a five-coordinate complex by a study of **LV** (Segal and Johnson, 1975) and for a six-coordinate complex (**LVI**) (Alt et al., 1974). The energy of activation for alkene rotation is in the range 30–80 kJ mol^{-1}.

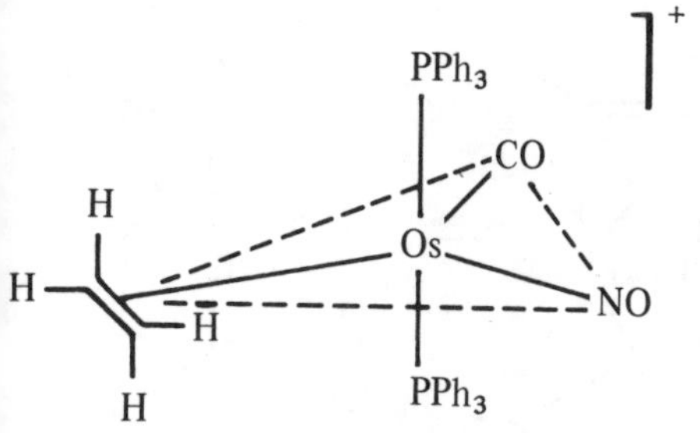

LV 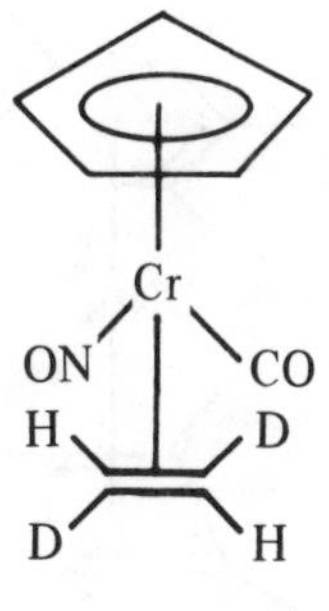   LVI

3.4.3 Polyene and Arene π-Complexes

The solution ^{1}H-n.m.r. spectrum of cyclooctatetraenetricarbonyliron (**XLIX**) consists of only one line at room temperature, although the solid state x-ray structure shows that the iron atom is only bonded to four of the eight carbon atoms. On cooling, the spectrum splits up as the protons become distinguishable on the n.m.r. time-scale. The ruthenium and osmium analogues show similar behaviour. Detailed studies including analysis of the ^{13}C-n.m.r. spectrum have established that the protons become equivalent at room temperature because the iron atom moves by 1,2-shifts around the ring (Cotton and Hunter, 1976).

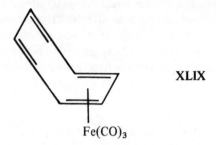

XLIX

Fe(CO)$_3$

The reader will appreciate that **XLIX** shows dynamic behaviour because the metal atom is only bonded to two of the four equivalent double bonds in the eight-membered ring. A similar situation occurs in some arene complexes such as *bis*(hexamethylbenzene)ruthenium, (HMB)$_2$Ru **(LVII)**. If the eighteen electron rule is followed, the Ru atom only requires 10 electrons from the two arene rings. This results in a η^6HMBRuη^4HMB sandwich structure, **LVII**, in the solid state.

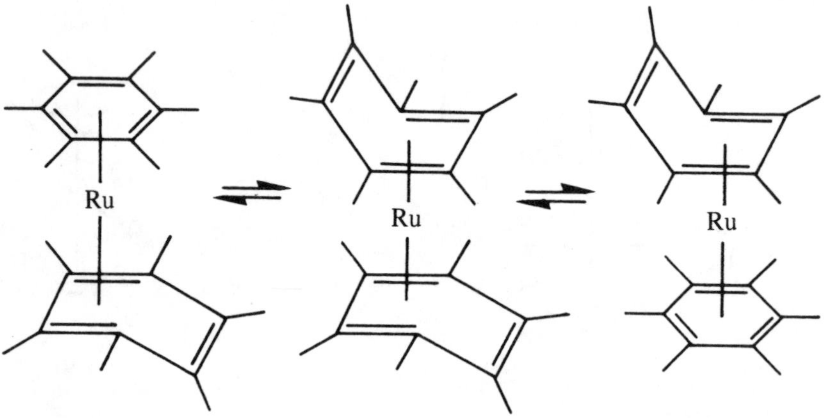

LVII, η^6Ruη^4	**LVIIa,** η^4Ruη^4	**LVIIb,** η^4Ruη^6

Scheme *ii*

A study of the temperature dependence of the ^{1}H-n.m.r. spectrum indicated that the site exchange occurs (scheme *ii*) via a η^4Ruη^4 intermediate, **LVIIa**, with 16 electrons. The ^{1}H-n.m.r. spectrum consists of one line at room temperature and separates out on cooling (Darensbourg and Muetterties, 1978).

3.4.4 Allyl Derivatives

The allyl derivatives of both main group and transition metals show dynamic behaviour in their n.m.r. spectra.

The n.m.r. spectrum of the allyl Grignard reagents shows only slight temperature dependence and has an AX_4 spectrum, indicating that all four methylene protons are equivalent. An explanation for this observation is that there is a dynamic equilibrium (equation 3.2) between **LVIIIa** and **LVIIIb**, with free rotation about the single C–C and C–Mg bonds (Hill, 1975).

$$(3.2)$$

LVIIIa   LVIIIb

The n.m.r. spectra of η^3-allyl complexes of transition metals show a large range of possibilities depending on the stoicheiometry of the compound. Much attention has been focused on the *syn/anti* site exchange, i.e. the equilibration of protons S and A on a η^3-allyl fragment (**LIX**). The phenomenon is, at first sight, strange in a π-bonded system, but can be explained by invoking a σ-bonded intermediate, **LX**, with free rotation about the single bonds (Vrieze, 1975).

LIX   LX

3.5 THERMOCHEMISTRY OF ORGANOMETALLIC COMPOUNDS

Knowledge of the thermochemistry of organometallic compounds has increased considerably in recent years. The mean bond disruption enthalpy, $\bar{D}$, is defined as the average enthalpy per M–C bond required to carry out the (hypothetical) disruption reaction shown in equation (3.3).

$$MR_n\ (g) \xrightarrow[\Delta H_{disrupt} = n\bar{D}]{} M\ (g) + nR\ (g) \tag{3.3}$$

The mean bond disruption enthalpy $(\bar{D})$ of a metal–carbon σ-bond of a non-transition element varies from about 150 kJ mol^{-1} for bismuth–carbon to about 380 kJ mol^{-1} for boron–carbon. A typical C–C bond has a $\bar{D}$ value of 350 kJ mol^{-1}. The heavier metals of a group usually form weaker metal–carbon bonds. The $\bar{D}$ values of metal–carbon σ-bonds in transition metal compounds vary over a narrower range from ~130 to ~250 kJ mol^{-1}. Mean bond disruption enthalpies have also been evaluated for some π systems. The $\bar{D}$ (M−Cp) in ferrocene is 351 kJ mol^{-1}, and in manganocene it is 266 kJ mol^{-1}. $\bar{D}$ (M−C$_6$H$_6$) in dibenzene chromium is 165 kJ mol^{-1} (Pilcher and Skinner, 1982).

Thermochemical measurements give only the mean bond disruption enthalpies. In a few cases the enthalpies of stepwise bond disruption have been estimated from mass-spectral and kinetic data. It is generally more difficult to disrupt the first bond compared with subsequent bonds. In dimethylmercury the first bond disruption enthalpy is 1.7 times the mean bond disruption enthalpy (scheme *iii*; Ashcroft and Beech, 1973).

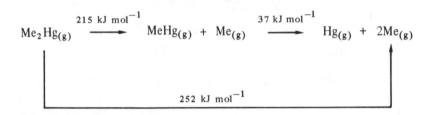

Scheme *iii*

REFERENCES

Albright, T. A., Hoffmann, R., Thibeault, J. C., and Thorn, D. L. (1979). *J. Am. chem. Soc.* **101**, 3801

Alt, H., Heberhold, M., Kreiter, C. G., and Strack, H. (1974). *J. organomet. Chem.* **77**, 353

Ashby, E. C., Laemmle, J., and Neumann, H. M. (1974). *Acc. chem. Res.* **7**, 273

Ashcroft, S. J., and Beech, G. (1973). *Inorganic Thermodynamics*. Van Nostrand Reinhold, London, p. 109

Badley, E. M., Chatt, J., Richards, R. L., and Sim, G. A. (1969). *Chem. Commun.* 1322

Baker, E. C., Halstead, G. W., and Raymond, K. N. (1976). *Struct. Bonding (Berlin)* **25**, 24

Baker, R. W., and Pauling, P. (1969). *Chem. Commun.* 745

Block, T. F., and Fenske, R. F. (1977). *J. Am. chem. Soc.* **99**, 4321

Birch, A. J., Fitton, H., Mason, R., Robertson, G. B., and Stangroom, J. E. (1966). *Chem. Commun.* 613

Bonny, A., Holmes-Smith, R. D., Hunter, G., and Stobart, S. R. (1982). *J. Am. chem. Soc.* **104**, 1855

Brock, C. P., and Ibers, J. A. (1975). *Acta crystallogr.* **31A**, 38

Brook, A. G., Nyburg, S. C., Abdesaken, F., Gutekunst, B., Gutekunst, G., Krishna, R., Kallury, M. R., Poon, Y. C., Chang, Y. M., and Wong-Ng, W., (1982). *J. Am. chem. Soc.* **104**, 5667

Bünder, W., and Weiss, E. (1975a). *J. organomet. Chem.* **92**, 65

Bünder, W., and Weiss, E. (1975b). *J. organomet. Chem.* **92**, 1
Bünder, W., and Weiss, E. (1978). *Z. Naturforsch. B* **33**, 1235
Cardin, D. J., Cetinkaya, B., Lappert, M. F., Manojlovic-Muir, L. J., and Muir, K. W. (1971). *Chem. Commun.* 400
Carrol, W. E., and Bau, R. (1978). *J. chem. Soc., chem. Commun.* 825
Casey, C. P., Burkhardt, T. J., Bunnell, C. A., and Calabrese, J. C. (1977). *J. Am. chem. Soc.* **99**, 2127
Churchill, M. R., and Mason, R. (1967). *Adv. organomet. Chem.* **5**, 93
Churchill, M. R., and Youngs, W. J. (1979). *Inorg. Chem.* **18**, 2454
Connolly, J. W., and Hoff, C. (1981). *Adv. organomet. Chem.* **19**, 123
Cotton, F. A. (1968). *Acc. chem. Res.* **1**, 257
Cotton, F. A. (1978). *Acc. chem. Res.* **11**, 225
Cotton, F. A., and Hunter, D. L. (1976). *J. Am. chem. Soc.* **98**, 1413
Cotton, F. A., and Millar, M. (1977). *J. Am. chem. Soc.* **99**, 7886
Cotton, F. A., Gage, L. D., Mertis, K., Shive, L. W., and Wilkinson, G. (1976). *J. Am. chem. Soc.* **98**, 6922
Cowan, D. O., Krieghoff, N. G., and Donnay, G. (1968). *Acta crystallogr.* **24B**, 287
Darensbourg, M. Y., and Muetterties, E. L. (1978). *J. Am. chem. Soc.* **100**, 7425
Davidson, P. J., Harris, D. H., and Lappert, M. F. (1976). *J. chem. Soc. Dalton Trans.* 2268
Davies, G. R., Jarvis, J. A. J., Kilbourn, B. T., and Pioli, A. J. R. (1971). *Chem. Commun.* 677
Dickens, B., and Lipscomb, W. N. (1962). *J. chem. Phys.* **37**, 2084
Dietrich, H. (1963). *Acta crystallogr.* **16**, 681
Drew, M. G. B., and Wilkins, J. D. (1973). *J. chem. Soc. Dalton Trans.* 1830
Dubey, R. J. (1976). *Acta crystallogr.* **32B**, 199
Dubler, E., Textor, M., Oswald, H.-R., and Salzer, A. (1974). *Angew. Chem., int. Edn Engl.* **13**, 135
Edwards, P., Mertis, K., Wilkinson, G., Hursthouse, M. B., and Malik, K. M. A. (1980). *J. chem. Soc. Dalton Trans.* 334
Efraty, A. (1977). *Chem. Rev.* **77**, 691
Fischer, E. O., and Schubert, U. (1975). *J. organomet. Chem.* **100**, 59
Gavens, P. D., Guy, J. J., Mays, M. J., and Sheldrick, G. M. (1971). *Acta crystallogr.* **33B**, 137
Gramlich, V., and Pfefferkorn, K. (1973). *J. organomet. Chem.* **61**, 247.
Greiser, T., Kopf, J., Thoennes, D., and Weiss, E. (1980). *J. organomet. Chem.* **191**, 1
Guest, M. F., Hillier, I. H., and Saunders, V. R. (1972). *J. organomet. Chem.* **44**, 59
Guest, M. F., Hillier, I. H., Higginson, B. R., and Lloyd, D. R. (1975). *Molec. Phys.* **29**, 113
Guggenberger, L. J., and Rundle, R. E. (1968). *J. Am. chem. Soc.* **90**, 5375
Haaland, A. (1979). *Acc. chem. Res.* **12**, 415

Hill, E. A. (1975). *J. organomet. Chem.* **93**, 123

Hope, H. and Power, P. P. (1983). *J. Am. chem. Soc.* **105**, 5321

Huffman, J. C., and Streib, W. E. (1971). *Chem. Commun.* 911

Ittel, S. D., and Ibers, J. A. (1976). *Adv. organomet. Chem.* **14**, 33

Jarvis, J. A. J., Kilbourn, B. T., Pearce, R., and Lappert, M. F. (1973). *J. chem. Soc. chem. Commun.* 475

Jensen, W. B. (1978). *Chem. Rev.* **78**, 1

Kaduk, J. A., Poulos, A. T., and Ibers, J. A. (1977). *J. organomet. Chem.* **127**, 245

Kettle, S. F. A., and Mason, R. (1966). *J. organomet. Chem.* **5**, 97

Keulen, E., and Jellinek, F. (1966). *J. organomet. Chem.* **5**, 490

Lefferts, J. J., Molloy, K. C., Hossain, M. B., Van der Hlem, D., and Zuckerman, J. J. (1982). *J. organomet. Chem.* **240**, 349

Leoni, P., Pasquali, M., and Ghilardi, C. A. (1983). *J. chem. Soc. chem. Commun.* 240

Love, R. A., Koetzle, T. F., Williams, G. J. B., Andrews, L. C., and Bau, R. (1975). *Inorg. Chem.* **14**, 2653

Mason, R., and Mingos, D. M. P. (1973). *J. organomet. Chem.* **50**, 53.

Mills, O. S., and Robinson, G. (1963). *Acta crystallogr.* **16**, 758

Mills, O. S., and Redhouse, A. O. (1968). *J. chem. Soc. (A)* 642

Muller, N., and Otermat, A. L. (1965). *Inorg. Chem.* **4**, 296

Oliver, J. P. (1977). *Adv. organomet. Chem.* **15**, 235

Pez, G. P., and Armor, J. N. (1981). *Adv. organomet. Chem.* **19**, 1

Pilcher, G., and Skinner, H. A. (1982). In *The Chemistry of the Metal–Carbon Bond*, Vol. 1 (F. R. Hartley and S. Patai, eds). Wiley, Chichester, p. 43

Piper, T. S., and Wilkinson, G. (1956). *J. inorg. nucl. Chem.* **3**, 104

Poller, R. C. (1979). In *Comprehensive Organic Chemistry*, Vol. 3 (D. H. R. Barton and W. D. Ollis, eds). Pergamon Press, Oxford, p. 1111

Preut, H., Haupt, H.-J., and Huber, F. (1973). *Z. anorg. allg. Chem.* **396**, 81

Robinson, E. A. (1981). *J. chem. Soc. Dalton Trans.* 2373

Rohmer, M. H., and Veillard, A. (1975). *Chem. Phys.* **11**, 349

Scovell, W. M., Kimura, B. Y., and Spiro, T. G. (1971). *J. coord. Chem.* **1**, 107

Segal, J. A., and Johnson, B. F. G. (1975). *J. chem. Soc. Dalton Trans.* 677

Seiler, P., and Dunitz, J. D. (1979). *Acta crystallogr.* **35B**, 1068

Seiler, P., and Dunitz, J. D. (1980). *Acta crystallogr.* **36B**, 2255

Smith, A. E. (1965). *Acta crystallogr.* **18**, 331

Thoennes, D., and Weiss, E. (1978). *Chem. Ber.* **111**, 3157

Uttech, R., and Dietrich, H. (1965). *Z. Krystallogr.* **122**, 60

Vrieze, K. (1975). In *Dynamic Nuclear Magnetic Resonance Spectroscopy* (L. M. Jackman and F. A. Cotton, eds). Academic Press, New York, p. 441

Wakefield, B. J. (1974). *The Chemistry of Organolithium Compounds*. Pergamon Press, Oxford, p. 14

Weiss, E. (1964). *J. organomet. Chem.* **2**, 314

Weiss, E., and Lucken, E. A. C. (1964). *J. organomet. Chem.* **2**, 197

Weiss, E., and Köster, H. (1977). *Chem. Ber.* **110**, 717

Zubieta, J. A., and Zuckerman, J. J. (1978). *Prog. inorg. Chem.* **24**, 251

GENERAL READING

Armstrong, D. R., and Perkins, P. G. (1981). Main group organometallic compounds. *Co-ord. Chem. Rev.* **38**, 139

Churchill, M. R. (1970). Structural studies on transition-metal complexes containing σ-bonded carbon atoms. In *Perspectives in Structural Chemistry*, Vol. 1 (J. D. Dunitz and J. A. Ibers, eds). Wiley, New York, p. 191

Connor, J. A. (1977). Thermochemical studies on organo-transition metal carbonyls and related compounds. *Top. curr. Chem.* **71**, 71

Davidson, P. J., Lappert, M. F., and Pearce, R. (1976). Metal hydrocarbyls, MR_n stoichiometry, structure, stabilities and thermal decomposition pathways. *Chem. Rev.* **76**, 219

Faller, J. W. (1977). Fluxional and non-rigid behaviour of transition metal organometallic π-complexes. *Adv. organomet. Chem.* **16**, 211

Green, J. C. (1981). Gas phase photoelectron spectra of d and f block organometallic compounds. *Struct. Bonding (Berlin)* **43**, 37

Halpern, J. (1982). Determination and significance of transition-metal–alkyl bond dissociation energies. *Acc. chem. Res.* **15**, 238

Hartley, F. R. (1973). Thermodynamic data for olefin and acetylene complexes of the transition metals. *Chem. Rev.* **73**, 163

Kuz'mida, L. G., Bokis, N. G., and Struchov, Ya. T. (1975). The structural chemistry of organic compounds of mercury and its analogues (Zn and Cd). *Russ. Chem. Rev. (Engl. Transl.)* **44**, 73

Maslowsky, E. (1980). Synthesis, structure and vibrational spectra of organomethyl compounds. *Chem. Soc. Rev.* **9**, 25

Muetterties, E. L., Bleeke, J. R., Wucherer, E. J., and Albright, T. A. (1982). Structural, stereochemical and electronic features of arene–metal complexes. *Chem. Rev.* **82**, 499

Schubert, U. (1984). Structural consequences of bonding in transition metal carbene complexes. *Co-ord. Chem. Rev.* **55**, 261

Tel'noi, V. I., and Rabinovitch, I. B. (1977). Thermochemistry of organic compounds of transition metals. *Russ. Chem. Rev. (Engl. Transl.)* **46**, 689

Tel'noi, V. I., and Rabinovitch, I. B. (1980). Thermochemistry of organic derivatives of non-transition elements. *Russ. Chem. Rev. (Engl. Transl.)* **49**, 603

Vrieze, K. and van Leeuwen, P. W. N. M. (1971). Studies of dynamic organometallic compounds of the transition metal by means of nuclear magnetic resonance. *Prog. inorg. Chem.* **14**, 1

4

ORGANOMETALLIC COMPOUNDS AS SOURCES OF CARBANIONS

Alkanes, with their strong, low polarity C–C and C–H bonds, rarely participate in ionic reactions unless some heteroatom is present. The common heteroatoms such as halogen, oxygen and nitrogen are all more electronegative than carbon. The substituted carbon atoms are therefore cationoid and susceptible to nucleophilic attack; hence the predominance of S_N reactions in aliphatic chemistry.

Metals are more electropositive than carbon and the resulting polarisation of the carbon–metal bond makes organometallic compounds convenient sources of species containing carbon atoms with integral (carbanion) or partial (carbanionoid) negative charge. We shall consider attack at these carbons by electrophilic reagents, then the use of organometallic compounds as nucleophilic reagents, and, finally, the use of organometallics as basic reagents.

Almost all the discussion in this chapter is confined to σ-bonded main group organometallic compounds. Most of the arguments used are also applicable to σ-bonded organic compounds of the transition metals, but these are notably fewer in number and unlikely to be used as carbanion sources since they offer no advantages over the more accessible main group organometallic compounds. For the π-bonded d-block metal derivatives, the unsaturated organic ligands are, in any case, electron-rich and the dualistic nature of the electron exchange between ligand and metal often leads to bonds of low polarity. These ligands then do not, necessarily, have enhanced anionic character and the reactions of their complexes usually depend upon other factors. For example, π-bonded aromatic groups may become enriched or deficient in electrons, depending upon the transition metal and its ligands leading to significant modifications in reactivity.

4.1 GENERATION AND STABILITY OF CARBANIONS

A carbanion can be generated by a dissociation of the type shown in equation (4.1), where the most likely possibilities for Y are hydrogen or a metal.

$$\underset{\diagup}{\overset{\diagup}{\diagdown}}C-Y \rightleftharpoons \underset{\diagup}{\overset{\diagup}{\diagdown}}C^- + Y^+ \qquad (Y = H \text{ or } M) \qquad (4.1)$$

The acidities of the carbon acids, $\overset{\diagdown}{\diagup}C-H$, reflect stabilities of the anions, $-\overset{\diagdown}{\diagup}C^-$, those with the highest acidity giving the most stable ionic metal derivatives. Thus the weakest carbon acids, the alkanes ($pK_a \sim 50$), only give fully ionised organometallic compounds with the heavier (i.e. the more electropositive) alkali metals. This is exemplified by butylsodium, $C_4H_9^-\,Na^+$, which has a high melting point, is insoluble in hydrocarbons (other solvents react), and extremely reactive to oxygen and water. It is also thermodynamically unstable and slowly decomposes *in vacuo* to butene and sodium hydride. A possible mechanism for this reaction is indicated in equation (4.2).

$$CH_3CH_2CH\underset{\underset{H}{\overset{|}{C}}}{\overset{\frown}{}}CH_2\,Na^+ \longrightarrow CH_3CH_2CH{=}CH_2 + NaH \qquad (4.2)$$

More stable anions, such as those derived from acetylenes ($pK_a \sim 24$) and cyclopentadiene (pK_a 18), give sodium derivatives of higher stability (**I** and **II**).

$$RC{\equiv}C^-Na^+$$

I	**II**	**III**

The remarkable stability of the cyclopentadiene anion is a consequence of the delocalization of the charge into an aromatic 6π system. The pK_a of pentamethylcyclopentadiene is 26, showing the reduction in acidity when hydrogens are replaced by electron donating methyl groups (Bordwell and Bausch, 1983). Proton and ^{7}Li-n.m.r. data (as well as other measurements) confirm that the cyclooctatetraenyl dianion in compound **III** gains stability from the 10π aromatic system, as predicted by Hückel's rule (Cox *et al.*, 1971).

Stabilisation can also occur when the charge is delocalised over one, or preferably two, neighbouring carbonyl groups as in the familiar examples of anions derived from diethyl malonate (**IV**), ethyl acetoacetate (**V**), and β-diketones such as acetylacetone (**VI**).

IV	**V**	**VI**

As discussed later in this chapter, anions are important nucleophilic reagents used, for example, to replace the halogen of a halogenoalkane by an organic group (equation 4.3).

$$R^- Na^+ + R'X \longrightarrow R-R' + NaX \qquad (4.3)$$

In these and other reactions, the tightness of binding between the carbanion and its metallic counter ion influences reactivity. Whether the $R^- M^+$ species exist as contact, or solvent separated, ion pairs depends on a number of factors. The smaller cations favour contact ion pairing, and as M^+ increases in size, solvent separation of the ions often becomes more important. Solvents, such as tetrahydrofuran, which are particularly effective at solvating cations, favour solvent separation, as does the addition of cation-complexing crown ethers. In nucleophilic substitutions, such as, for example, alkylation of the reactive methylene groups in compounds such as **IV**, **V** and **VI**, the free carbanions react more rapidly than the ion pairs (Lowry and Richardson, 1981, p. 470).

Carbonium ions undergo many rearrangements, and in this respect have a much richer chemistry than the carbanions. Nevertheless, some significant rearrangements are shown by anions derived from organometallic compounds and we begin our discussion by considering those involving migrations of groups on to neighbouring atoms, that is 1,2-shifts. Thus, 1,1,1,2-tetraphenylethane can be converted to the corresponding 1,1,2,2-compound by treatment with phenylsodium (equations 4.4–4.6).

$$Ph_3 CCH_2 Ph + Ph^- Na^+ \longrightarrow [Ph_3 C\overline{C}HPh\ Na^+] + PhH \qquad (4.4)$$

$$[Ph_3 C\overline{C}HPh\ \overset{+}{Na}] \xrightarrow{\text{Spontaneously}} Ph_2 \overline{C}CHPh_2\ Na^+ \qquad (4.5)$$

$$Ph_2 \overline{C}CHPh_2\ Na^+ + ROH \longrightarrow Ph_2 CHCHPh_2 + RONa \qquad (4.6)$$

The view that this type of 1,2-phenyl shift occurs via a carbanion is supported by the observation that treatment of 2,2,2-triphenylethyl chloride with potassium or sodium at $-66\,^\circ$C gave the rearrangement product but, at this temperature, the less completely ionised organolithium compound did not rearrange (Grovenstein and Williams, 1961; Zimmerman and Zweig, 1961; scheme *i*). Note that in these rearrangements the phenyl group migrates as a cation, $C_6 H_5^+$. Orbital symmetry considerations predict that 1,2-migrations of unsaturated groups, such as phenyl or vinyl, can occur, but the corresponding migrations of alkyl groups are symmetry forbidden (Grovenstein, 1978). Accordingly, we find that rearrangements involving movement of an alkyl group from one carbon to a neighbouring carbon atom are not observed in saturated acyclic systems. However, carbanion rearrangements involving 1,2-shifts of alkyl groups from heteroatoms to carbon are known as, for example, in the Wittig rearrangement of benzyl and allyl ethers (schemes *ii* and *iii*).

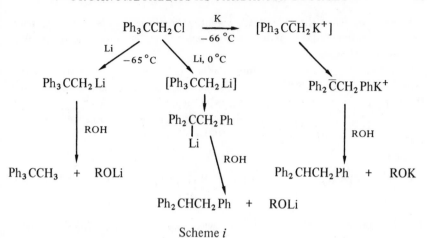

Scheme *i*

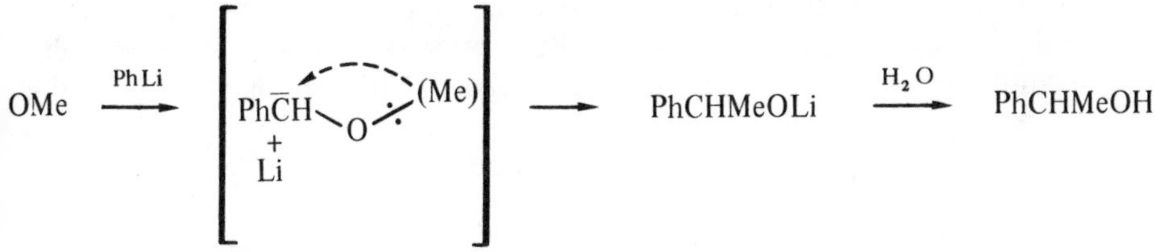

OMe $\xrightarrow{\text{PhLi}}$ $\left[\begin{array}{c} \text{Ph}\bar{\text{C}}\text{H} \\ + \\ \text{Li} \end{array} \text{O} \cdots (\text{Me}) \right]$ $\longrightarrow$ PhCHMeOLi $\xrightarrow{\text{H}_2\text{O}}$ PhCHMeOH

Scheme *ii*

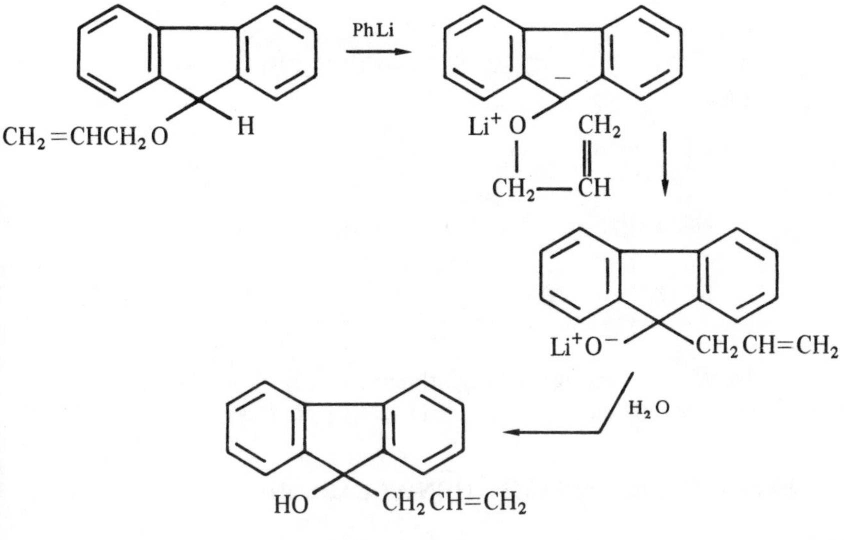

Scheme *iii*

There is evidence that the Wittig and the closely related Stevens rearrangements may occur at least under some circumstances by homolytic rather than heterolytic pathways.

It is possible that certain organometallic species which contain strained ring systems and rearrange to open chain compounds may do so via the anions. Thus ^{1}H-n.m.r. study of the freshly-prepared Grignard reagent from cyclopropylmethyl chloride and magnesium contained 99 per cent or more of the but-3-enylmagnesium chloride (Silver et al., 1960; equation 4.7).

$$\begin{array}{c} CH_2 \\ | \quad \diagdown \\ \quad \quad CHCH_2MgCl \longrightarrow CH_2{=}CHCH_2CH_2MgCl \qquad\qquad (4.7) \\ | \quad \diagup \\ CH_2 \end{array}$$

The corresponding lithium compound was prepared from cyclopropylmethyl iodide by metal–halogen exchange at $-70\,^\circ$C. The proportion of rearranged product (shown by isolation of the benzaldehyde adducts) was minimised by using a mixture of 10 parts of light petroleum to one of ether as solvent, rather than pure ether, as well as reducing the reaction time (scheme iv). Essentially quantitative yields of the unrearranged product (**VII**) were obtained using the

$$\begin{array}{c} CH_2 \\ | \quad \diagdown \\ \quad \quad CHCH_2I \quad + \quad RLi \\ | \quad \diagup \\ CH_2 \end{array}$$

$$\begin{array}{c} CH_2 \\ | \quad \diagdown \\ \quad \quad CHCH_2Li \rightleftharpoons CH_2{=}CHCH_2CH_2Li \\ | \quad \diagup \\ CH_2 \end{array}$$

| 1. PhCHO | 1. PhCHO |
| 2. H$_2$O | 2. H$_2$O |

$$PhCH(OH)CH_2CH\begin{array}{c} \diagup CH_2 \\ | \\ \diagdown CH_2 \end{array} \qquad PhCH(OH)CH_2CH_2CH{=}CH_2$$

VII

Scheme iv

mixed solvent, and allowing sec-butyllithium (= RLi) to react for only 2 min at $-70\,^\circ$C before adding the benzaldehyde (Lansbury et al., 1964).

4.2 ELECTROPHILIC SUBSTITUTION AT CARBON

Most organometallic compounds are susceptible to attack at the carbon–metal bond by halogens or protic reagents (equations 4.8 and 4.9).

$$X_2 + M-C \Longleftarrow \longrightarrow M-X + X-C \Longleftarrow \qquad (X_2 = Cl_2, Br_2, I_2) \qquad (4.8)$$

$$H-Z + M-C \Longleftarrow \longrightarrow M-Z + H-C \Longleftarrow \qquad (4.9)$$

Regarding the latter, there is a wide spectrum of reactivity, from tetraalkyl-germanium compounds, which are unattacked by aqueous mineral acids, to the organic derivatives of lithium, which must be carefully protected from atmospheric moisture which rapidly decomposes them (equation (4.9); M = Li, Z = OH). Clearly, reactivity is related to the polarity of the M–C bond, and the electro-negativity value of the metal is a useful guide to this. Attempts to quantify the relationship between chemical reactivity and bond polarity should, however, be resisted because of inherent imprecision in the measurement of electronegativity. This results in uncertainties in the electronegativity values for elements and for the bond polarities calculated from them. As we shall see, reactions such as those shown in equations (4.8) and (4.9) may occur by more than one mechanism and generalisations must be treated with caution.

We are familiar with the fact that bimolecular nucleophilic substitutions at carbon proceed with complete inversion of configuration. In contrast, the electro-philic analogues, the S_E2 reactions, commonly, though not invariably, proceed with retention of configuration. This difference can be explained by orbital symmetry considerations whereby favourable interactions will occur between the highest occupied molecular orbital (HOMO) of one molecular species and the lowest unoccupied molecular orbital (LUMO) of another, providing that these have comparable energies and are of the same symmetry. A simplified but illuminating picture of the orbital symmetry differences in S_N2 and S_E2 pro-cesses is shown in figure 4.1 (Fleming, 1976; Anh and Minot, 1980). We see that frontside attack in an S_N2 process does not occur, because the favourable over-lap between the HOMO of the nucleophile and the larger lobe of the substrate's LUMO is compensated by an unfavourable, out of phase overlap. Rearside approach of the nucleophile, however, involves only bonding overlaps. In the S_E2 case it is the LUMO of the reagent which overlaps with the HOMO of the substrate and now both rearside and frontside approaches are symmetry allowed.

In practice S_E reactions may not occur in a concerted manner and are often complicated by nucleophilic assistance via coordination of the reagent or some other species to the metal atom.

Studies in this area were, at first, hampered by difficulties in obtaining optically active compounds having chiral carbon atoms joined to metals. Much early work was with the relatively stable organomercury compounds. Many examples of electrophilic substitutions with retention of configuration at carbon atoms bound to mercury have now been identified. Some of these reactions proceed via open transition states (S_E2) as in equation (4.10) (Charman et al., 1959), and others involve cyclic transition states (S_Ei = internal electrophilic substitution) illustrated by the bromide-ion-catalysed exchange shown in equation (4.11) (Charman et al., 1961).

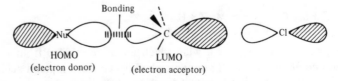

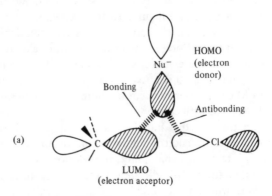

(a)

Frontside attack, not allowed

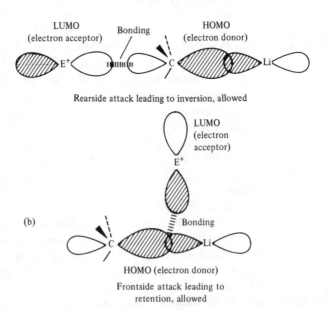

(b)

Figure 4.1 (a) Frontier orbitals for S_N2 reaction; (b) frontier orbitals for S_E2 reaction. The open and hatched lobes denote opposite signs of the wavefunction.

$$[EtCHMe]_2Hg \; + \; HgBr_2 \; \rightleftharpoons \; EtCH\begin{array}{c} \nwarrow HgCHMeEt \\ | \quad \searrow HgBr \\ Me \quad \searrow Br \end{array} \qquad (4.10)$$

$$2EtCHMeHgBr$$

$$EtCHMeHgBr \; + \; {}^{203}HgBr_2 \; + \; Br^- \; \rightleftharpoons \; EtCH\begin{array}{c} Br \\ \nwarrow Hg \searrow \\ \diagup \quad Br \\ | \; {}^{203}Hg \diagup \\ Me \quad Br \end{array} \qquad (4.11)$$

$$EtCHMe^{203}HgBr \; + \; HgBr_2 \; + \; Br^- \longleftarrow \rfloor$$

In the experiment summarised in equation (4.10) the optical activity of the starting material was derived from only one of the secondary butyl groups, the other being racemic. Since the optically active group is diluted to half its original concentration in this reaction, substitution with 100 per cent retention should give a product with the specific rotation halved; this was what was observed. The rate of mercury exchange in the reaction starting with s-butylmercuric bromide (equation 4.11) was monitored by measurement of the uptake of the radioactive ^{203}Hg label, while the optical activity was maintained again showing that substitution proceeded with retention of configuration.

More recently, it has been possible to find examples of S_E2 reactions which can be made to occur with either retention or inversion by relatively minor changes in reaction conditions, or in the structure of the substrate (Fukuto and Jensen, 1983). Thus optically active s-butyltrineopentyltin reacts with bromine in methanol (containing bromide ion to suppress ionization of the organotin bromide formed) to give 100 per cent inversion (equation (4.12); here and below, sBu* indicates optically active secondary butyl).

$$Np_3SnsBu* + Br_2 \xrightarrow[\text{NaBr}]{\text{MeOH}} sBu*Br + Np_3SnBr \qquad (4.12)$$
$$100 \text{ per cent inversion}$$

The conditions were deliberately chosen to favour this stereochemical outcome. Thus, the large neopentyl groups inhibited frontside attack; indeed if these were replaced by smaller groups such as isopropyl, then, under the same conditions, frontside attack became dominant and the net result was 12 per cent retention (equation 4.13).

$$iPr_3SnsBu* + Br_2 \xrightarrow[\text{NaBr}]{\text{MeOH}} sBu*Br + iPr_3SnBr \qquad (4.13)$$
$$12 \text{ per cent retention}$$

It can be shown that, in the transition state, charge separation is less for substitution with retention than with inversion, so that polar solvents favour the latter. Accordingly, if in equation (4.12) the solvent is changed from methanol to carbon tetrachloride (containing a suitable inhibitor to suppress the intervention of radical reactions) then, instead of 100 per cent inversion, 89 per cent retention is observed. Fewer studies have been reported on compounds of the transition metals but bromine cleavage of an Fe$-$C* bond occurred with 95 per cent inversion at the chiral C* atom (Bock *et al.*, 1974).

Brominolysis of *trans*-methylcyclopropyllithium gave principally the isomer with retained configuration, but in this case substantial amounts of the inverted product were also obtained (equation (4.14); Applequist and Peterson, 1961).

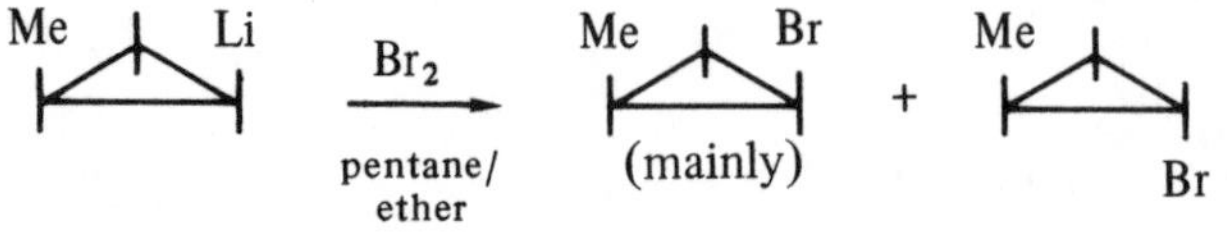

$$(4.14)$$

The dominant factor here may be the difficulty in inverting a cyclopropyl ring since similar brominolyses of 2-norbornyllithium (Applequist and Chmurny, 1967) and 4-*t*-butylcyclohexyllithium (Glaze *et al.*, 1969) showed predominant inversion accompanied by some retention. As with the organotin compounds, the reactions appear to be poised between the two stereochemical pathways.

Rates of cleavage of aryl groups from metal atoms are generally greater than those for alkyl because of resonance stabilisation of the transition state. The mechanism of substitution by protons of metals attached to benzene rings has been discussed by Eaborn (1975). These protodemetallations (equation 4.15) should be compared with simple aromatic substitutions such as that indicated in equation (4.16).

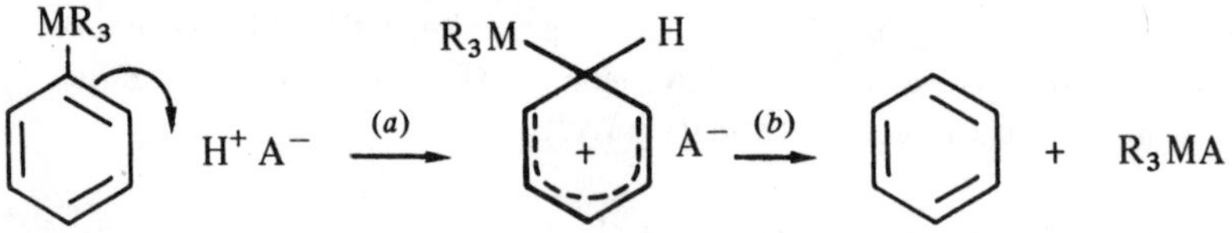

(M = Si, Ge, Sn, Pb)   (4.15)

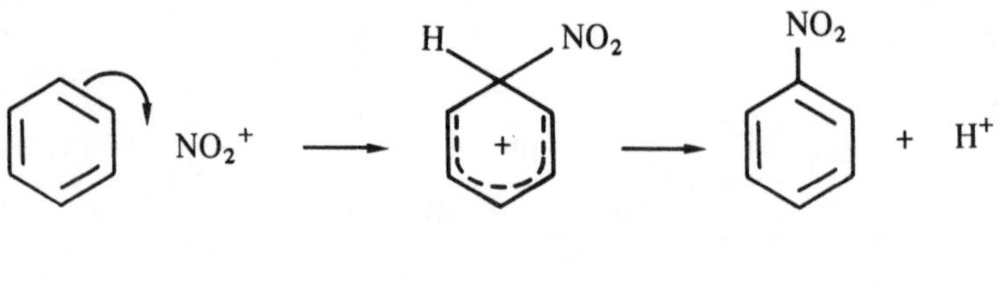

$$(4.16)$$

It was shown that the bond-forming step ((a) in equation 4.15) was rate determining since, when various aryl-MR_3 species (M = Si, Ge, Sn, Pb) were cleaved with hydrogen chloride in aqueous dioxan, the reaction was always slower when the water was replaced by D_2O. (In reaction (4.16) also, the first step is rate determining.) The rate of cleavage of the Ar—MEt_3 bond by aqueous methanolic perchloric acid increases in the sequence (M =) Si, 1; Ge, 36; Sn, 3.5×10^5; Pb, 2×10^8. This dramatic increase in rate was ascribed to the enhanced ability of the heavier elements to stabilise the Wheland intermediate by hyperconjugative electron release from M, as in the canonical form, **VIII**.

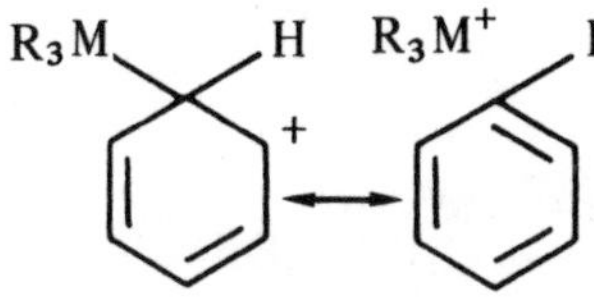

VIII

Because the organometallic substrates are more difficult to obtain, fewer kinetic studies have been made of electrophilic substitutions, nevertheless, it is clear that unimolecular mechanisms are less common than is the case with nucleophilic substitutions. As with S_N1 reactions, S_E1 processes are favoured by enhanced ion stability and strongly solvating solvents. An example is the conversion of protonated 4-pyridylmethylmercuric chloride to the 4-methylpyridinium ion in aqueous perchloric acid in the presence of chloride ions (Dodd and Johnson, 1969). A major pathway for this reaction is shown in scheme ν. As expected, when mercuric chloride was added the reaction was slowed, presumably by suppression of the rate-determining dissociation step.

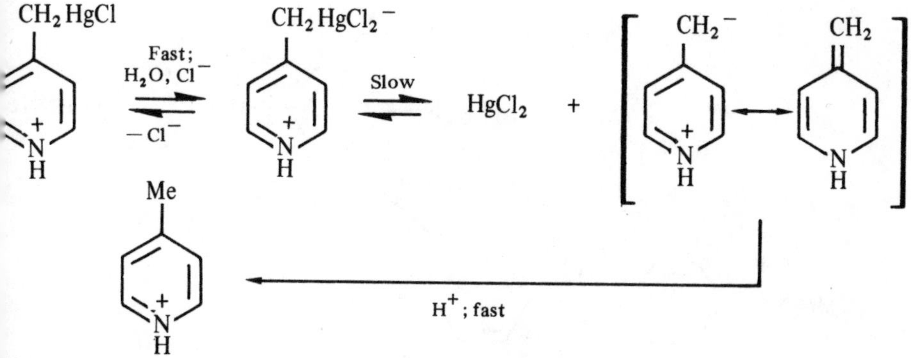

Scheme ν

4.3 CARBANIONS AS NUCLEOPHILIC REAGENTS

We now consider reactions in which the carbanionoid groups in organometallic compounds function as nucleophilic reagents. The principal areas of interest are nucleophilic substitutions at electron deficient carbon atoms and at metal atoms.

The familiar production of a tertiary alcohol from Grignard reagent and ketone involves nucleophilic attack at carbonyl carbon with assistance from co-ordination of oxygen to magnesium illustrated in a simplified form by scheme vi.

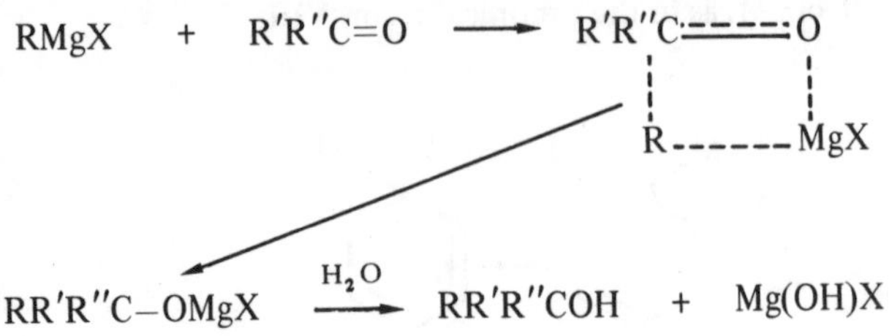

$$RMgX \quad + \quad R'R''C{=}O \quad \longrightarrow \quad R'R''C{\cdots\cdots}O$$

$$R{\cdots\cdots}MgX$$

$$RR'R''C{-}OMgX \quad \xrightarrow{H_2O} \quad RR'R''COH \quad + \quad Mg(OH)X$$

Scheme vi

This reaction is discussed more fully on p. 162, but we note here that the detailed mechanism is more complicated and that there are a number of side reactions which, under certain circumstances, may become major reaction pathways.

Gilman's colour test (Gilman and Schulze, 1925), for the detection of the Grignard reagent and other organometallic species, relies on addition of the compound across the carbonyl group in Michler's ketone (**IX**), followed by hydrolysis and oxidation with iodine to give a diphenylmethane dye (**X**), as shown in scheme vii. A small number of σ-bonded transition metal alkyls, such as MeTiCl₃,

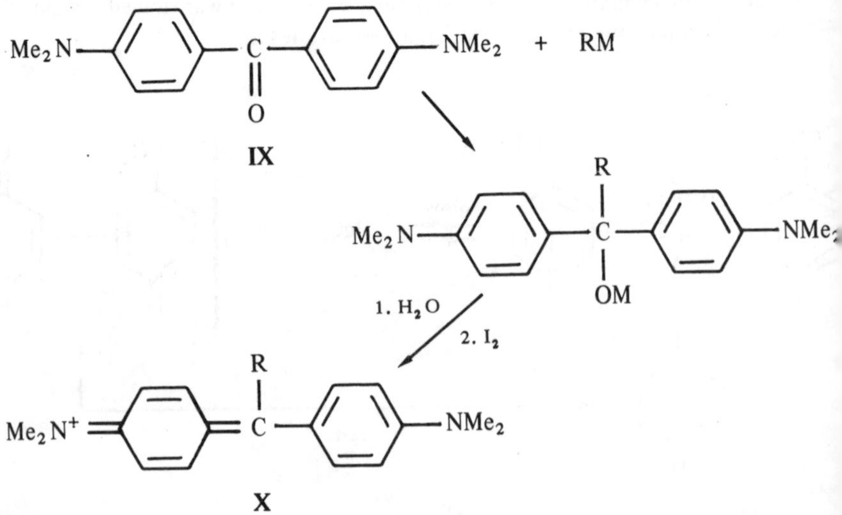

Scheme vii

do not react, but most binary metal alkyls give a positive reaction (Schrock and Parshall, 1976).

Pyridine reacts with organolithium compounds to give the 2-substituted derivatives (equation 4.17).

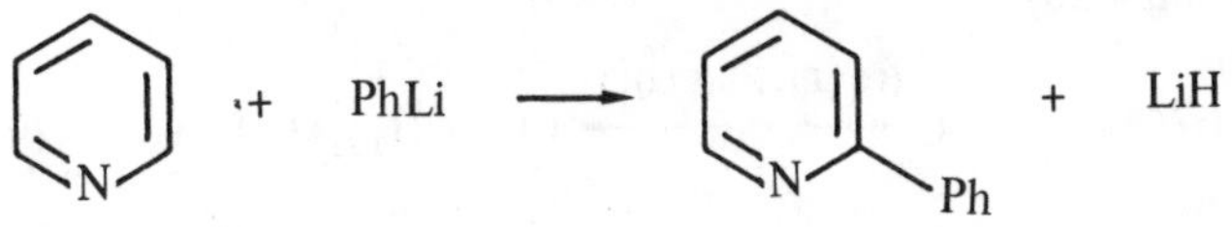

$$+ \quad PhLi \longrightarrow \qquad + \quad LiH \qquad (4.17)$$

The strictly *ortho* orientation and unusual elimination of hydride ion suggests that this is not a simple nucleophilic substitution. It has been shown (Giam and Stout, 1969) that there is an initial addition across the azomethine double bond to give the adduct (**XI**), which is decomposed by heating to the 2-phenylpyridine. However, the similarity of **XI** to the normal Meisenheimer intermediate, invoked in aromatic nucleophilic substitutions, is considerable, the negative charge is delocalised and carbonation gives 2-phenylpyridine-5-carboxylic acid (**XII**), as shown in scheme *viii* (Doyle and Yates, 1970).

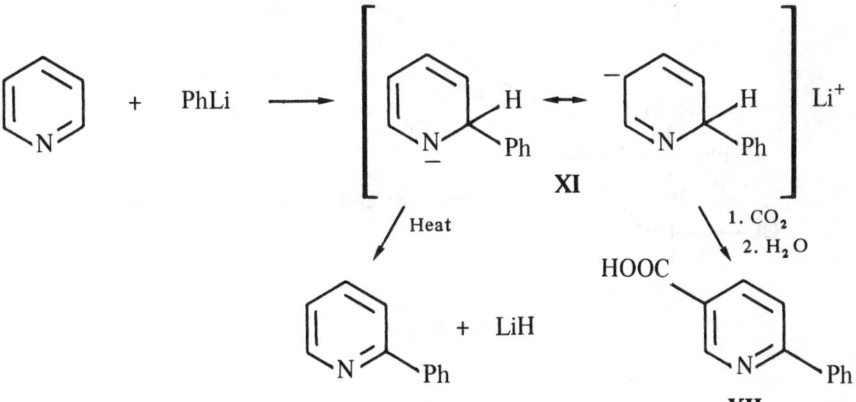

Scheme *viii*

It is difficult to use organolithium or Grignard reagents to obtain a ketone from an acyl chloride by nucleophilic displacement of the chlorine because the reagents also add to the carbonyl group, giving mixtures of products. The methods used to overcome this problem provide a useful illustration of some of the ways in which organometallic chemistry has developed. The traditional approach was to add a stoicheiometric quantity of cadmium chloride to the Grignard solution, thereby generating the organocadmium compound which does not add to the carbonyl group (equations 4.18 and 4.19).

$$BuMgCl + CdCl_2 \longrightarrow BuCdCl + MgCl_2 \qquad (4.18)$$

$$BuCdCl + CH_3COCl \longrightarrow CH_3COBu + CdCl_2 \qquad (4.19)$$

More recently, acyl halides have been transformed to ketones in high yields by using lithium dialkylcuprates, R_2CuLi (p. 187) in place of the organocadmium reagent. Alternatively, an organolithium compound in the presence of a stoicheiometric amount of a rhodium(I) complex (which can be recovered unchanged) is employed (equation 4.20).

$$MeLi + CH_3(CH_2)_{10}COCl \xrightarrow{(Ph_3P)_2Rh(CO)Cl} CH_3(CH_2)_{10}COCH_3 \qquad (4.20)$$

86 per cent

The reaction can also be carried out with a Grignard reagent and the proposed mechanism involves an initial alkylation of rhodium with subsequent transfer of methyl from rhodium to the acyl group via an oxidative addition/reductive elimination sequence (scheme ix; Hegedus et al., 1973). An example of the use of a more selective organometallic reagent for the same purpose is that of the vinylsilane shown in equation (4.21) (Fleming and Pearce, 1975).

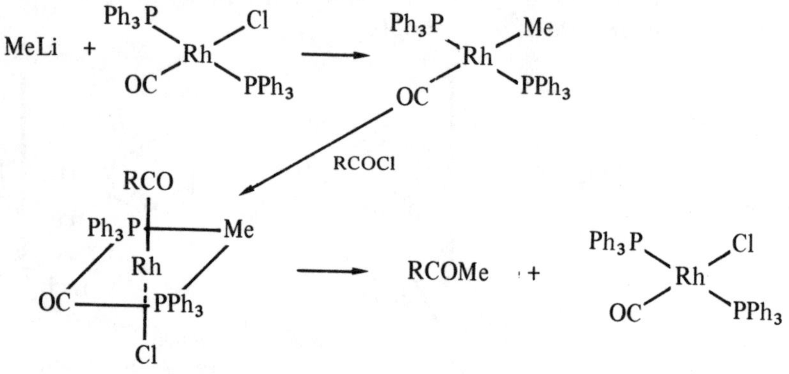

Scheme ix

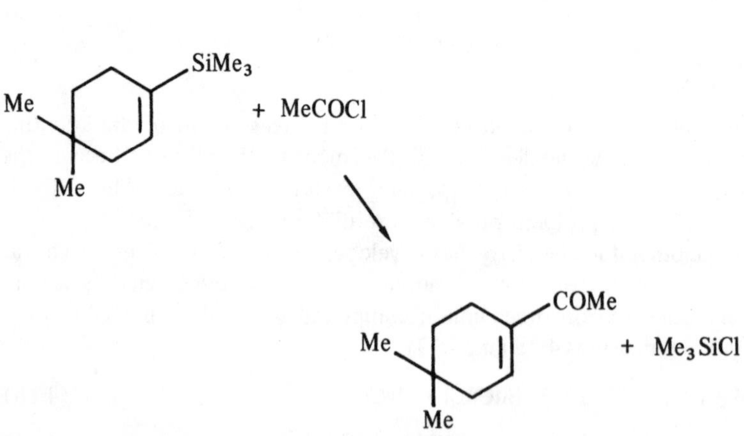

$$(4.21)$$

Finally, as discussed on p. 174, lithium tetraalkylborates react with acyl halides to give good yields of ketones (equation 4.22).

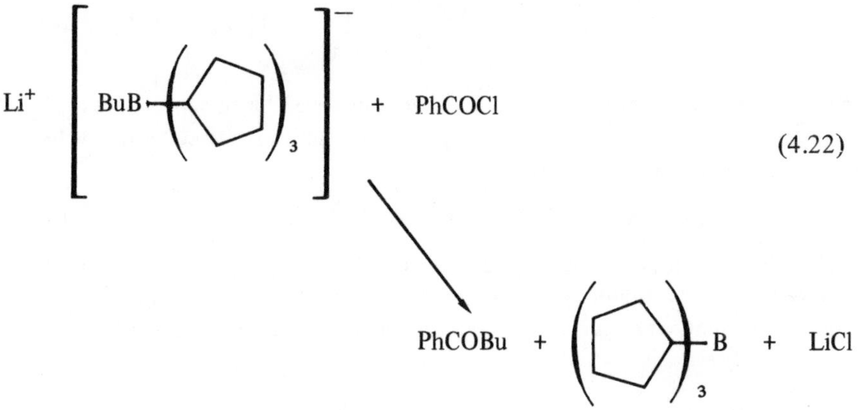

$$Li^+ \left[BuB \left(\!\!\!\! \diagdown\!\!\!\diagup \right)_3 \right]^- + PhCOCl \qquad (4.22)$$

$$PhCOBu + \left(\!\!\!\! \diagdown\!\!\!\diagup \right)_3 \!\!-\!\! B + LiCl$$

Nucleophilic displacement by carbanion of halogen from alkyl and aryl halides is the overall process in a number of important methods of making carbon–carbon bonds, though other mechanisms may intervene (equation 4.23).

$$R-M + R'-X \longrightarrow R-R' + M-X \qquad (4.23)$$

Examples include the Wurtz reaction (M = Na, R = R′ = alkyl), the Ullmann reaction (M = Cu, R = R′ = aryl) and reactions in which R–M is a lithium dialkylcuprate and R′–X may be an alkyl or aryl halide. These reactions are discussed in chapter 8.

Carbanions attack certain metal carbonyl compounds at the carbon atoms transforming them to acyl derivatives (equations 4.24 and 4.25).

$$Fe(CO)_5 + RLi \longrightarrow Li[RCOFe(CO)_4] \qquad (4.24)$$

$$Ni(CO)_4 + RLi \longrightarrow Li[RCONi(CO)_3] \qquad (4.25)$$

The products are sources of acyl carbanions and by treatment with protons or cationoid carbon give, respectively, aldehydes or ketones (equation 4.26).

$$RCHO \xleftarrow{\;H^+\;} Li[RCOFe(CO)_4] \xrightarrow{\;EtBr\;} RCOEt \qquad (4.26)$$

Metals bound to electronegative elements such as halogen are susceptible to attack by carbanionic carbon atoms. These reactions (equation 4.27) are of great importance in the preparation of organometallic compounds and are discussed more fully in chapter 2.

$$R-M + M'-X \longrightarrow R-M' + M-X \qquad (4.27)$$

Here we note that reactions summarised by equation (4.27) will, like all reactions, be thermodynamically favoured when the stabilities of the products are greater than those of the reactants. This is the case in, for example, the preparation of 2-furanyltrimethylsilane (equation 4.28).

$$\text{furyl-Li} + Me_3SiCl \longrightarrow \text{furyl-SiMe}_3 + LiCl \qquad (4.28)$$

In less favourable cases the reaction can sometimes be driven to the right by precipitation or distillation of the required product or, as in equation (4.29) by complexation of a by-product.

$$4Bu_3Al + 3SnCl_4 + 4R_3N \longrightarrow 3Bu_4Sn + 4AlCl_3.R_3N \qquad (4.29)$$

When iron(II) chloride was reacted with certain pentadienyl anions, the products were the $bis(\eta^5$-pentadienyl)iron compounds analogous to ferrocene, except that the five-membered rings are open (equation (4.30); Wilson $et\ al.$, 1980).

$$FeCl_2 + \text{[pentadienyl]}^- \; M^+ \xrightarrow{\text{(M = Li, K)}} \text{Fe(pentadienyl)}_2 + 2MCl \qquad (4.30)$$

4.4 METALLATION AND DETERMINATION OF ACID STRENGTHS

Metallation is the direct replacement of hydrogen by metal to give a new carbon-metal bond. The reagents used may be inorganic or organometallic and the reactions are discussed more fully in chapter 2. In the present context we are concerned with metallations which occur by protolysis of an existing carbon-metal bond with a carbon acid (equation 4.31).

$$R-M + R'H \rightleftharpoons R'-M + RH \qquad (4.31)$$

These reactions are of importance in determining the strengths of weak carbon acids since the position of equilibrium is related to the acid strengths of RH and R'H. For example, McEwen (1936) developed his acidity scale for weak carbon acids by the use of alkali metal derivates in ether and measuring the position of equilibrium colorimetrically. The method has been extended and improved by subsequent workers. Acidities have been determined using organocaesium metallations in cyclohexylamine, even though there is little dissociation in this solvent, and the term ion-pair acidity has been coined. More recently, an acidity scale has been developed based on DMSO which, although a stronger acid than cyclohexylamine and therefore more restricted for measuring the weaker acids, ensures dissociation. It has been found possible to relate these acidity scales to pK_a values measured in water (Lowry and Richardson, 1981, p. 250).

The same reaction is used preparatively, particularly for lithium derivatives. Many aromatic compounds are successfully lithiated with an alkyllithium species, the relatively acidic aromatic hydrogens ensuring good yields (equation 4.32).

$$\text{OMe} \qquad \qquad \text{OMe, Li} \qquad \qquad \text{+ BuLi} \longrightarrow \qquad \qquad \text{+ BuH} \qquad (4.32)$$

As discussed on p. 25, the metal usually enters the ring *ortho* to a substituent. Strong bases are required in the preparation of ylides (equation 4.33).

$$E^+ - CHR_2 + B^- \longrightarrow (\overset{+}{E} - \overset{-}{C}R_2 \longleftrightarrow E = CR_2) + BH \qquad (4.33)$$

A common use of ylides is in the Wittig reaction, when an organolithium compound is often used as the base (scheme x). In certain cases the method of

$$Ph_3P + CH_3CH_2Br \longrightarrow Ph_3\overset{+}{P}CH_2CH_3Br^-$$

$$Ph_3\overset{+}{P}CH_2CH_3Br^- + BuLi \longrightarrow [Ph_3\overset{+}{P}-\overset{-}{C}HCH_3 \longleftrightarrow Ph_3P=CHCH_3]$$

$$+ BuH + LiBr$$

$$Ph_3P=CHCH_3 + RR'C=O \longrightarrow RR'C=CHCH_3 + Ph_3PO$$

Scheme x

generation of the ylide may affect the outcome of a Wittig reaction (Wakefield, 1974).

In this section we have emphasised the basic properties of organometallic compounds with reference to carbon acids. In practice, protolysis of carbon-metal bonds, using stronger acids, is an important step in most procedures involving organometallic reagents and, as noted earlier, the choice of protic species depends on the lability of the bond. Some illustrative examples are given in equations (4.34)-(4.36).

$$EtMgBr + H_2O \longrightarrow EtH + MgBr(OH) \qquad (4.34)$$

$$R-B\big< + CH_3COOH \longrightarrow RH + CH_3COOB\big< \qquad (4.35)$$

$$\text{(4.36)}$$

CH₃COOH

HCl

Me₃Si

REFERENCES

Anh, N. T., and Minot, C. (1980). *J. Am. chem. Soc.* **102**, 103

Applequist, D. E., and Chmurny, G. N. (1967). *J. Am. chem. Soc.* **89**, 875

Applequist, D. E., and Peterson, A. H. (1961). *J. Am. chem. Soc.* **83**, 862

Bock, P. L., Boschetto, D. J., Rasmussan, J. R., Demers, J. P., and Whitesides, G. M. (1974). *J. Am. chem. Soc.* **96**, 2814

Bordwell, F. G., and Bausch, M. J. (1983). *J. Am. chem. Soc.* **105**, 6188

Charman, H. B., Hughes, E. D., and Ingold, C. K. (1959). *J. chem. Soc.* 2530

Charman, H. B., Hughes, E. D., Ingold, C. K., and Volger, H. C. (1961) *J. chem. Soc.* 1142

Cox, R. H., Terry, H. W., and Harrison, L. W. (1971). *Tetrahedron Lett.* 4815

Dodd, D., and Johnson, M. D. (1969). *J. chem. Soc. (B)* 1071

Doyle, P., and Yates, R. R. J. (1970). *Tetrahedron Lett.* 3371

Eaborn, C. (1975). *J. organomet. Chem.* **100**, 43

Fleming, I. (1976). *Frontier Orbitals and Organic Chemical Reactions*. Wiley, Chichester, p. 74

Fleming, I., and Pearce, A. (1975). *J. chem. Soc. chem. Commun.* 633

Fukuto, J. M., and Jensen, F. R. (1983). *Acc. chem. Res.* **16**, 177

Giam, C. S., and Stout, J. L. (1969). *Chem. Commun.* 142

Gilman, H., and Schulze, F. (1925). *J. Am. chem. Soc.* **47**, 2002

Glaze, W. H., Selman, C. M., Ball, A. L., and Bray, L. E. (1969). *J. org. Chem.* **34**, 641

Grovenstein, E. (1978). *Angew. Chem. int. Edn Engl.* **17**, 313

Grovenstein, E., and Williams, L. P. (1961). *J. Am. chem. Soc.* **83**, 412; 2537

Hegedus, L. S., Lo, S. M., and Bloss, D. E. (1973). *J. Am. chem. Soc.* **95**, 3040

Lansbury, P. T., Pattison, V. A., Clement, W. A., and Sidler, J. D. (1964). *J. Am. chem. Soc.* **86**, 2247

Lowry, T. H., and Richardson, K. S. (1981). *Mechanism and Theory in Organic Chemistry*, 2nd edn. Harper and Row, New York

McEwen, W. K. (1936). *J. Am. chem. Soc.* **58**, 1124

Schrock, R. R., and Parshall, G. W. (1976). *Chem. Rev.* **76**, 263

Silver, M. S., Schafer, P. R., Nordlander, J. E., Rüchardt, C., and Roberts, J. D. (1960). *J. Am. chem. Soc.* **82**, 2646

Wakefield, B. J. (1974). *The Chemistry of Organolithium Compounds*. Pergamon Press, Oxford, p. 184

Wilson, D. R., Dilullo, A. A., and Ernst, R. D. (1980). *J. Am. chem. Soc.* **102**, 5928

Zimmerman, H. E., and Zweig, A. (1961). *J. Am. chem. Soc.* **83**, 1196

GENERAL READING

Bates, R. B., and Ogle, C. A. (1983). *Carbanion Chemistry*. Springer, Berlin
Buncel, E., and Durst, T. (eds) (1980). *Comprehensive Carbanion Chemistry*, Part A. Elsevier, Amsterdam
Cram, D. J. (1965). *Fundamentals of Carbanion Chemistry*. Academic Press, New York
Flood, T. C. (1981). Stereochemistry of reactions of transition metal–carbon sigma bonds. *Top. Stereochem.* **12**, 37.
Lowry, T. H., and Richardson, K. S. (1981). *Mechanism and Theory in Organic Chemistry*, 2nd edn. Harper and Row, New York

5

REACTIONS OF ORGANIC GROUPS BONDED TO METALS IN WHICH THE METAL – CARBON BOND IS RETAINED

The reactivity of organic groups bonded to metals is an interesting study in itself, but it becomes particularly important when the ramifications to organic synthesis are considered. At its simplest, metal complex formation can be used to block a reactive site while comparatively harsh conditions are used at another part of the molecule. In more sophisticated cases, the bonding of an organic group to a metal can be used to modify reactivity and regiospecificity. The reactions described in this chapter may be used to modify the organic group in an organometallic reagent which is used in organic synthesis.

5.1 σ-BONDED COMPOUNDS

As we have seen from previous chapters, the metal–carbon bond is a reactive point in an organometallic compound. However, it is possible to carry out reactions on the organic groups which do not involve metal–carbon bond cleavage. Transition metal σ-bonded compounds provide few examples of this type of reaction, although some are known. For example, the σ-cyclopentadienyl platinum compound, **I**, undergoes a Diels–Alder reaction with hexafluorobutyne (equation (5.1); Clark *et al.*, 1975).

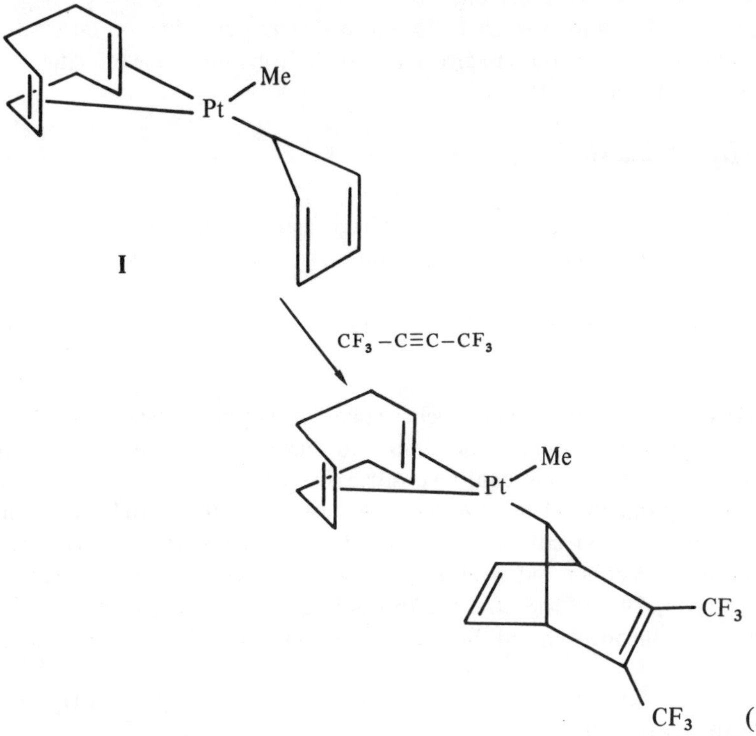

$$\text{CF}_3-\text{C}\equiv\text{C}-\text{CF}_3$$

(5.1)

There is also a considerable amount of data on cobalt alkyls which have been studied in an attempt to understand the behaviour of vitamin B_{12} coenzyme (Tsutsui and Courtney, 1977). In a study of a model system it was found that **II** and **III** interconvert in chloroform solution to an equilibrium mixture in a way which is similar to the 1,2-rearrangements found in biological systems (equation (5.2); Bury *et al.*, 1978).

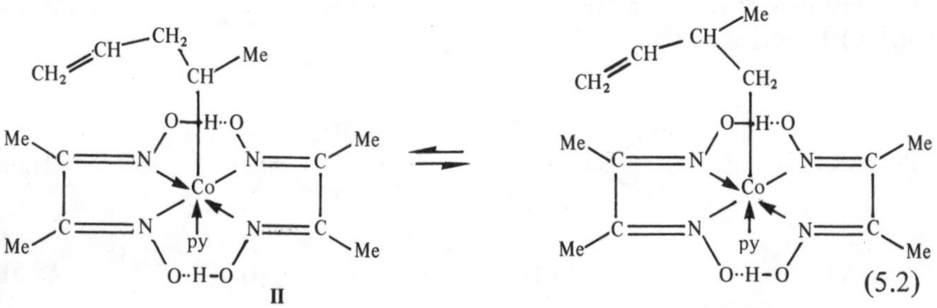

(5.2)

5.1.1 Main Group Compounds with Saturated Organic Groups

The alkyl derivatives of the Group IVb elements provide an interesting comparative series. The metal–carbon bond strength falls from silicon to lead, and so there are far more examples of reactions occurring without metal–carbon cleavage for silicon than the other elements.

The methyl groups of tetramethylsilane undergo chlorination on treatment with chlorine and irradiation to give chloromethyltrimethylsilane (equation (5.3); Whitmore and Sommer, 1946).

$$Me_4Si \xrightarrow[h\nu]{Cl_2} Me_3SiCH_2-Cl \qquad (5.3)$$

Both silicon and germanium alkyls undergo a similar reaction with sulphuryl chloride and dibenzoyl peroxide (equation (5.4); Vyazankin et al., 1963).

$$Et_4M \xrightarrow[(M=Si,Ge)]{SO_2Cl_2,\,(PhCOO)_2} Et_3M-\underset{\underset{Cl}{|}}{C}H-Me \qquad (5.4)$$

The α-halo compounds generally behave like the analogous organic alkyl halides. However, the β-haloalkyl silicon compounds have an enhanced reactivity in reactions whose mechanisms require a positive charge on the β-carbon. For example, β-chloroethyltrimethylsilane, **IV**, has been estimated to be $10^{9.5}$ times more reactive than methyl chloride in S_N1 solvolysis (Cook et al., 1970). This is known as the β-effect and it also occurs with germanium compounds (Jarvie, 1970; Colvin, 1981). The effect is thought to arise because of an interaction between the silicon atom and the β-carbon, as shown in **V** (see p. 178).

$$Me_3SiCH_2-CH_2-Cl \qquad\qquad Me_3Si\overset{\diagup CH_2}{\underset{\diagdown CH_2^+}{|}}$$

IV   **V**

Reactions of saturated organotin compounds in which the tin–carbon bond is retained are rarer than with silicon, and this is more typical of the majority of σ-bonded organometallic compounds. Dichlorocarbene generated from the mercury derivative, **VI**, attacks the ring of dimethylstannacyclohexane, **VII**, to give **VIII** (equation (5.5); Seyferth et al., 1970).

$$PhHgCCl_2Br \;+\; \underset{Me}{\overset{Me}{>}}Sn\!\!\bigcirc \longrightarrow \underset{Me}{\overset{Me}{>}}Sn\!\!\bigcirc_{CCl_2H} \;+\; PhHgBr$$

VI   **VII**   **VIII** CCl_2H (5.5)

5.1.2 Main Group Compounds with Unsaturated Organic Groups

A considerable chemistry has been developed using silicon, germanium and tin compounds containing double bonds. Electrophilic addition to double bonds can be carried out under controlled conditions. However, even with silicon, which is the most favoured case, silicon–carbon cleavage occurs under fairly mild conditions.

Owing to the β-effect mentioned above, silicon influences the regiospecificity of electrophilic addition. In vinyltrimethylsilane, **IX**, because of the β-effect, the β-carbon atom has enhanced electrophilic character. This results in the addition of hydrogen bromide occurring in anti-Markownikoff orientation (see equation (5.6); Sommer *et al.*, 1954; Seyferth, 1962).

$$Me_3Si-CH=CH_2 + HBr \longrightarrow Me_3Si-CH_2-CH_2Br \qquad (5.6)$$

IX

However, the allyl analogue, **X**, undergoes Markownikoff addition (equation (5.7); Sommer *et al.*, 1948).

$$Me_3Si-CH_2-CH=CH_2 + HCl \longrightarrow Me_3Si-CH_2-CHCl-Me \qquad (5.7)$$

X

The reactions of allyl groups bonded to silicon have been extensively used in organic synthesis (see p. 179; Colvin, 1981; Fleming, 1981).

Silyl alkynes are moderately stable to electrophilic cleavage and free-radical addition of hydrogen bromide to trimethylsilylacetylene is possible (equation (5.8); Komarov and Yarosh, 1971).

$$Me_3Si-C{\equiv}C-H \xrightarrow[\text{(PhCOO)}_2]{\text{HBr}} \begin{array}{c} Me_3Si \\ \diagdown \\ C \end{array} = \begin{array}{c} H \\ \diagup \\ C \end{array} \begin{array}{c} \text{94 per cent } \textit{trans} \\ \text{6 per cent } \textit{cis} \end{array} \qquad (5.8)$$

The alkyne–silicon linkage is very readily cleaved by nucleophilic reagents.

Reactions may be carried out on unsaturated groups bonded to germanium, but the tendency for germanium–carbon cleavage is greater than with silicon. Addition of hydrogen bromide to vinyl groups is again anti-Markownikoff (equation (5.9); Mazerolles and Lesbre, 1959).

$$Bu_3Ge-CH=CH_2 + HBr \xrightarrow{-80°C} Bu_3GeCH_2CH_2Br \qquad (5.9)$$

In a comparative study of silicon and germanium vinyl compounds, it was found that the germanium compound was slightly more reactive in the Diels–Alder reaction with 2,3-dimethylbutadiene (equation (5.10); Minot *et al.*, 1976).

$$\underset{CH_2}{\overset{\displaystyle Cl_3M\diagdown}{\underset{\|}{CH}}} + \overset{\displaystyle}{\diagup\!\!\diagdown}\underset{(M = Si,\ Ge)}{\longrightarrow} \overset{\displaystyle Cl_3M\diagdown}{\diagup\!\!\diagdown} \qquad (5.10)$$

Alkyne groups bonded to germanium may be brominated without cleavage of the germanium carbon bond (equation (5.11); Davidsohn and Henry, 1966).

$$Ge(C{\equiv}CH)_4 + 4Br_2 \xrightarrow{CCl_4} Ge(CBr{=}CHBr)_4 \qquad (5.11)$$

The tin–carbon bond is very susceptible to cleavage by electrophilic reagents, and most successful addition reactions involve free radicals. For example, allyl-triethyltin, **XI**, reacts with thiophenol in the presence of dibenzoyl peroxide to give **XII**, the anti-Markownikoff orientation, being a result of the free radical mechanism (equation (5.12); Ayrey *et al.*, 1972).

$$Et_3Sn{-}CH_2{-}CH{=}CH_2 + PhSH \xrightarrow[h\nu]{(PhCOO)_2} Et_3Sn{-}CH_2CH_2CH_2SPh \quad (5.12)$$

XI   **XII**

Under carefully controlled conditions ionic reactions can be carried out successfully without tin–carbon cleavage, for example the oxidation of **XIII** to **XIV** using very dilute potassium permanganate (equation (5.13); Murphy and Poller, 1982).

$$Ph_3Sn{-}CH_2{-}CH{=}CH_2 \xrightarrow[H_2O/tBuOH]{Very\ dilute\ KMnO_4} \underset{\underset{OH\ \ OH}{|\ \ \ |}}{Ph_3Sn{-}CH_2{-}CH{-}CH_2}$$

XIII   **XIV**   (5.13)

Acetylenic groups bonded to tin and lead are very readily cleaved by ionic reagents, but cycloaddition reactions are known, for example the Diels–Alder reaction (equation (5.14); Seyferth and Evnin, 1967).

$$\underset{Me_3Sn}{\overset{Me_3Sn}{\underset{|}{\overset{|}{\underset{C}{\overset{C}{\interleave}}}}}} + \underset{Cl}{\overset{Cl}{\diagup\!\!\diagdown}} \xrightarrow[25\ h]{Bu_2O,\ reflux} \qquad (5.14)$$

5.1.3 Main Group Compounds with Aryl Groups

The metal–carbon bond in aryl derivatives of the Group IV elements is readily cleaved by electrophiles (see p. 88). Thus an electrophile E^+ will replace the metal substituent in the arene ring, the counterion, X^-, becoming attached to the metal (equation 5.15).

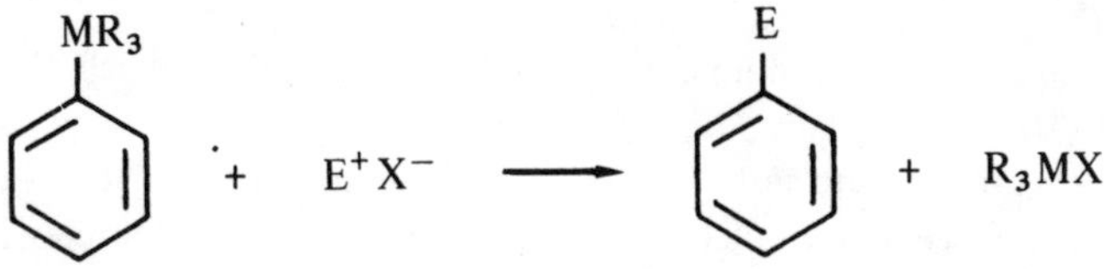

$$\text{(5.15)}$$

In *m*-methoxyphenyltrimethylsilane, electrophilic substitution occurs without loss of the silicon substituent to give **XV** (equation (5.16); Eaborn and Webster, 1960).

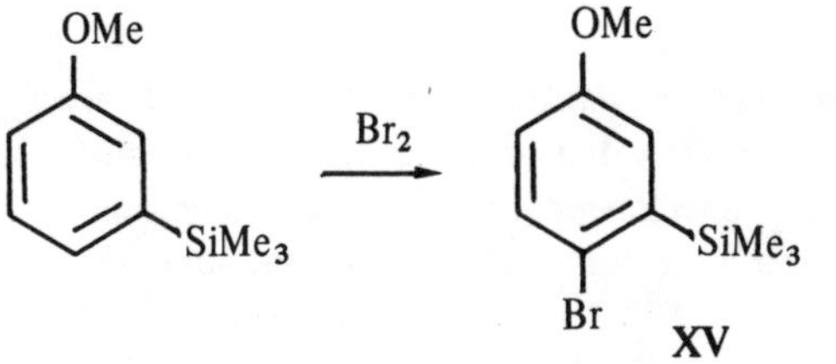

$$\text{(5.16)}$$

When Group Vb aryl compounds are considered we find that the bond is now able to withstand much more vigorous conditions. Both benzene arsonic acid, **XVIa**, and benzene stibonic acid, **XVIb**, may be nitrated using nitric and sulphuric acid mixture (equation 5.17).

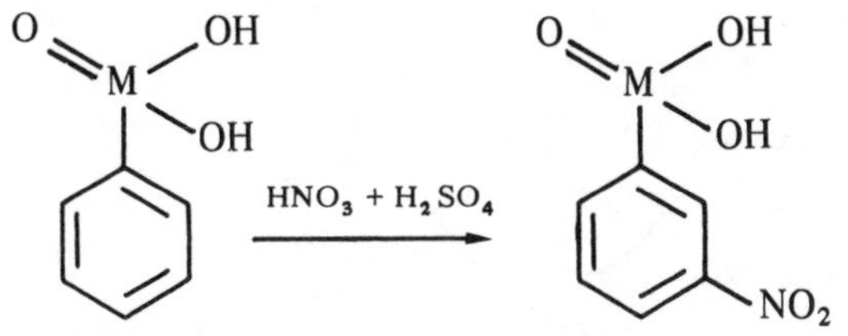

$$\text{(5.17)}$$

XVIa, M = As
XVIb, M = Sb

The nitro group enters *meta* in both cases, but vigorous conditions are required to bring about the nitration in the arsonic acid case (Schmidt, 1920; Freedman and Doak, 1959).

5.2 π-BONDED COMPLEXES

Many π-complexes of the transition metals possess aromatic character. In the case of benzene and η^5-cyclopentadienyl complexes this is predicted by the Hückel rule which requires $(4n + 2)$ π-electrons for aromatic properties. Cyclobutadiene can be stabilised by coordination to transition metals and these complexes also undergo reactions typical of aromatic systems; for example, electrophilic substitution. Molecular orbital calculations show that there are effectively six π-electrons in the cyclobutadiene–metal system and it is said to be metalloaromatic (Bursten and Fenske, 1979).

Another important feature in the chemistry of π-complexes is their reaction with nucleophiles. In general a nucleophile may attack a metal π-complex at the ligand or the metal atom, and there has been much controversy over the last 30 years as to which site is favoured. Some of these questions have now been settled and in general the metal is not directly involved, although its presence markedly alters the reactivity compared with the free organic ligand (Pauson, 1977, 1980).

5.2.1 Ferrocene

Ferrocene undergoes Friedel–Crafts acylation very readily and in much of its organic chemistry it resembles an activated benzene such as anisole. Ferrocene, **XVII**, undergoes acylation 10^6 times faster than benzene. Using acetyl chloride and aluminium chloride the diacetyl compound, **XVIII**, is obtained (equation 5.18). There is only one isomer of **XVIII** because of free rotation of the cyclopentadienyl rings.

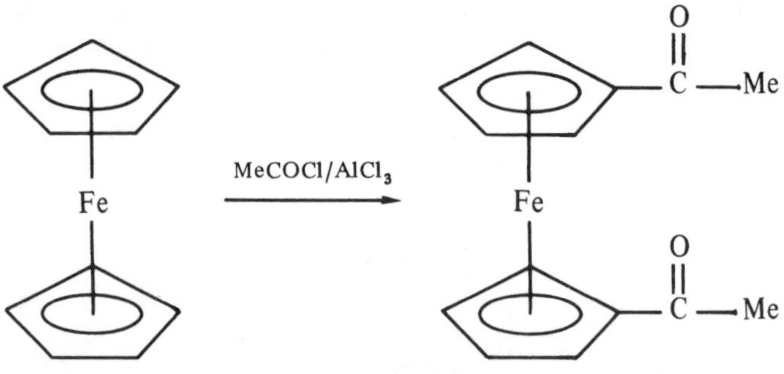

$$\text{(5.18)}$$

$$\text{XVII} \qquad \text{XVIII}$$

When disubstitution occurs, as in this case, the second acetyl group enters the second ring because the first acetyl group deactivates the ring to which it is bonded. Under mild conditions the monosubstitution product, **XIX**, may be obtained (equation 5.19).

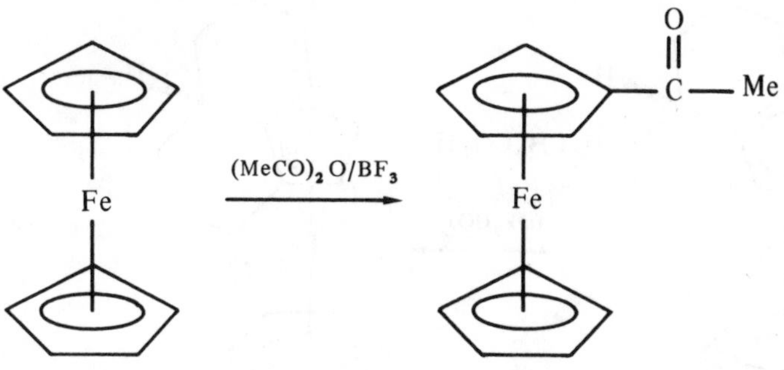

$$(5.19)$$

XVII   **XIX**

There is a considerable range of reagents which may be used to carry out electrophilic substitution on ferrocene. However, the electrophile must not be oxidising because oxidation to the ferricinium cation, Cp_2Fe^+, gives a positively charged species which is not attacked by electrophiles. As a result, nitration cannot be carried out and sulphonation requires the use of the sulphur trioxide: pyridine adduct or chlorosulphonic acid (Dublitz and Rinehardt, 1969).

The mechanism of electrophilic substitution of ferrocene has been the subject of much debate since the reaction was discovered. The modern view is that electrophilic substitution does not involve direct participation of the iron atom. This was established by Rosenblum and Abbate (1966), who devised a very elegant experiment to demonstrate the point. The experiment involved the cyclisation of two isomeric ferrocene carboxylic acids, **XX** (the *exo* isomer), and **XX** (the *endo* isomer), to give **XXI** (*exo*) and **XXI** (*endo*), respectively (equations 5.20a and 5.20b).

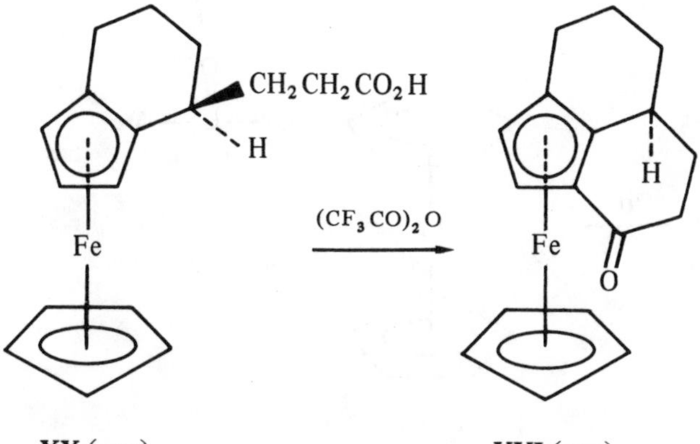

$$(5.20a)$$

XX (*exo*)   **XXI** (*exo*)

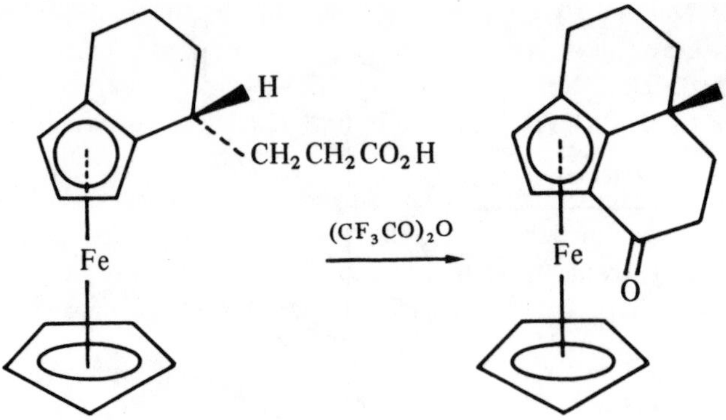

(5.20b)

XX (*endo*) **XXI** (*endo*)

It was found that the *exo* isomer, in which it is more difficult for the side chain to reach to the iron atom, cyclised faster than the *endo* isomer. If the iron atom were the initial site of attack for the electrophilic acylium ion, then the *endo* isomer should have cyclised more quickly than the *exo*.

Ferrocene derivatives also undergo nucleophilic substitution reactions. The mechanism of the reaction has not been established and it should be noted that

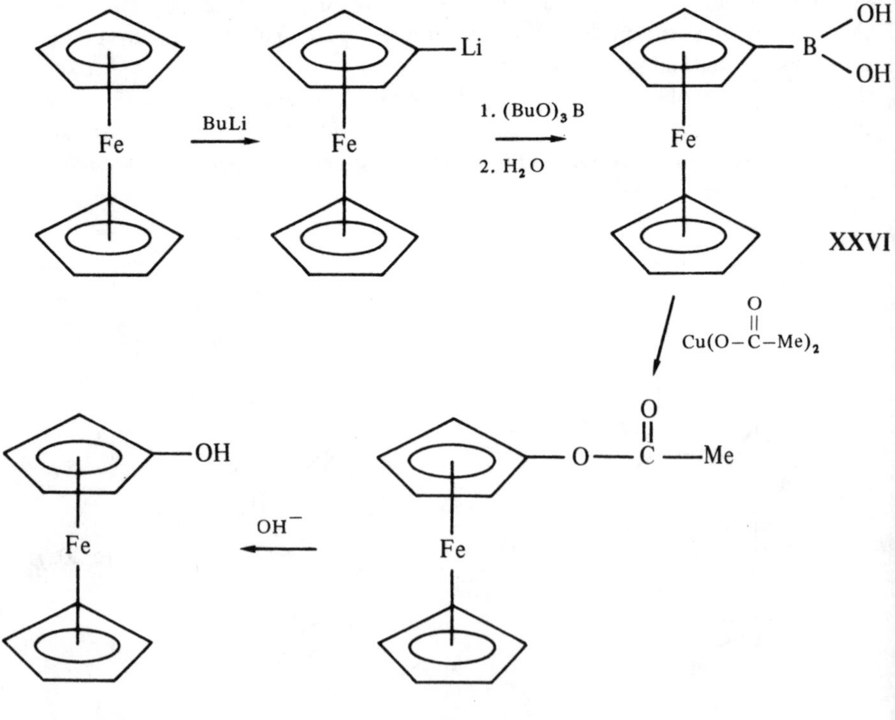

XXVI

XXVIII **XXVII**

Scheme *i*

the reactions are catalysed by copper(II) and may not be simple displacement reactions. Scheme *i* also illustrates the metallation of ferrocene and its conversion to the intermediate ferrocene boronic acid (**XXVI**) which undergoes nucleophilic substitution to give **XXVII** which on hydrolysis yields hydroxyferrocene, **XXVIII**.

Substituent effects in ferrocenes have been investigated, but regiospecificity is rather lower than found in substituted benzenes. The acetylation of monomethylferrocene, **XXII**, by acetic anhydride and boron trifluoride gave **XXIII**, **XXIV** and **XXV** in the ratio of 1:1.2:1.3 (equation (5.21); Slocum and Ernst, 1970).

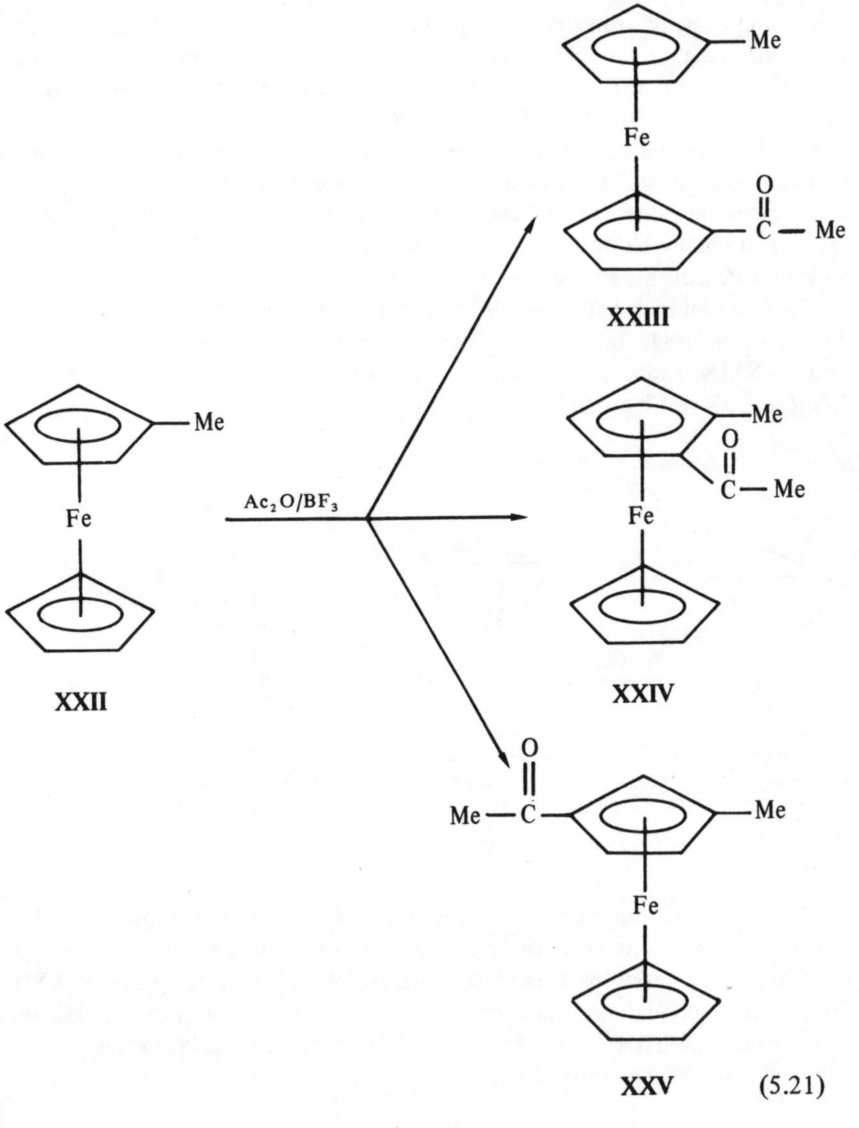

(5.21)

Electrophilic substitution using H^+ or D^+ (such as protodesilylation or H/D exchange) was thought to be a special case and involve direct participation of metal electrons. This now seems unlikely. Under the conditions of most reactions, a π-complex between the proton and the cyclopentadienyl ring is the most probable intermediate, although n.m.r. studies show that protonation at the iron atom does occur at high acid concentration (Cerichelli et al., 1977).

5.2.2 η^5-Cyclopentadienyl Compounds other than Ferrocene

Although ferrocene is by far the most widely studied of all η^5-cyclopentadienyl compounds, many other compounds have been investigated. A comparative study showed that the rate of reaction towards acetyl chloride/aluminium trichloride decreases in the sequence $Cp_2Fe >$ anisole $> Cp_2Ru > CpMn(CO)_3 > Cp_2Os \sim C_6H_6 > CpRe(CO)_3$ (Fischer et al., 1967).

Cobaltocene and nickelocene do not appear in this list because they do not undergo electrophilic substitution, although they do react with electrophiles. Cobaltocene has one electron above the ideal configuration of ferrocene. As a result it is readily oxidised to form the very stable cobalticinium ion, Cp_2Co^+, which is isoelectronic with ferrocene.

The reaction of cobaltocene with organic chlorides is particularly interesting. The reaction leads to disproportionation of the Co(II), to the Co(I) compound, **XXIX**, and cobalticinium chloride, **XXX** (equation (5.22); Fischer and Herberich, 1961; Churchill, 1965).

$$\text{(5.22)}$$

XXIX   **XXX**

Disproportionation reactions of this type have been used to produce other substituted cyclopentadiene complexes such as **XXXI** (scheme ii) which undergo ring expansion reactions to give cationic cyclohexadienyl complexes, **XXXII**. These, in turn, undergo nucleophilic addition to give neutral cyclohexadiene complexes, **XXXIII** (Herberich et al., 1970; Herberich and Schwarzer, 1970; Herberich and Michelbrink, 1970).

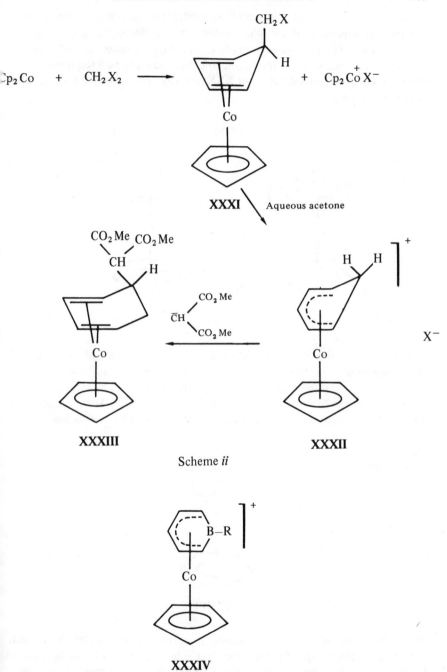

Scheme *ii*

When alkyl boron halides are employed in place of CH_2X_2, a borabenzene complex, **XXXIV**, is formed (Herberich *et al.*, 1979).

The stereochemistry of addition of nucleophiles to the cobalticinium ion was, like the mechanism of electrophilic substitution of ferrocene, a controversial issue (Pauson, 1980). It is now generally accepted that in most cases additions of this type occur directly on the ligand to give an *exo* product (equation 5.23). The x-ray structure determination of **XXXV** clearly demonstrated this (Churchill and Mason, 1964).

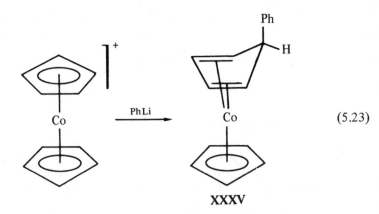

$$\qquad\qquad\qquad\qquad\qquad\qquad\qquad\qquad\qquad (5.23)$$

XXXV

The electron complement of nickelocene is two above that of ferrocene and this causes a marked difference in reactivity. The reactions of nickelocene can be understood by considering it as an η^5,η^3 system, with an uncoordinated double bond in one of the five membered rings, as shown in **XXXVI**. Although structure **XXXVI** offers a convenient rationalisation of the reactivity of nickelocene, it

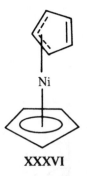

XXXVI

should be borne in mind that structurally nickelocene is a sandwich molecule with D_{5d} symmetry (p. 64). Thus the double bond in **XXXVI** may be reduced by sodium amalgam and ethanol to give **XXXVII**, and it undergoes a [2+2] cyclo-addition with tetrafluoroethene to give **XXXVIII** (McBride *et al.*, 1962). Both **XXXVII** and **XXXVIII** are diamagnetic, whereas nickelocene is paramagnetic.

XXXVII   **XXXVIII**

5.2.3 η^6-Arene Complexes

Dibenzene chromium is oxidised under the conditions of the Friedel–Crafts reaction, but tricarbonylarenechromium complexes do undergo electrophilic substitution. The ratio of *ortho:meta:para* isomers produced by electrophilic substitution of substituted arene complexes differs from that of the free ligand. However, these differences in reactivity towards electrophiles have been little used in organic synthesis in comparison with the substantial amount of work with nucleophiles and bases (Jackson and Jennings, 1969).

The complexation of one benzene ring of diphenyl to a tricarbonylchromium group restricts the rotation about the bond connecting the two rings (Eyer *et al.*, 1981).

The most thoroughly investigated area of the chemistry of coordinated arenes is the reactivity towards nucleophiles and bases. The reaction can proceed in five different ways depending on the reagent and substrate (scheme *iii*).

The reaction of tricarbonylchlorobenzenechromium (**XLI**, X = Cl) with methoxide ion was reported in 1959 (Nicholls and Whiting, 1959). It was found that the chlorine atom was displaced much more readily than in free chlorobenzene. This observation suggested that the tricarbonylchromium group is electron withdrawing, and so increases the susceptibility of the chlorine to nucleophilic displacement. The effect of coordinating chlorobenzene to tricarbonylchromium is about the same as introducing a *para*-nitro group into chlorobenzene. This was supported by measurements of the dissociation constants of the benzoic acids **XLV–XLVII**.

The electron withdrawing effect of the chromium tricarbonyl group has also been established from studies of the hydrolysis of benzoate esters (Klopman and Calderazzo, 1967) and benzyl halides (Ceccon and Sartori, 1973).

The reaction of phenyllithium with tricarbonylbenzenechromium followed by alkylation leads to a carbene complex, **XXXIX**, by attack on a carbon monoxide group (Fischer *et al.*, 1976). The electron withdrawing effect of the tricarbonyl-

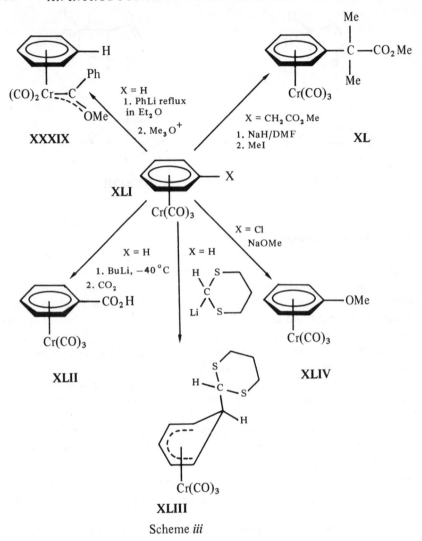

Scheme *iii*

chromium group increases the acidity of the side chain such that dialkylation to produce **XL** can be carried out using very mild conditions (Simmoneaux and Jaouen, 1979). Metallation of benzene coordinated to tricarbonylchromium occurs on reaction with butyllithium at $-40°C$, as shown by carbonation to give **XLII** (Rausch et al., 1979). Metallation of dibenzenechromium has also been carried out.

The final mode of reaction to consider for tricarbonylarenechromium complexes is nucleophilic addition to give cyclohexadienyl derivatives. This has been proved by an x-ray structure determination in the case of **XLIII** (Semmelhack et al., 1979). This reaction, followed by decomplexation using iodine, has applications in organic synthesis (see p. 208).

Tropylium complexes of transition metals may be synthesised by hydride ion abstraction from the corresponding cycloheptatriene complexes (equation (5.24); Dauben and Honnen, 1958). Nucleophilic addition to tropylium complexes generally gives substituted cycloheptatriene complexes, although there are some complicating factors (Pauson, 1980).

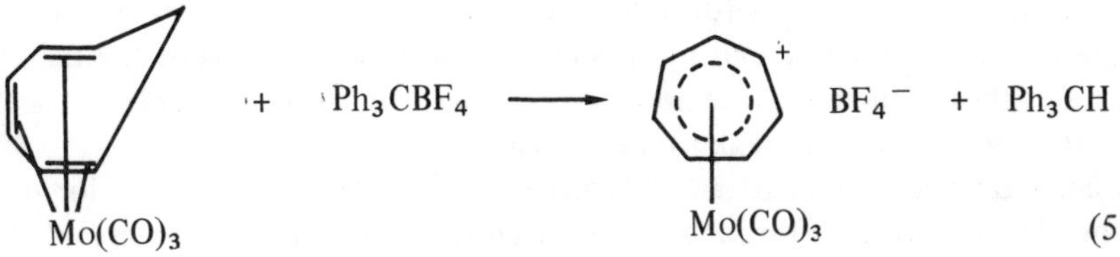

$$+ \quad Ph_3CBF_4 \longrightarrow \quad BF_4^- \quad + \quad Ph_3CH \tag{5.24}$$

One other series of arene complexes has been studied extensively and this is the cyclopentadienyl(arene)iron(II) cations. Nucleophilic addition occurs preferentially on the six-membered ring in these complexes (equation (5.25); Pauson, 1980).

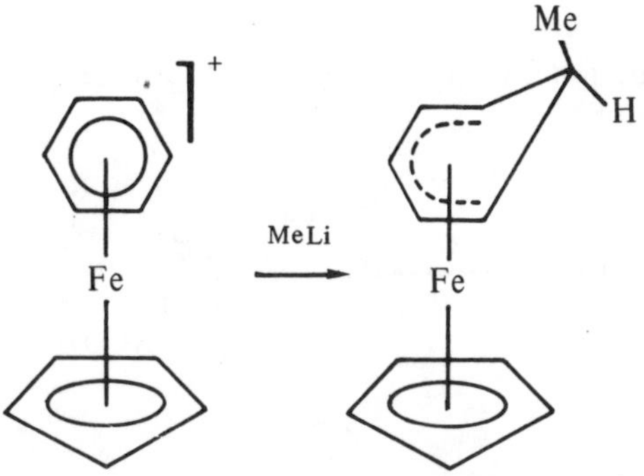

$$\xrightarrow{MeLi} \tag{5.25}$$

When halogen is present on either ring it can be displaced by methoxide ion to give the methoxy derivative (equation 5.26).

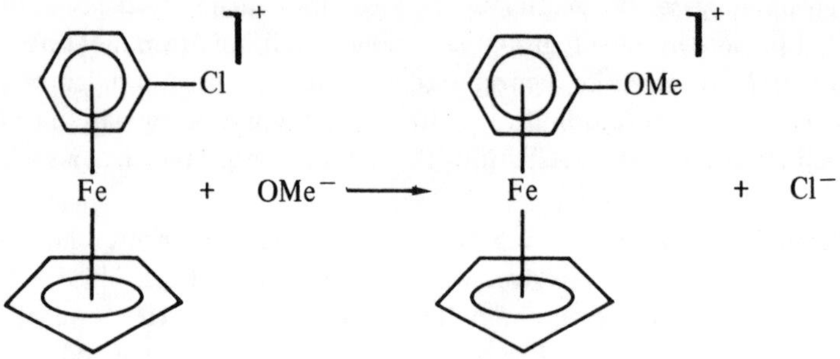

$$\text{(5.26)}$$

Comparative studies show that chlorine in the benzene ring is about 10^3 times more reactive than chlorine in the cyclopentadienyl ring (Nesmeyanov et al., 1968).

5.2.4 Tricarbonyl(diene)iron Complexes

The chemistry of tricarbonyl(diene)iron complexes, including cyclobutadiene complexes, has been developed to produce a broad understanding of the field. As a result these compounds are now used in some sophisticated organic syntheses (Pettit, 1975; Birch and Jenkins, 1976; Pearson, 1980).

A common tendency in tricarbonyl(diene)iron complexes is the formation of cations. These may be formed by hydride abstraction from a diene complex using triphenylmethyl fluoroborate (equation 5.27).

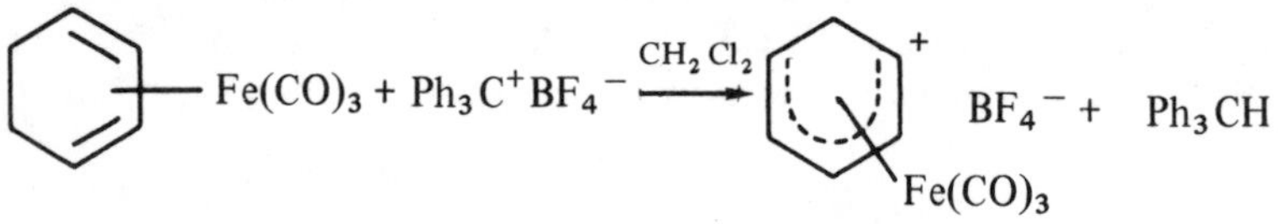

$$\text{(5.27)}$$

An alternative preparation is by protonation (equations 5.28 and 5.29).

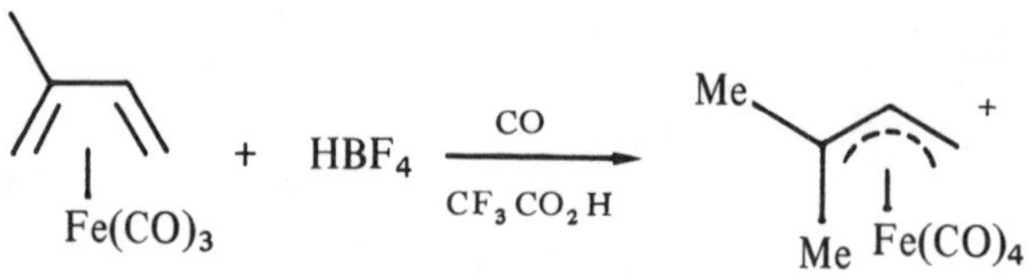

$$\text{(5.28)}$$

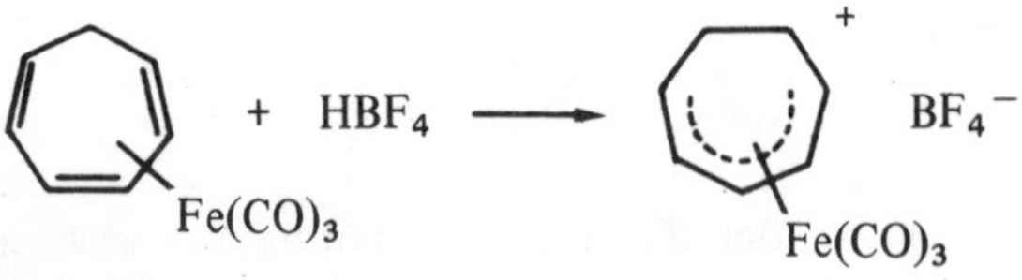

$$\text{(5.29)}$$

XLVIII

The reaction in equation (5.28) requires carbon monoxide to give a compound which obeys the eighteen electron rule, and in equation (5.29) the starting material, **XLVIII**, has only two of its three double bonds coordinated to the iron atom. These cations are susceptible to nucleophilic attack and behave as stabilised carbonium ions, although the nomenclature adopted regards them as Fe(II) derivatives with anionic ligands. The nucleophilic addition occurs at the end of the dienyl group (equation 5.30).

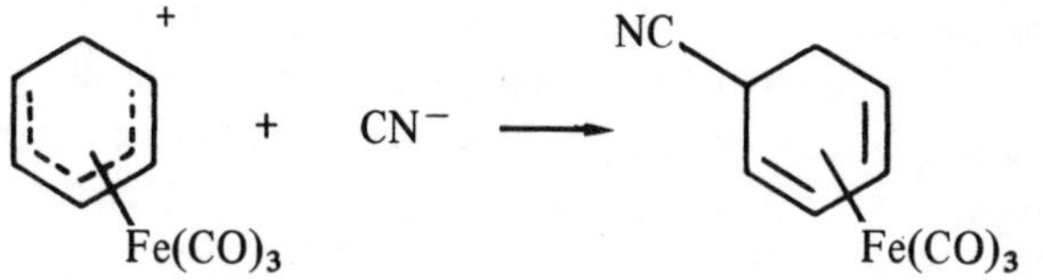

$$ \text{(5.30)} $$

The dienyl cations will also react with activated arenes to bring about electrophilic substitution (equation 5.31).

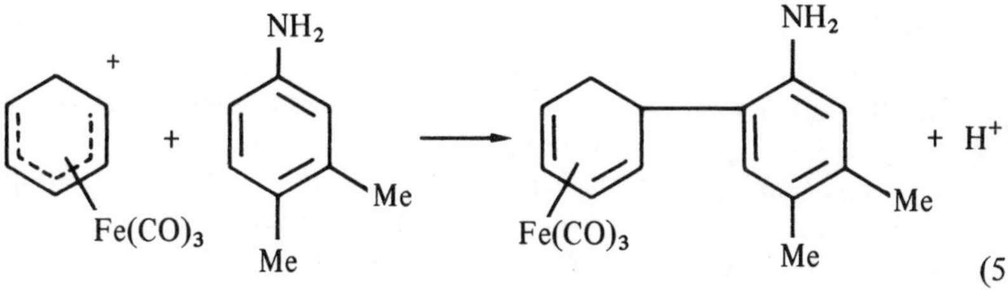

$$ \text{(5.31)} $$

5.2.5 Diene Complexes of Pd(II) and Pt(II)

Diene complexes of Pd(II) and Pt(II) undergo nucleophilic addition at the ligand (equation 5.32).

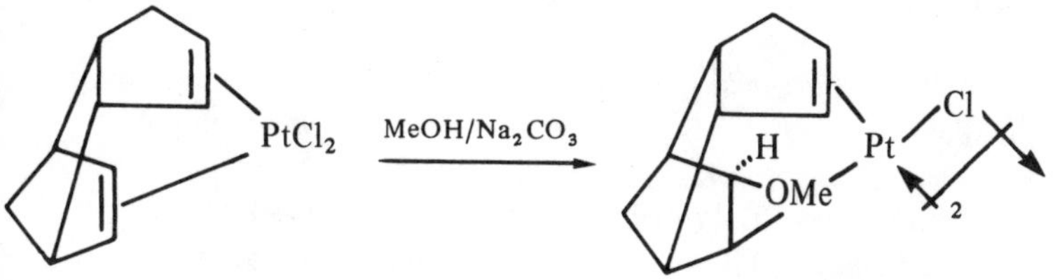

$$ \text{(5.32)} $$

XLIX

The dimeric product (**XLIX**) is bonded to the ligand via one σ- and one π-bond. An x-ray structure showed that the MeO group is *exo* and this implies that the attack occurs directly on the ligand without precoordination at the metal (Whitla *et al.*, 1966).

5.2.6 Generalisations Concerning Nucleophilic Addition to π-complexes

We have quoted a number of examples of nucleophilic addition during this chapter. In general it appears that this occurs directly on the ligand, without pre-coordination at the metal.

Another point to consider is which π-bonded ligand will be preferentially attacked in competitive situations. An early series dealt only with cyclic ligands (Efraty and Maitlis, 1967). More recently it has been suggested that two main criteria determine the reactivity. Ligands are classified as to whether (a) the hapto number is even or odd; (b) the π-electronic system is cyclically or acyclically conjugated. The cyclically conjugated or 'closed' π-systems are ligands such as η^6-benzene or η^5-cyclopentadienyl. The acyclically conjugated systems or 'open' π-systems are ligands such as η^3-allyl or η^5-cyclohexadienyl.

Using this classification it is found that generally the reactivity is even > odd and open > closed (Davies $et\ al.$, 1978).

REFERENCES

Ayrey, G., Brasington, R. D., and Poller, R. C. (1972). $J.\ organomet.\ Chem.$ **35**, 105

Birch, A. J., and Jenkins, I. D. (1976). In $Transition\ Metal\ Organometallics\ in\ Organic\ Synthesis$, Vol. 1 (H. Alper, ed.). Academic Press, New York, p. 1

Bursten, B. E., and Fenske, R. F. (1979). $Inorg.\ Chem.$ **18**, 1760

Bury, A., Ashcroft, M. R., and Johnson, M. D. (1978). $J.\ Am.\ chem.\ Soc.$ **100**, 3217

Ceccon, A., and Sartori, S. (1973). $J.\ organomet.\ Chem.$ **50**, 161

Cerichelli, G., Illuminati, G., Ortaggit, G., and Guilani, A. M. (1977). $J.\ organomet.\ Chem.$ **127**, 357

Churchill, M. R. (1965). $J.\ organomet.\ Chem.$ **4**, 258

Churchill, M. R., and Mason, R. (1964). $Proc.\ R.\ Soc.\ A$ **279**, 191

Clark, H. C., Ibbott, D. G., Payne, N. C., and Shaver, A. (1975). $J.\ Am.\ chem.\ Soc.$ **97**, 3555

Colvin, E. W. (1981). $Silicon\ in\ Organic\ Synthesis$. Butterworth, London, p. 15

Cook, M. A., Eaborn, C., and Walton, D. R. M. (1970). $J.\ organomet.\ Chem.$ **24**, 301

Dauben, H. J., and Honnen, L. R. (1958). $J.\ Am.\ chem.\ Soc.$ **80**, 5570

Davidsohn, W., and Henry, M. C. (1966). $J.\ organomet.\ Chem.$ **5**, 29

Davies, S. G., Green, M. L. H., and Mingos, D. M. P. (1978). $Tetrahedron$ **34**, 3047

Dublitz, D. E., and Rinehart, K. L. (1969). $Org.\ React.$ **17**, 1

Eaborn, C., and Webster, D. E. (1960). $J.\ chem.\ Soc.$ 179

Efraty, A., and Maitlis, P. M. (1967). $J.\ Am.\ chem.\ Soc.$ **89**, 3744

Eyer, M., Schlögl, K., and Schölm, R. (1981). $Tetrahedron$ **37**, 4239

Fischer, E. O., and Herberich, G. E. (1961). $Chem.\ Ber.$ **94**, 1517

Fischer, E. O., Von Foerster, M., Kreiter, C. G., and Schwarzhans, K. E. (1967). *J. organomet. Chem.* **7**, 113

Fischer, E. O., Stückler, P., Beck, H.-J., and Kreissl, F. R. (1976). *Chem. Ber.* **109**, 3089

Fleming, I. (1981). *Chem. Soc. Rev.* **10**, 83

Freedman, L. D., and Doak, G. O. (1959). *J. org. Chem.* **24**, 1590

Herberich, G. E. and Michelbrink, R. (1970). *Chem. Ber.* **103**, 3615

Herberich, G. E., and Schwarzer, J. (1970). *Chem. Ber.* **103**, 2016

Herberich, G. E., Engelke, C., and Pahlmann, W. (1979). *Chem. Ber.* **112**, 607

Herberich, G. E., Greiss, G., and Heil, F. (1970). *J. organomet. Chem.* **22**, 273

Jackson, W. R., and Jennings, W. B. (1969). *J. chem. Soc. (B)* 1221

Jarvie, A. W. P. (1970). *Organomet. Chem. Rev. A* **6**, 153

Klopman, G., and Calderazzo, F. (1967). *Inorg. Chem.* **6**, 977

Komarov, N. V., and Yarosh, O. G. (1971). *Bull. Acad. Sci. USSR Div. chem. Sci.* **20**, 1476

McBride, D. W., Pruett, R. L., Pitcher, E., and Stone, F. G. A. (1962). *J. Am. chem. Soc.* **84**, 497

Mazerolles, P., and Lesbre, M. (1959). *C. r. hebd. Seanc. Acad. Sci., Paris* **248**, 2018

Minot, C., Laportiere, A., and Dubac, J. (1976). *Tetrahedron* **32**, 1523

Murphy, J., and Poller, R. C. (1982). *Chem. Ind. (Lond.)* 724

Nesmeyanov, A. N., Volkenau, N. A., Isaeva, L. S., Bolesova, I. N. (1968). *Dokl. Chem. (Engl. Transl.)* **183**, 1042

Nicholls, B., and Whiting, M. C. (1959). *J. chem. Soc.* 551

Pauson, P. L. (1977). *Pure appl. Chem.* **49**, 839

Pauson, P. L. (1980). *J. organomet. Chem.* **200**, 207

Pearson, A. J. (1980). *Acc. chem. Res.* **13**, 463

Pettit, R. (1975). *J. organomet. Chem.* **100**, 205

Rausch, M. D., Moser, G. A., and Lee, W. A. (1979). *Synth. React. inorg. metal-org. Chem.* **9**, 357

Rosenblum, M., and Abbate, F. W. (1966). *J. Am. chem. Soc.* **88**, 4178

Schmidt, H. (1920). *Justus Leibigs Annln Chem.* **421**, 174

Semmelhack, M. F., Hall, H. T., Farina, R., Yoshifugi, M., Clark, G., Bargar, T., Hirotsu, K., and Clardy, J. (1979). *J. Am. chem. Soc.* **101**, 3535

Seyferth, D. (1962). *Prog. inorg. Chem.* **3**, 129

Seyferth, D., and Evnin, A. B. (1967). *J. Am. chem. Soc.* **89**, 1468

Seyferth, D., Washburne, C. C., Attridge, C. J., and Yamamoto, K. (1970). *J. Am. chem. Soc.* **92**, 4405

Simmoneaux, G., and Jaouen, G. (1979). *Tetrahedron* **35**, 2249

Slocum, D. W., and Ernst, C. R. (1970). *Organomet. Chem. Rev. A* **6**, 337

Sommer, L. H., Tyler, L. J., and Whitmore, F. C. (1948). *J. Am. chem. Soc.* **70**, 2872

Sommer, L. H., Bailey, D. L., Goldberg, G. M., Buck, C. E., Bye, T. S., Evans, F. J., and Whitmore, F. C. (1954). *J. Am. chem. Soc.* **76**, 1613

Tsutsui, M., and Courtney, A. (1977). *Adv. organomet. Chem.* **16**, 241

Vyazankin, N. S., Gladyshev, E. N., and Razuvaev, G. A. (1963). *Dokl. Chem. (Engl. Transl.)* **153**, 878

Whitla, W. A., Powell, H. M., and Venanzi, L. M. (1966). *Chem. Commun.* 310

Whitmore, F. C., and Sommer, L. H. (1946). *J. Am. chem. Soc.* **68**, 481

GENERAL READING

Barnett, K. W. (1974). The chemistry of nickelocene. *J. organomet. Chem.* **78**, 139

Deganello, G. (1979). *Transition Metal Complexes of Cyclic Polyolefins*. Academic Press, New York

Efraty, A. (1977). Cyclobutadienemetal complexes. *Chem. Rev.* **77**, 691

Leonova, E. V., and Kochetkova, N. S. (1973). Chemical reactions of cobalto-cene and nickelocene. *Russ. Chem. Rev. (Engl. Transl.)* **42**, 278

Rosenblum, M. (1965). *Chemistry of the Iron Group Metallocenes: Ferrocene, Ruthenocene and Osmocene*. Wiley, New York

Semmelhack, M., Clark, G. R., Garcia, J. L., Harrison, J. J., Thebtaranonth, Y., Wulf, W., and Yamashita, A. (1981). Addition of carbon nucleophiles to arene-chromium complexes. *Tetrahedron* **37**, 3957

Sheats, J. E. (1979). The chemistry of cobaltocene, cobalticinium salts and other cobalt sandwich compounds. *J. organomet. Chem. Library* **7**, 461

6

THE COORDINATION
CHEMISTRY OF
ORGANOMETALLIC
COMPOUNDS

6.1 GENERAL SURVEY

In this chapter we shall be mainly concerned with interactions of the Lewis acid–base type. Most of these interactions will involve the organometallic compound acting as a Lewis acid and receiving electrons from some donor species. It is also possible, in a formal sense, to regard an organometallic compound RM as arising from combination of the base R^- and acid M^+. These, and related, ideas have considerable impact on structure and reactivity in organometallic chemistry. The aim of this chapter is, primarily, to show that coordination is a concept which can be used to unify and systematise much material that occurs elsewhere in this book.

The continuing validity of Lewis's original definitions was emphasised in a valuable review in which various attempts to extend and quantify the original concepts are discussed (Jensen, 1978). It is clear that Lewis acids (electron pair acceptors) are of many types. In an early and useful classification, metal ions were shown to belong to one or other of two categories (Ahrland *et al.*, 1958). Stated simply, those metals which formed most stable complexes with ligands containing first row elements as donor atoms were labelled class 'a'. If, however, the complex with a second row (or heavier) element was more stable the metal ion was class 'b'. This classification was extended by Pearson (1963, 1968), whereby class 'a' acceptors became hard acids and class 'b' soft acids with a similar division of donors into hard and soft bases. Reactions between hard acids and hard bases, or between soft acids and soft bases, are favoured. In the original classification, some organometallic species were included and this list has subsequently been extended (Table 6.1). Thus, we would expect that the adduct

Table 6.1   Classification of Lewis acids and bases

Hard acids	*Soft acids*
$BeMe_2$, $AlMe_3$	$MeHg^+$, $GaMe_3$, $TlMe_3$
Me_2Sn^{2+}, $MeSn^{3+}$	RTe^+
Li^+, Si^{4+}, Sn^{4+}	M^0, Pt^{2+}, Cu^+, Ag^+, Au^+, Hg^{2+}
Hard bases	*Soft bases*
NH_3, RNH_2, H_2O, OH^-, ROH	R^-, C_2H_4, C_6H_6, CO
RO^-, R_2O, $MeCOO^-$, CO_3^{2-}	CN^-, R_3P, R_3As, R_2S
NO_3^-, PO_4^{3-}, SO_4^{2-}, F^-, Cl^-	RSH, RS^-, I^-

formed from trimethylaluminium and the hard base trimethylamine would be more stable than that formed from the softer trimethylphosphine and this is the case. It also follows that alkenes will complex with, for example, Fe(0) or Pt^{2+}, which are classified as soft acids. However, if carbanions are to be regarded as soft bases, then this HSAB concept predicts that alkylsilver compounds will be more stable than tetraalkylsilanes, and this is not so. Thus, although the HSAB concept is undoubtedly useful, it has limitations and is essentially qualitative. More quantitative approaches to Lewis acidity are discussed by Jensen (1978).

We have, so far, mentioned two aspects of coordination chemistry. One of these is that the formation of an organometallic compound can be regarded as a Lewis acid–base interaction, as in equations (6.1) and (6.2).

$$R^- + [MX_n]^+ \longrightarrow RMX_n \qquad (6.1)$$

$$CH_2{=}CH_2 + MX_n \longrightarrow \begin{array}{c} CH_2 \\ \| \\ CH_2 \end{array}\!\!{-}MX_{n-1} + X \qquad (6.2)$$

The second aspect is the ability of an existing organometallic compound to react with Lewis bases, as expressed generally in equation (6.3).

$$RM + D: \longrightarrow RM \leftarrow D \qquad (6.3)$$

In equations (6.1)–(6.3) the metal is the acid, but this may not be true for the Group V elements which bear lone pairs, thereby enabling their organic derivatives to function as Lewis bases. Using enthalpy of adduct formation measurements, it was shown that the relative base strengths were $Me_3P > Me_3As > Me_3Sb$ for reaction (6.4) (Mente *et al.*, 1975).

$$Me_3M + BX_3 \xrightarrow{\text{(M = P, As, Sb; X = Cl, Br)}} Me_3M \rightarrow BX_3 \qquad (6.4)$$

There are two other much more general mechanisms which are closely related to one another, whereby organometallic compounds can function as electron donors. Thus, when a dialkylmercury compound is added to a solution of tetracyanoethene, TCNE, transient colours due to charge transfer complexes are seen. The colours disappear as the alkene inserts into the mercury–carbon bond (scheme *i*; Kochi, 1980). Similarly, tetraethyltin reacts with TCNE in the dark at room temperature to give a 95 per cent yield of the insertion product (equation 6.5).

$$R_2Hg \;+\; TCNE \;\rightleftharpoons\; [R_2Hg, TCNE] \;\rightleftharpoons\; [R_2Hg^{\cdot+} \, TCNE^{\cdot-}]$$

$$\text{charge transfer} \qquad\qquad \text{electron transfer}$$
$$\text{complex} \qquad\qquad\qquad \text{complex}$$

$$\Updownarrow$$

$$RHgC(CN)_2 C(CN)_2 R$$

$$TCNE = (CN)_2 C{=}C(CN)_2$$

Scheme *i*

$$Et_4Sn + TCNE \longrightarrow Et_3SnC(CN)_2 C(CN)_2 Et \qquad (6.5)$$

This general area is discussed later in this chapter but, from scheme *i*, we see that the organometallic compound is a donor species with transfer of negative charge to the TCNE preceding the transfer of an entire electron.

Despite these charge transfer complexes and the donor properties of the Group V elements referred to above, most of the phenomena in this chapter are concerned with organometallic compounds as Lewis acids in which electrons are transferred, in pairs, to the metal.

6.2 EFFECTS ON STRUCTURE

Coordination is the most common process whereby stability is conferred on what would otherwise be unstable compounds. We shall examine, firstly, intra- and inter-molecular coordination in which lone pairs already present in the molecule coordinate to the metal, a process referred to as auto-complexation. We shall then discuss adduct formation in which the solvent or some other donor molecule complexes with the organometallic species.

This section is mainly confined to σ-bonded compounds. One reason is that, in π-bonded organometallic compounds, it is not always easy or profitable to separate the coordination and organometallic chemistry. The alkene or arene is often an alternative to other π-bonding ligands such as carbon monoxide and organophosphines. Thus the starting material and the product in the synthesis of a π-bonded organotransition metal compound are often closely related complexes. In contrast, the typical σ-bonded organometallic compound is a coordinatively unsaturated Lewis acid (though it may be stabilised by complexation with the solvent or other σ-donor species).

6.2.1 Auto-complexation

Auto-complexation is pronounced in Group IVb, particularly with organic derivatives of the heavier elements. Generally, for tin and lead compounds of the types R_nMX_{4-n} (M = Sn, Pb; R = alkyl or aryl) polymeric structures with bridging X groups result, providing lone pairs are available in X. For $n = 3$, the normal structure is I, as exemplified by trimethyltin chloride. The coordinative

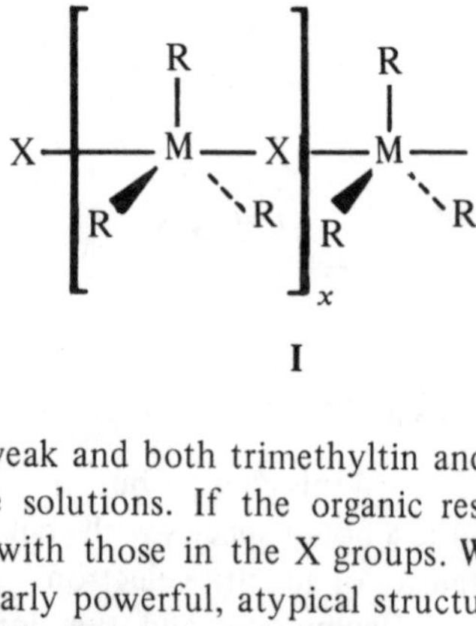

I

bridging bonds are weak and both trimethyltin and trimethyllead chlorides are monomeric in dilute solutions. If the organic residues contain donor atoms, these may compete with those in the X groups. Where this intramolecular coordination is particularly powerful, atypical structures may result as in the ionic organotin bromide, II (van Koten et al., 1978).

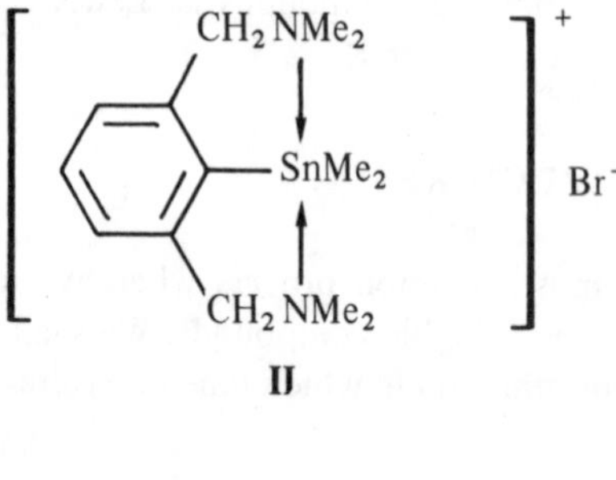

II

Bridging structures abound in Group IIIb compounds, generally not involving Lewis acid-base interactions but, instead, hydrogen or carbon bridges and three-centre molecular orbitals. However, where there is a choice, as in dimethyl-aluminium chloride, then chlorine bridging by electron pair donation is preferred.

In Group IIa, the coordination number at magnesium in the Grignard reagent, RMgX, is raised from two to four by one or more of three methods, namely, bridging halogen, bridging carbon, or coordination of solvent (or other donor) molecules (p. 45). The fact that Hg^{2+} and $MeHg^+$ are classed as soft acids may be the explanation why many monoalkyl (or monoaryl) mercury compounds, RHgX, resist bridging by 'hard' chlorine or oxygen donors and exist as linear or near linear monomers.

In the Group Ia metal alkyls there are no X groups to form bridges and, in any case, the typical structure is ionic, R^-M^+, except for the lithium derivatives which associate via carbon bridging (p. 43). Addition of donor solvents or additives increases the reactivity of organolithium compounds by full or partial depolymerisation and by increasing the polarity of the lithium-carbon bond (p. 26).

6.2.2 Adducts of Organometallic Compounds

Adduct formation is well established in Group IVb organometallics providing the metal retains one or more electronegative groups such as halogen. Thus, tetra-alkyltin compounds show no perceptible Lewis acid properties but, interestingly, tetramethyltitanium which decomposes above $-60\,^{\circ}C$ is stabilised by complex formation and the dipyridine adduct, $Me_4Ti.2py$, is stable up to $25\,^{\circ}C$ (Thiele, 1972).

Oligomeric or polymeric structures derived from the metals of Groups II–IV can usually be converted to monomeric adducts by the addition of sufficiently powerful donors, as exemplified in equations (6.6) and (6.7).

$$\text{(Et}_6\text{Al}_2\text{Et)} + 2Me_3N \longrightarrow 2Et_3Al.Me_3N \quad (6.6)$$

$$\left[\text{Me}_x\text{Sn—Cl}\right]_x + x\,\text{pyridine} \longrightarrow x\,Me_3SnCl.py \quad (6.7)$$

The donor species may be the solvent and the presence of a donor solvent will usually improve the yields in the direct synthesis of an organometallic compound from the metal and an organic halide. In some cases, such as that of the Grignard reagent, its presence is crucial.

As we discussed on p. 19, many alkyltransition metal compounds are unstable, decomposing to the metal hydride by expulsion of an alkene (scheme *ii*).

$$M-CH_2CH_2R \rightleftharpoons \left[\begin{array}{c} H-----CHR \\ \vdots \quad \| \\ M-----CH_2 \end{array} \right]^{\ddagger} \rightleftharpoons \begin{array}{c} H \\ | \\ M \end{array} \underset{CH_2}{\overset{CHR}{\|}} \rightleftharpoons M-H + RCH=CH_2$$

III

Scheme *ii*

Since the transition state (**III**) involves an increase in the coordination number of the metal by transient bonding to both α-carbon and β-hydrogen, stabilisation may be conferred by filling all the coordination sites at the metal. Thus the six-coordinate ethylrhodium cation, **IV**, is stable whereas attempts to make four-coordinate alkylrhodium gave unstable products (Wilkinson, 1974). (The high reactivity of α-hydrogens, as well as β-hydrogens, in σ-alkyltransition metal compounds is referred to later on p. 147.

$$\left[\begin{array}{c} Et \\ H_3N \diagdown \quad | \quad \diagup NH_3 \\ Rh \\ H_3N \diagup \quad | \quad \diagdown NH_3 \\ NH_3 \end{array} \right]^{2+}$$

IV

6.3 EFFECTS ON REACTIVITY

There are many qualitative and quantitative ways in which coordination to a metal atom may influence reactivity. The transition metals in particular, with their multiplicity of valency states, show considerable scope for oxidation and reduction reactions occurring within the coordination sphere.

6.3.1 Oxidative Addition and Reductive Elimination

In our discussion of the eighteen electron rule (p. 5) we noted that, although most organotransition metal compounds have 18 electrons in the valence shell of the metal, certain metals, notably rhodium, nickel, palladium and platinum, often have only 16 electrons. Compounds of metals whose valence shells contain less than 18 electrons are coordinatively unsaturated and are likely candidates for participation in oxidative addition reactions.

Oxidative addition is commonly accompanied by an increase of the coordination number, as in the reaction of an acyl halide with the methylrhodium complex, **V** (equation 6.8).

$$\text{(6.8)}$$

In reaction (6.8), not only has Rh(I) been oxidised to Rh(III), but the coordination number has increased from four to six. (Reaction (6.8) was proposed to explain the catalysed alkylation of acyl halides with organolithium compounds (p. 92); note that proposed intermediates such as **V** have not been isolated and it follows from observations made in the previous section that they would be expected to be very labile.)

In other cases of oxidative addition, prior or simultaneous dissociation of ligands may free coordination positions as in equation (6.9), where the coordination is unchanged as the metal is oxidised from Pd(0) to Pd(II).

$$\text{Pd(PPh}_3)_4 + \text{PhBr} \longrightarrow \text{PhPd(PPh}_3)_2\text{Br} + 2\text{PPh}_3 \qquad \text{(6.9)}$$

Oxidative additions can be used to produce reactive intermediates in which σ-bonded organic groups react with other ligands coordinated to the metal, the product being ejected in a reductive elimination process. Thus from the product, **VI**, of reaction (6.8) a ketone is obtained together with the Rh(I) species, **VII**, which is the effective catalyst (equation 6.10). Compound **VII** is then methylated to regenerate **V** in a continuing cycle.

$$\text{(6.10)}$$

In practice, this sequence of coordination of a substrate molecule by oxidative addition followed by reaction with another species within the coordination sphere with subsequent reductive elimination of the product is the basis of many catalytic cycles. A typical example is the industrially important conversion of methanol to acetic acid, which is catalysed by rhodium complexes (scheme *iii*).

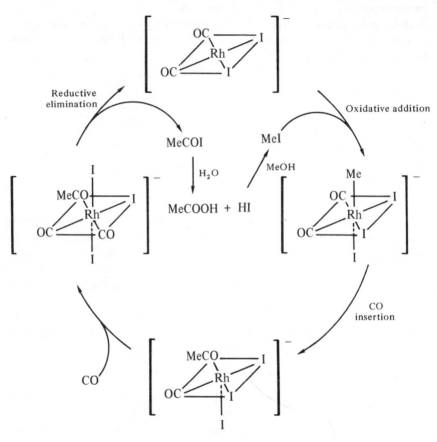

Scheme *iii*

Other examples of oxidative addition–reductive elimination catalytic cycles are given in chapter 9. These include the hydrogenation of alkenes catalysed by *tris*(triphenylphosphine)rhodium(I) chloride (p. 216) where, paradoxically, Rh(I) is oxidised to Rh(III) by oxidative addition of hydrogen.

Reductive eliminations occurring outside catalytic cycles are rare because the low coordination number products are so reactive. An example is the formation of biaryls from diaryl*bis*(triphenylphosphine)platinum(II) complexes (equation (6.11); Braterman *et al.*, 1977).

$$Ph_3P \diagdown \diagup Ph$$
$$\qquad Pt \qquad \xrightarrow{\text{heat}} \quad PhPh \quad + \quad [Pt(PPh_3)_2]$$
$$Ph_3P \diagup \diagdown Ph \qquad\qquad\qquad\qquad\qquad \text{undergoes} \qquad (6.11)$$
$$\qquad\qquad\qquad\qquad\qquad\qquad\qquad \text{further decomposition}$$

Biaryls (and alkanes) are also formed by reductive elimination from reactive organometallic intermediates which have been oxidised by electron transfer processes (p. 131).

6.3.2 Migratory Insertion Reactions

In scheme *iii*, carbon monoxide was inserted into the methyl–rhodium bond. These insertion processes, occurring within the coordination sphere, are key steps in catalytic cycles. The mechanism of one particular carbonyl insertion, that occurring in methylmanganese pentacarbonyl (equation 6.12). has been studied in detail (Calderazzo, 1977).

$$MeMn(CO)_5 + CO \longrightarrow MeCOMn(CO)_5 \qquad (6.12)$$

The carbon monoxide inserted comes from the five groups already bound and not from the newly entering CO molecule since, when the latter was radioactively labelled, no label was found in the acetyl group (equation 6.13).

$$MeMn(CO)_5 + {}^{14}CO \longrightarrow MeCOMn(CO)_4\ {}^{14}CO \qquad (6.13)$$

By use of labelled acetyl chloride it was possible to synthesise a specimen of methylmanganese pentacarbonyl (**VIII**) containing only one carbonyl group carrying a ${}^{13}C$ label, this group being *cis* to the methyl residue (scheme *iv*; Noack

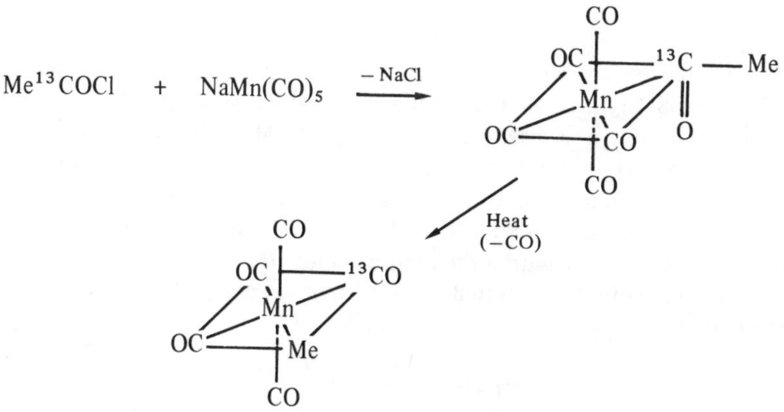

$$Me^{13}COCl \quad + \quad NaMn(CO)_5 \quad \xrightarrow{-NaCl} \quad$$

VIII

Scheme *iv*

and Calderazzo, 1967). When **VIII** was treated with unlabelled carbon monoxide three products were obtained. Of the acetylmanganese pentacarbonyl formed, 25 per cent was labelled in the acetyl group, 50 per cent had the acetyl group *cis* to the ${}^{13}CO$ group, and 25 per cent *trans* (equation 6.14). Noack and Calderazzo explained the distribution of products by proposing that the methyl group migrates as shown. The alternative mechanism in which a carbonyl group inserts into the Me–Mn bond would not give any *trans* product and can, therefore, be discounted. It is considered that carbonylation of octahedral alkylmetal compounds generally occurs by migration of the alkyl group; theoretical studies support this view (Berke and Hoffmann, 1978), and a similar mechanism applies in square planar complexes (Anderson and Cross, 1984). These reactions are best named migratory insertion reactions.

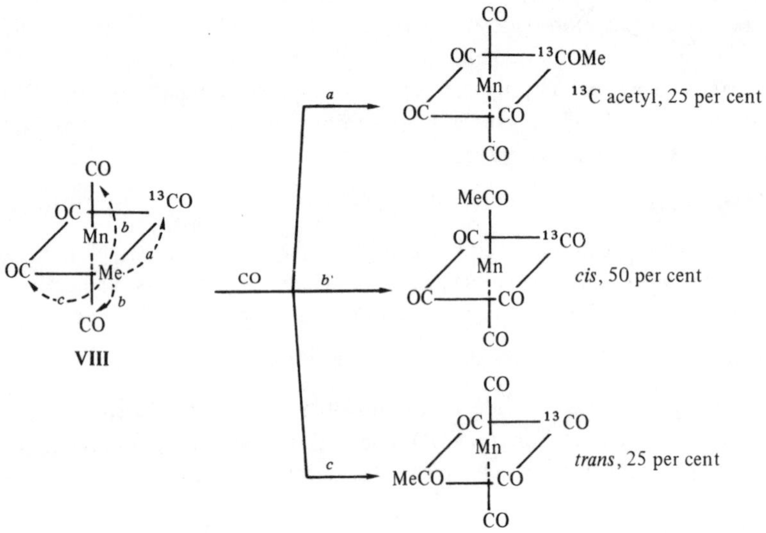

$$(6.14)$$

Closely related are insertions of RNC, $R_2C:$, $\diagup C{=}C\diagdown$, $-C{\equiv}C-$, CO_2, O_2, SO_2, S_x and $SnCl_2$ into M$-$C as well as other M$-$X bonds (Calderazzo, 1977). Of these reactions, the insertion of alkenes into M$-$C bonds is important in polymerisation, while insertion of alkenes into M$-$H bonds is a standard method of preparing organometallic compounds. A well authenticated example of the insertion of an alkene into a metal-hydrogen bond, and the reverse reaction, is provided by the ethenehydridorhodium complex, **IX**, which, in solution, exists in equilibrium with the solvated ethylrhodium complex, **X** (equation (6.15), Werner and Feser, 1979).

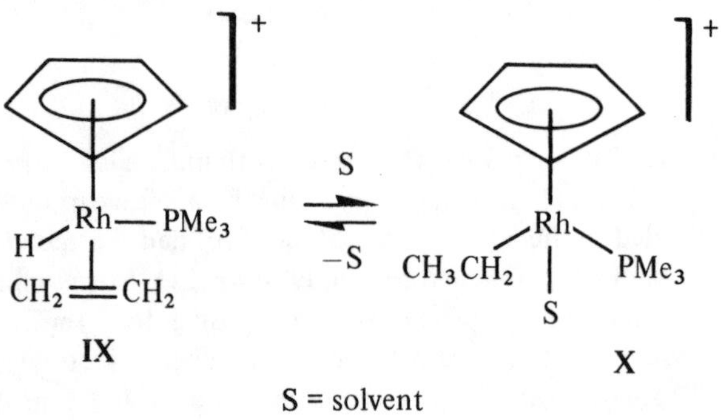

$$(6.15)$$

S = solvent

If, in compound **IX**, the ethene residue is replaced by carbon monoxide to give η^5-C_5H_5RhHCOPMe$_3$, no insertion of CO into the rhodium-hydrogen bond is observed. In fact, despite many determined attempts the insertion of carbon monoxide into metal-hydrogen bonds has not yet been demonstrated. Indeed, the expected products from such a reaction, formylmetal derivatives, are unstable; they spontaneously decompose to give the metal hydride and carbon monoxide (equation 6.16; Gladysz, 1982).

$$M-\underset{\underset{O}{\|}}{C}-H \longrightarrow M-H + CO \qquad (6.16)$$

6.3.3 Effects on Electrophilic Substitutions

In chapter 4 the cleavage of carbon-metal bonds by halogens or hydrogen halides was seen to proceed via electrophilic substitution at carbon. These S_E reactions are complicated by the possibility of coordination of the reagent, or solvent, or other donor species, at the neighbouring metal atom.

In organic chemistry, the use of a Lewis acid (such as iron(III) bromide) to polarise a bromine molecule, producing relatively positive bromine atoms, able to effect electrophilic substitution in aromatic systems, is well established. An organometallic substrate, on the other hand, may simultaneously present an anionic carbon, vulnerable to electrophilic substitution, as well as a metal atom which has sufficient Lewis acidity to polarise the halogen. This is depicted, in a general way, in scheme ν.

$$\overset{\delta -}{R}-\overset{\delta +}{M} \quad + \quad Br-Br \quad \longrightarrow \quad \left[\begin{array}{cc} R^{\delta -}\text{-----}M \\ | \qquad\quad | \\ Br^{\delta +}\text{-----}Br \end{array}\right]^{\ddagger} \quad \textbf{XI}$$

$$R-Br \ + \ M-Br$$

Scheme ν

In early studies of S_E2 reactions of chiral organometallic compounds, closed transition states such as **XI** were postulated when retention of configuration was observed. It is now realised that orbital symmetry considerations allow both inversion and retention of configuration (p. 85), so that the latter can be observed with open transition states. Nevertheless, there is clear evidence for cyclic transition states, of the type **XI**, as in the bromide ion-catalysed exchange between chiral 2-butylmercury bromide and [203]HgBr$_2$ (p. 87). Also, it is likely that, under certain conditions, bromine cleavage of alkyl groups from tin, in the absence of polar solvents, proceeds through transition states such as **XI** in which the reagent is coordinated to the metal (Chambers and Jensen, 1978). These reactions are complicated by a number of factors which include steric effects related to the size of the alkyl groups, the intervention of radical mechanisms, and coordination of polar solvent molecules to the tin atom.

As discussed on p. 25, in the lithiation of anisole by butyllithium, which can be regarded as an aromatic electrophilic substitution, lithium enters the ring *ortho* to the methoxyl group. One factor in the regiospecificity of this reaction was shown to be coordination of methoxyl oxygen to lithium whereby the metal was directed on to the *ortho* position, though other, electronic, influences are also involved.

6.3.4 Reactions Proceeding by Charge- and Electron-transfer Mechanisms

When tetraethyltin is added to a suspension of *tris*(phenanthroline)iron(III) perchlorate, $(phen)_3 Fe^{3+}(ClO_4)_3^-$, in deuteroacetonitrile, the ^{1}H-n.m.r. signal due to the tetraethyltin is immediately replaced by that of triethyltin perchlorate (equation (6.17); Wong and Kochi, 1979).

$$Et_4 Sn + 2(phen)_3 Fe^{3+}(ClO_4)_3^- \longrightarrow Et_3 SnClO_4 + 2(phen)_3 Fe^{2+}(ClO_4)_2^- + [EtClO$$

reacts

further

(6.17)

Similar reactions occur with alkyl-lead or -mercury compounds and the iron complex can be replaced with hexachloroiridate(IV). Kochi (1980) has shown that these ready carbon–metal bond cleavages are preceded by electron transfer from a carbon–metal bonding orbital (scheme *vi*). There is ample evidence to support

$$R_4 Sn + Fe(III) \xrightarrow{\text{Slow}} R_4 Sn^{\bullet +} + Fe(II)$$

$$R_4 Sn^{\bullet +} \xrightarrow{\text{Fast}} R_3 Sn^+ + R^\bullet$$

$$R^\bullet + Fe(III) \xrightarrow{\text{Fast}} [R^+] + Fe(II)$$

Scheme *vi*

this mechanism. Thus, the rate of cleavage is related to the ionisation potential of the organometallic compounds which, in turn, for a given metal, depends upon the nature of the organic groups. For groups attached to tin the ease of cleavage in these reactions is tBu > iPr > Et > Me.

Cleavage of carbon–metal bonds following electron transfer seems to be fairly general and certainly includes transition metal compounds, as exemplified in scheme *vii* (Almemark and Åkermark, 1978).

$$(o\text{-}MeC_6H_4)_2Ni(II)(PEt_3)_2 \xrightarrow[\substack{\text{anodic} \\ \text{oxidation}}]{e^-} (o\text{-}MeC_6H_4)_2Ni(III)(PEt_3)_2{}^+$$

$$o\text{-}MeC_6H_4 - C_6H_4Me\text{-}o \quad + \quad Ni(I)(PEt_3)_2^+$$

Scheme *vii*

The anionic methylgold complexes, lithium dimethylaurate(I) and tetramethyl-aurate(III), are stable, but the paramagnetic oxidation products dimethylgold(II) and tetramethylgold(IV) spontaneously decompose by reductive elimination of methyl groups to give ethane (scheme *viii*).

$$Me_4Au(III)^- \xrightarrow[(-e)]{O_2} Me_4Au(IV) \longrightarrow 2MeMe + Au(0)$$

Scheme *viii*

Tin–tin and germanium–germanium bonds are cleaved by a variety of reagents which include one-electron oxidants. Thus hexaalkyl and hexaaryl distannanes react with *tris*(phenanthroline)iron(III) ions as in equation (6.18).

$$R_3SnSnR_3 + 2(phen)_3Fe^{3+} \xrightarrow{k} 2R_3Sn^+ + 2(phen)_3Fe^{2+} \tag{6.18}$$

When the rate constants, k, for various groups R were compared, it was found that the relative rates were $Bu_6Sn_2 > Me_6Sn_2 > Me_3SnSnPh_3 > Ph_6Sn_2$. The hexamethyl compound reacted approximately 7×10^{12} times faster than the phenyl analogue, in agreement with the electron donor properties of these compounds and in direct contradiction to predictions made from bond energy considerations. This supports the view that the rate-determining step is an electron transfer from the distannane to the oxidant. The insertion of tetra-cyanoethene (TCNE) into carbon–metal bonds was referred to at the beginning of this chapter and we note here that insertion into tin–tin bonds can also occur (scheme *ix*; Kochi, 1978). The likely structure of the paramagnetic cleavage

$$n_2 + TCNE \longrightarrow [Bu_6Sn_2^{\cdot+}TCNE^{\cdot-}] \longrightarrow Bu_3SnTCNE^\cdot + Bu_3Sn^\cdot$$

XII

$$Bu_3Sn^\cdot + TCNE \longrightarrow Bu_3SnTCNE^\cdot$$

XII

Scheme *ix*

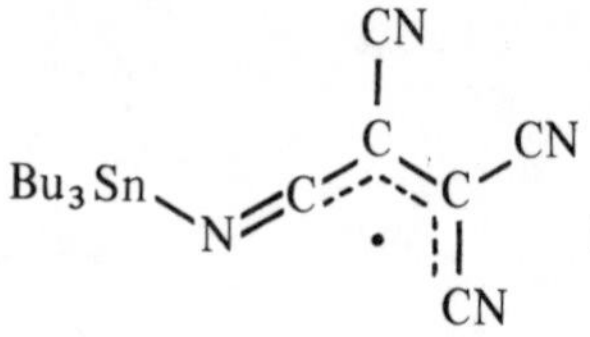

$$
\begin{array}{c}
CN \\
| \\
C \\
Bu_3Sn\diagdown \qquad \diagup C \diagdown C \diagdown CN \\
N \mathbin{\text{\Large $=$}} C \quad \bullet \; \| \\
CN
\end{array}
\qquad \textbf{XIIa}
$$

product, **XII**, is shown in **XIIa** (Krusic *et al.*, 1975). It is stabilised by delocalisation of the lone electron over the cyano groups as well as the alkene skeleton and belongs to that very exclusive group of radicals which are stable enough to be isolated in pure form. The hexaphenyldistannane reacts differently as do the hexalkyl-silicon and -lead analogues. However, dimanganese decacarbonyl undergoes a similar oxidative homolytic cleavage of the metal–metal bond (equation 6.19).

$$
Mn_2(CO)_{10} \xrightarrow{\text{TCNE}} (CO)_5 MnTCNE\cdot \tag{6.19}
$$

It was shown by Kochi (1978, p. 504) that a number of reactions between organometallic compounds and organic halides proceed via electron transfer processes. Thus, in the formation of cross-coupled products from Grignard reagents and allyl halides (equation 6.20), there was evidence for the participation of radicals, indicating that this is not a simple S_N2-type displacement of Br^- by R^-. An e.s.r. signal was observed during the reaction and, when the reagent was

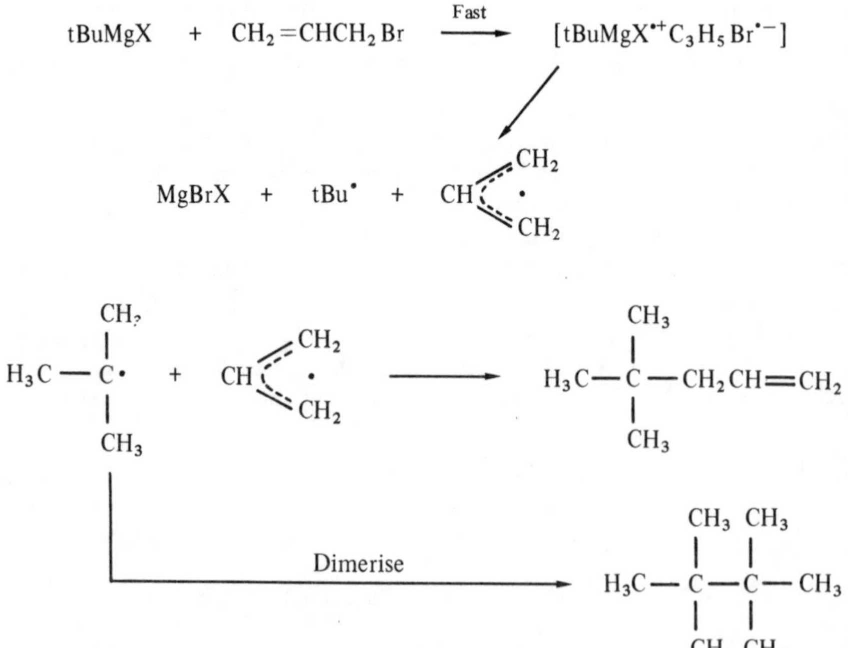

$$
tBuMgX \;+\; CH_2{=}CHCH_2Br \xrightarrow{\text{Fast}} [tBuMgX^{\bullet+}C_3H_5Br^{\bullet-}]
$$

$$
MgBrX \;+\; tBu^{\bullet} \;+\; CH\!\!\diagdown_{\!\!CH_2}^{\!\!CH_2}\;\bullet
$$

$$
\begin{array}{c}
CH_? \\
| \\
H_3C - C\bullet \\
| \\
CH_3
\end{array}
\;+\; CH\!\!\diagdown_{\!\!CH_2}^{\!\!CH_2}\;\bullet \;\longrightarrow\;
\begin{array}{c}
CH_3 \\
| \\
H_3C - C - CH_2CH{=}CH_2 \\
| \\
CH_3
\end{array}
$$

$$
\xrightarrow{\text{Dimerise}} \quad
\begin{array}{c}
CH_3\; CH_3 \\
|\quad\; | \\
H_3C - C - C - CH_3 \\
|\quad\; | \\
CH_3\; CH_3
\end{array}
$$

Scheme *x*

t-butylmagnesium chloride, appreciable amounts of 2,2,3,3-tetramethylbutane were produced (Gough and Dixon, 1968).

$$RMgBr + CH_2=CHCH_2Br \longrightarrow CH_2=CHCH_2R + MgBr_2 \qquad (6.20)$$

All these factors are accommodated if the reaction occurred as in scheme x.

In a somewhat similar manner, di-t-butylmercury decomposes spontaneously in the presence of carbon tetrachloride. The principal products are t-butyl chloride, isobutylene, chloroform and metallic mercury and these are thought to be formed by the reactions shown in scheme xi (Nugent and Kochi, 1977).

$$tBu_2Hg + CCl_4 \xrightarrow{\text{Slow}} [tBu_2Hg^{\bullet+} CCl_4^{\bullet-}] \xrightarrow{\text{Fast}} tBuHg^+ + tBu^\bullet + Cl_3C^\bullet + Cl^-$$

$$tBuHg^+ + Cl^- \longrightarrow tBuHgCl$$

$$Cl_3C^\bullet + tBu_2Hg \longrightarrow Cl_3CH + Me_2C=CH_2 + tBuHg^\bullet$$

$$tBuHg^\bullet \longrightarrow tBu^\bullet + Hg$$

$$tBu^\bullet + CCl_4 \longrightarrow tBuCl + Cl_3C^\bullet$$

Scheme xi

6.3.5 Homolytic Reactions

Increased interest is being shown in the hitherto neglected area of free radical organometallic chemistry. Details of reaction mechanisms are sparse but it appears that coordination effects are less important than is the case with polar reactions. Nevertheless, it is convenient at this point to make a few observations on homolytic organometallic reactions (Lappert and Lednor, 1976).

It should by now be clear that metals generally form stronger bonds to the more electronegative elements than they do to carbon. This means that carbon-metal bonds can often be cleaved by treatment with electronegative radicals such as $RO^\bullet$, $ROO^\bullet$, $RS^\bullet$, $R_2N^\bullet$. Examples are shown in equations (6.21) and (6.22).

$$tBuO^\bullet + Bu_3B \longrightarrow tBuOBBu_2 + Bu^\bullet \qquad (6.21)$$

$$\overline{CO(CH_2)_4CON}^\bullet + Bu_4Sn \longrightarrow \overline{CO(CH_2)_4CON}SnBu_3 + Bu^\bullet \qquad (6.22)$$

Attack by an alkylperoxy radical is a special case since this is one of the propagation steps in the autoxidation of an organometallic compound (scheme xii). Autoxidation is a very common reaction in organometallic chemistry. It may occur inadvertently when, for example, an organozinc compound spontaneously ignites on exposure to the atmosphere. A more useful example is the Alfol process for the industrial preparation of long chain alcohols (scheme $xiii$).

$$R^{\bullet} + O_2 \longrightarrow ROO^{\bullet}$$

$$ROO^{\bullet} + RM \longrightarrow ROOM + R^{\bullet}$$

Scheme *xii*

$$R-Al\langle \xrightarrow{O_2} ROOAl\langle \xrightarrow{RAl\langle} ROAl\langle \xrightarrow{H_2O} ROH + HOAl\langle$$

Scheme *xiii*

REFERENCES

Ahrland, S., Chatt, J., and Davies, N. R. (1958). *Quart. Rev. (Lond.)* **12**, 265

Almemark, M., and Åkermark, B. (1978). *J. chem. Soc. chem. Commun.* 66

Anderson, G. K., and Cross, R. J. (1984). *Acc. chem. Res.* **17**, 67

Berke, H., and Hoffmann, R. (1978). *J. Am. chem. Soc.* **100**, 7224

Braterman, P. S., Cross, R. J., and Young, G. B. (1977). *J. chem. Soc. Dalton Trans.* 1892

Calderazzo, F. (1977). *Angew. Chem. int. Edn Engl.* **16**, 299

Chambers, R. L., and Jensen, F. R. (1978). In *Aspects of Mechanism and Organometallic Chemistry* (J. H. Brewster, ed.). Plenum Press, New York, p. 169

Gladysz, J. A. (1982). *Adv. organomet. Chem.* **20**, 1

Gough, R. G., and Dixon, J. A. (1968). *J. org. Chem.* **33**, 2148

Jensen, W. B. (1978). *Chem. Rev.* **78**, 1

Kochi, J. K. (1978). *Organometallic Mechanisms and Catalysis.* Academic Press, New York

Kochi, J. K. (1980). *Pure appl. Chem.* **52**, 571

Krusic, P. J., Stoklosa, H., Manzer, L. E., and Meakin, P. (1975). *J. Am. chem. Soc.* **97**, 667

Lappert, M. F., and Lednor, P. W. (1976). *Adv. organomet. Chem.* **14**, 345

Mente, D. C., Mills, J. L. and Mitchell, R. E. (1975). *Inorg. Chem.* **14**, 123

Noack, K., and Calderazzo, F. (1967). *J. organomet. Chem.* **10**, 101

Nugent, W. A., and Kochi, J. K. (1977). *J. organomet. Chem.* **124**, 37

Pearson, R. G. (1963). *J. Am. chem. Soc.* **85**, 3533

Pearson, R. G. (1968). *J. chem. Educ.* **45**, 581

Thiele, K.-H. (1972). *Pure appl. Chem.* **30**, 575

van Koten, G., Jastrzebski, J. T. B. H., Noltes, J. G., Spek, A. L., and Schoone, J. C. (1978). *J. organomet. Chem.* **148**, 233

Werner, H., and Feser, R. (1979). *Angew. Chem. int. Edn Engl.* **18**, 157

Wilkinson, G. (1974). *Science* **185**, 109

Wong, C. L., and Kochi, J. K. (1979). *J. Am. chem. Soc.* **101**, 5593

GENERAL READING

Daub, G. W. (1977). Oxidatively induced cleavage of transition metal carbon bonds. *Prog. inorg. Chem.* **22**, 409

Henrici-Olivé, G., and Olivé, S. (1977). *Coordination and Catalysis.* Verlag Chemie, Weinheim

Jensen, W. B. (1978). The Lewis acid–base definitions: a status report. *Chem. Rev.* **78**, 1

Kochi, J. K. (1978). *Organometallic Mechanisms and Catalysis.* Academic Press, New York; see particularly part III

Pearson, R. G. (1968). Hard and soft acids and bases, HSAB, Part I. *J. chem. Educ.* **45**, 581

7

CARBENE CHEMISTRY AND ORGANOMETALLIC COMPOUNDS

Carbenes are neutral divalent carbon species with six electrons in their valence shell. A free carbene, $:CR_2$, is electron deficient, and has only a short life in a free state, but their high reactivity makes them useful reagents in organic chemistry (Kirmse, 1971).

The stabilisation of the carbene by bonding to a transition metal was first reported in 1964 by Fischer and Maasböl (1964). They reported the tungsten carbene complex, **I**. Ten years later Schrock (1974) reported the preparation of the tantalum compound, **II**, which, superficially, is very similar. We have men-

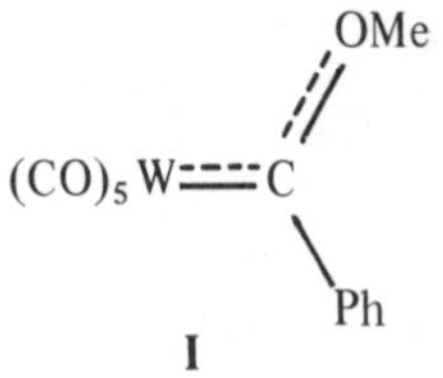

$$(CO)_5 \, W \!\!=\!\!\!= C \!\! \begin{array}{c} \nearrow^{OMe} \\ \searrow_{Ph} \end{array} \qquad\qquad (Me_3CCH_2)_3 \, Ta \!\!=\!\!\!= C \!\! \begin{array}{c} \nearrow^{CMe_3} \\ \searrow_{H} \end{array}$$

       I 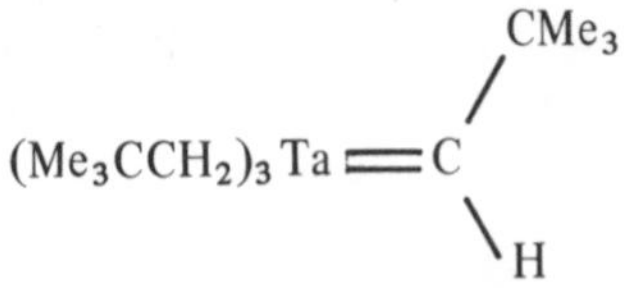   **II**

tioned these two compounds at this stage because they have very different reactivities.

Complexes, such as **I**, in which the metal atom is in a low oxidation state, can be used as a source of the neutral carbene in organic chemistry and give alkenes by dimerisation of the carbene ligand on heating. The electron density of the carbene carbon atom of such complexes is low. As a result, the carbene carbon atom is electrophilic, and in many cases becomes a site for nucleophilic attack. When the metal is in a higher oxidation state, as in **II**, the compound behaves rather like an ylide, that is, $R_3 Ta^+ - CH_2^-$. This means that the carbene carbon has a high electron density, anionic character and is nucleophilic; the opposite of the tungsten compound. Calculations suggest that the carbene carbon atom of the tantalum complex has a high electron density because of good overlap with the tantalum d-orbitals (Goddard *et al.*, 1980).

The nomenclature of carbene complexes is a vexed issue.† Carbenes are unstable neutral six-electron species, and it could be argued that in **I** a carbene has been stabilised by complexation to tungsten, and such compounds are usually called carbene complexes. Schrock prefers to describe **II** as an alkylidene complex which is formally derived from the eight-electron species CHR^{2-}. The difference between these two descriptions is neither trivial nor semantic as it has implications for the oxidation state of the metal. Thus the tungsten complex, **I**, is considered to be a complex of tungsten(0), whereas the oxidation state of tantalum in **II** is tantalum(V) because of the doubly negatively charged alkylidene group and the three singly negatively charged alkyl groups.

An additional distinction needs to be made between the types of substituents in carbene complexes. There are two classes: (a) complexes which have carbon or hydrogen substituents on the carbene carbon atom (non-hetero carbenes) and (b) complexes which have a heteroatom (often oxygen or nitrogen) as substituent (hetero carbenes). We shall not deal with compounds containing two metal atoms linked by CR_2, although the appearance of two recent reviews demonstrates the growing interest in these compounds (Herrmann, 1982a; Holton *et al.*, 1983).

7.1 CARBENE COMPLEXES DERIVED FROM METAL CARBONYLS

The preparation of transition metal carbene complexes from metal carbonyls was described in chapter 2. The basic reaction is shown in scheme *i*.

The reaction gives heterocarbenes in which the carbene carbon atom bears an oxygen atom, the electrons of which can donate into the empty p-orbital on the carbene carbon. The reaction works well with the Group VI and Group VII carbonyls (Fischer, 1976).

---

†A similar dilemma occurs between nitrene and imido complexes (Nugent and Haymore, 1980).

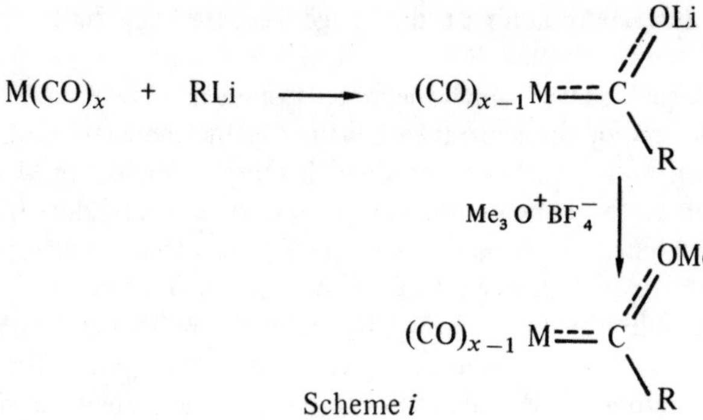

$$M(CO)_x \;+\; RLi \;\longrightarrow\; (CO)_{x-1}M \equiv C \diagup\!\!\!\!\overset{OLi}{\diagdown R}$$

$$\downarrow Me_3O^+BF_4^-$$

$$(CO)_{x-1}\,M \equiv C \diagup\!\!\!\!\overset{OMe}{\diagdown R}$$

Scheme *i*

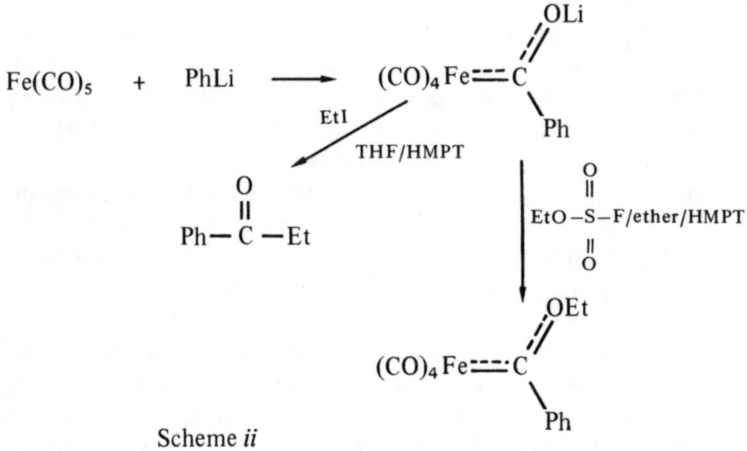

$$Fe(CO)_5 \;+\; PhLi \;\longrightarrow\; (CO)_4Fe \equiv C \diagup\!\!\!\!\overset{OLi}{\diagdown Ph}$$

EtI

THF/HMPT

$$\overset{O}{\overset{\|}{Ph - C - Et}}$$

$$\overset{O}{\underset{O}{\overset{\|}{EtO - S - F}}}/ether/HMPT$$

$$(CO)_4Fe \equiv C \diagup\!\!\!\!\overset{OEt}{\diagdown Ph}$$

Scheme *ii*

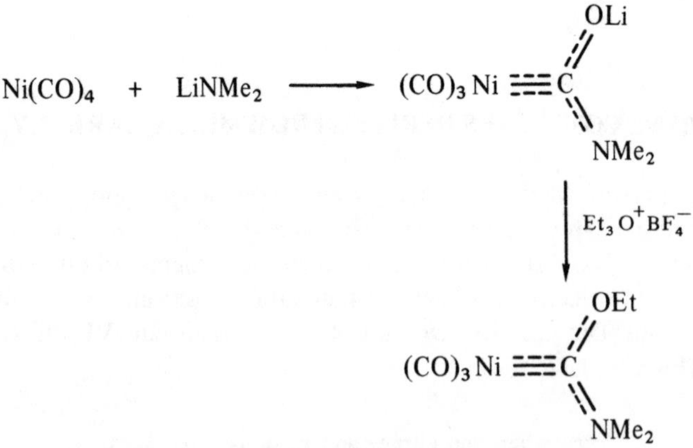

$$Ni(CO)_4 \;+\; LiNMe_2 \;\longrightarrow\; (CO)_3Ni \equiv C \diagup\!\!\!\!\overset{OLi}{\diagdown NMe_2}$$

$$\downarrow Et_3O^+BF_4^-$$

$$(CO)_3Ni \equiv C \diagup\!\!\!\!\overset{OEt}{\diagdown NMe_2}$$

Scheme *iii*

The behaviour of the acylmetallate anions derived from iron pentacarbonyl and organolithium compounds is a complicated subject. They may be converted to carbene complexes by alkylation using ethyl fluorosulphonate in ether/HMPT. However, with ethyl iodide as alkylating agent in THF/HMPT, ethyl phenyl ketone is obtained, in a similar way to the reaction of the sodium analogue in Collman's ketone synthesis (p. 195) (scheme *ii*; Semmelhack and Tamura, 1983).

No carbene complex has been isolated from $Ni(CO)_4$ using an organolithium reagent followed by alkylation, but carbenes stabilised by two heteroatoms using lithium dimethylamide in the first stage have been prepared (scheme *iii*; Fischer *et al*., 1972).

The dinuclear cobalt carbonyl $Co_2(CO)_8$ does not give a carbene complex on reaction with phenyllithium followed by alkylation with triethyloxonium fluoroborate, instead an alkylidynetricobalt nonacarbonyl cluster, **III**, is produced (Fischer *et al*., 1983). Metal cluster carbonyls only give carbene complexes in a few cases (Jensen and Kaesz, 1983).

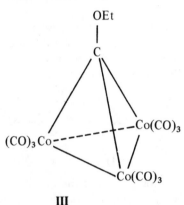

III

7.2 ROUTES TO HETEROCARBENE COMPLEXES NOT INVOLVING METAL CARBONYLS

Several routes to heterocarbene complexes which do not involve metal carbonyls have been developed. We shall mention five of these.

(a) A reaction not involving a carbonyl but electronically very similar to the methods described in section 7.1 is the nucleophilic addition of an alcohol (or amine) to a coordinated isocyanide (equation (7.1); Badley *et al*., 1971).

$$\begin{array}{c} Cl \\ \diagdown \\ Pt \\ \diagup \hspace{0.3cm} \diagdown \\ Cl \hspace{1.1cm} C \\ \hspace{1.5cm} \diagdown \\ \hspace{1.7cm} N \text{---} Ph \end{array} \quad \begin{array}{c} PEt_3 \end{array} \quad + \text{ EtOH} \longrightarrow \quad \begin{array}{c} Cl \hspace{1.2cm} PEt_3 \hspace{0.3cm} H \\ \diagdown \hspace{0.6cm} \diagup \hspace{1cm} | \\ Pt \hspace{1.5cm} \\ \diagup \hspace{0.4cm} \diagdown \hspace{0.3cm} C \text{---} N \text{---} Ph \\ Cl \hspace{2cm} \\ \hspace{1.5cm} OEt \end{array} \qquad (7.1)$$

(b) Nucleophilic attack by alcohols on cationic alkyne complexes of platinum also gives rise to carbene complexes. The reaction is carried out by generating the alkyne complex *in situ* (equation 7.2).

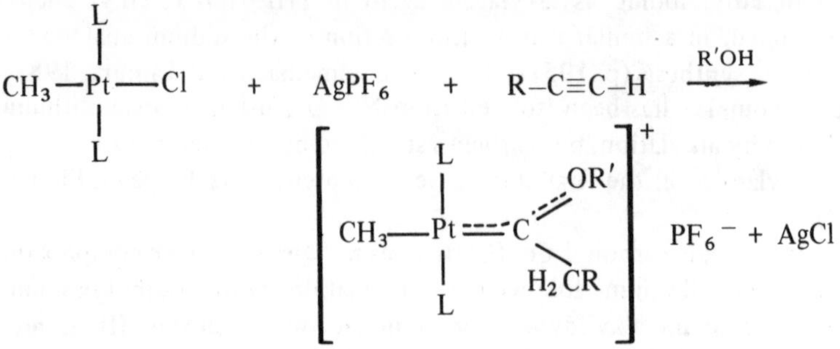

$$
\underset{\underset{L}{|}}{\overset{\overset{L}{|}}{CH_3-Pt-Cl}} \;+\; AgPF_6 \;+\; R-C\!\equiv\!C-H \;\xrightarrow{R'OH}
$$

$$
\left[\underset{\underset{L}{|}}{\overset{\overset{L}{|}}{CH_3-Pt=\!=\!C}}\!\!\begin{array}{l} \nearrow OR' \\ \searrow H_2CR \end{array} \right]^{+} \; PF_6^- \;+\; AgCl \qquad (7.2)
$$

These compounds can also be considered as stabilised carbonium ions (Chisholm *et al.*, 1975).

(c) The electron donation from the heteroatoms in carbene complexes is similar to the electronic effects present in some cationic species, such as the dialkyl chloromethylene iminium ion, **IV**. In this case the electron deficient carbon in **IVb** can accept electron density from nitrogen. The salts of such ions

$$
\underset{\mathbf{IVa}}{\overset{R}{\underset{R'}{\diagup}}N\!\!=\!\!C\overset{\diagup Cl}{\underset{\diagdown H}{}}} \qquad \longleftrightarrow \qquad \underset{\mathbf{IVb}}{\overset{R}{\underset{R'}{\diagup}}\overset{..}{N}\!-\!\overset{+}{C}\overset{\diagup Cl}{\underset{\diagdown H}{}}}
$$

$$
\mathbf{IV}
$$

react with low oxidation state compounds of the transition metals as shown in scheme *iv*. This sequence is a three-fragment oxidative addition reaction, and can

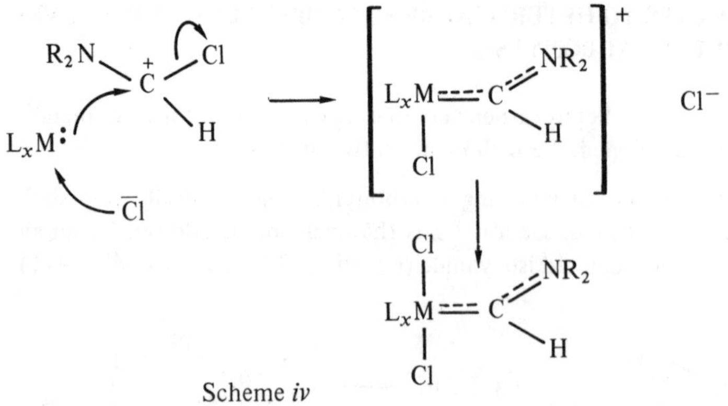

Scheme *iv*

be used with several of the platinum metals; for example rhodium (equation (7.3); Lappert, 1975).

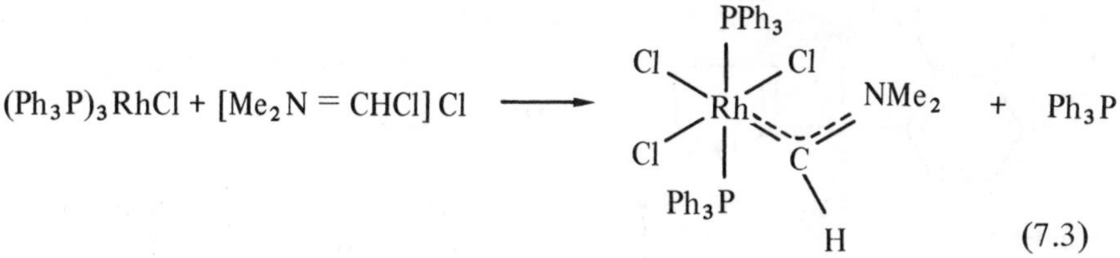

$$(Ph_3P)_3RhCl + [Me_2N = CHCl]Cl \longrightarrow$$

$$+ \quad Ph_3P \qquad (7.3)$$

(d) When an alkene has four nitrogen substituents as in **V**, it becomes so rich in electron density that it takes on special reactivity characteristics (Hoffmann, 1968). The C=C bond is much weaker than the double bond in ethene and the ionisation potential of 6 eV is very low for an organic molecule. The high electron density in **V** is available for coordination to a transition metal and it is found that **V** has some properties similar to a tertiary phosphine. Thus **V** can be used in a bridge splitting reaction with the platinum compound **VI**. The result of this reaction is that not only the chloride bridge of **VI**, but also the C=C of **V** is cleaved to give the heterocarbene complex **VII** (equation (7.4); Lappert, 1975).

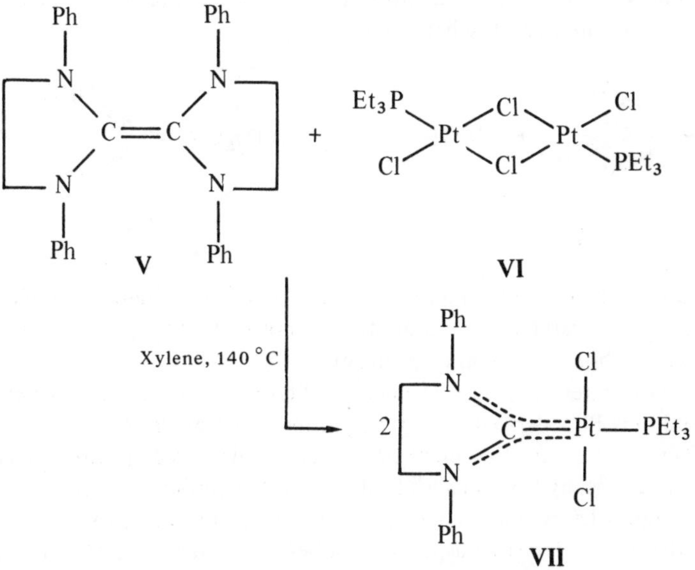

$$(7.4)$$

(e) The protonation of a σ-alkenyl nickel compound has been used to prepare a carbene complex (equation 7.5). The resulting cationic complex, **VIII**, has only one heteroatom bonded to the carbene carbon, in contrast to the complexes derived from nickel carbonyl in scheme *iii* (Miki *et al.*, 1982).

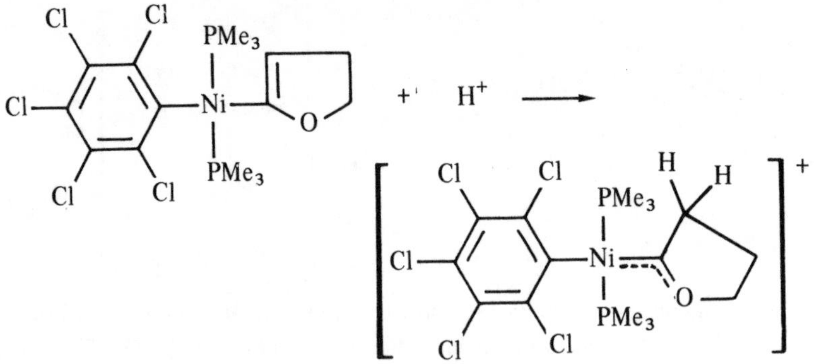

$$(7.5)$$

VIII

7.3 REACTIONS OF HETEROCARBENE COMPLEXES

The reactivity of heterocarbene complexes of low valent transition metals has been extensively studied. As explained in the introduction to this chapter, the complexes undergo nucleophilic substitution reactions; for example, with amines (equation (7.6); Connor and Fischer, 1969).

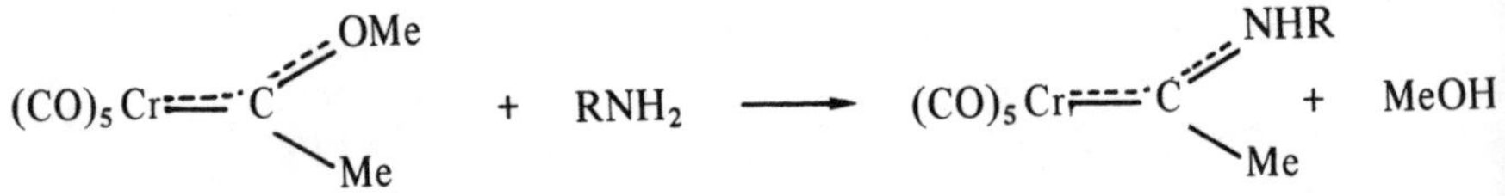

$$(7.6)$$

The reaction is similar to the aminolysis of esters to form amides. The electrophilic nature of the carbene carbon atom is supported by the low field ^{13}C shift of the carbene carbon in the n.m.r. spectrum.

A particularly interesting case of nucleophilic displacement occurs when the attacking anion is Ph^- derived from phenyllithium (Casey et al., 1977; Fischer et al., 1977). In this case the intermediate can be identified spectroscopically, and on reaction with hydrogen chloride at $-78\,^{\circ}C$ the methoxy group is lost and a carbene complex IX, without a heteroatom is obtained (scheme v).

One of the most interesting aspects of carbene complexes concerns their use as a source of carbenes in organic reactions. It is found that in a number of reactions, the organic products correspond to the transfer of the substituted carbene to the substrate. However, this is a complicated area and the products vary with the conditions. The elucidation of the olefin metathesis reaction in terms of a metallacyclobutane intermediate has shed light on some reactions of carbene complexes.

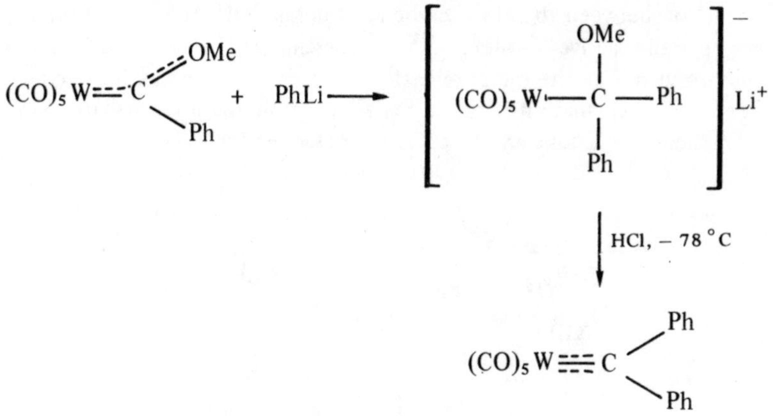

Scheme *v*

Phenylmethoxycarbenechromium pentacarbonyl, **X**, reacts with ethyl vinyl ether to give a cyclopropane product, **XI**, if the reaction is carried out under carbon monoxide pressure. However, in the absence of carbon monoxide, α-methoxystyrene, **XII**, is obtained (scheme *vi*; Fischer and Dötz, 1972).

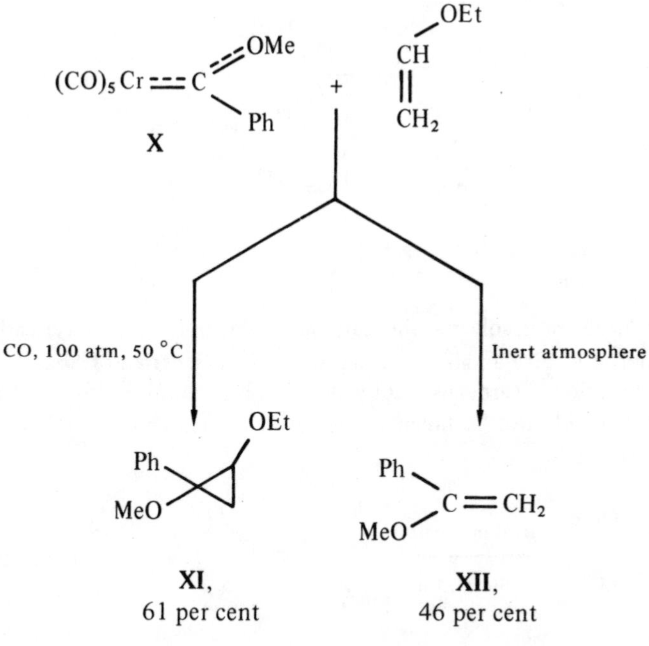

Scheme *vi*

The reaction between the chiral carbene complex, **XIII**, and diethyl fumarate gave an optically active product, **XV**, suggesting that carbon–carbon bond formation occurred at the metal and that no free carbene was involved. The most reasonable explanation is that a metallacyclobutane intermediate, **XIV**, is involved (scheme *vii*; Cooke and Fischer, 1973; see section 7.7).

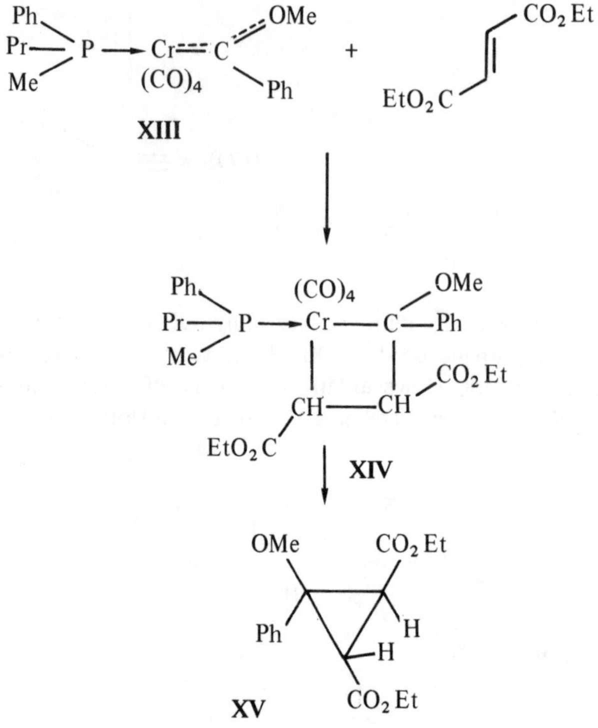

Scheme *vii*

On heating in an inert solvent, the carbene fragments of carbonyl carbene complexes dimerise to give an alkene (equation (7.7); Fischer and Plabst, 1974). The simple dimerisation of carbenes is not generally observed in organic chemistry and this reaction should not be taken as indicative of free carbenes (Casey and Anderson, 1975).

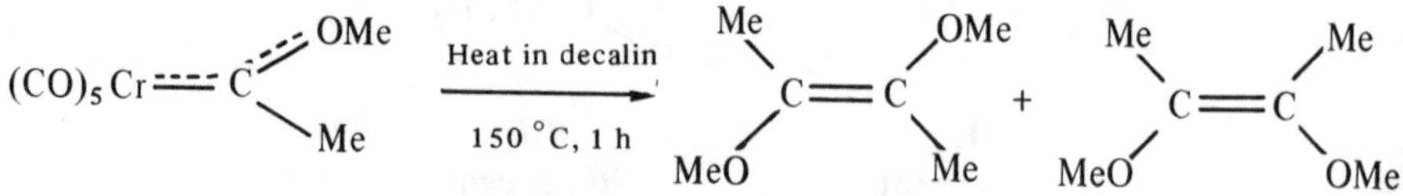

$$(7.7)$$

The reaction of chromium phenyl methoxy carbene complexes with alkynes has proved to be particularly fruitful from a synthetic standpoint. A co-cyclisation reaction occurs to give a naphthalene derivative initially with $Cr(CO)_3$ π-bonded to the substituted ring, the new ring being formed from the carbene, alkyne and one carbon monoxide group. On heating the $Cr(CO)_3$ group migrates to the unsubstituted ring (scheme *viii*). A large number of naphthalene derivatives

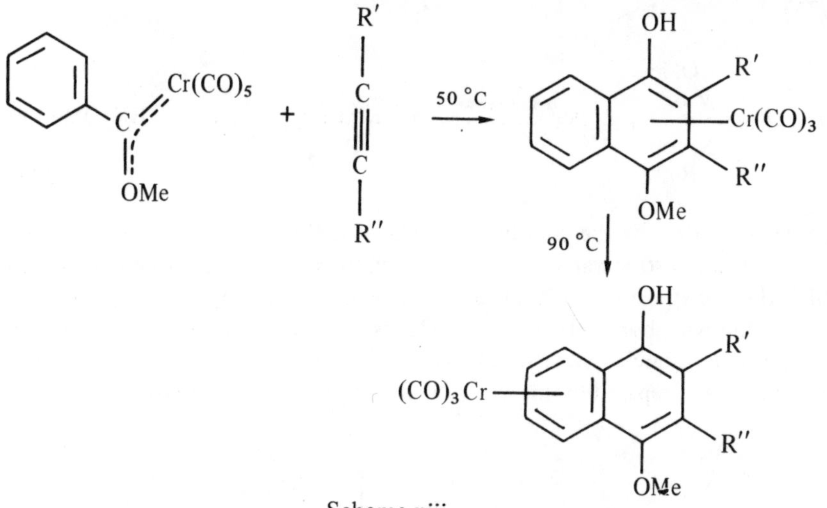

Scheme *viii*

have been prepared in this way (Dötz *et al.*, 1983). An extension of this work is the synthesis of compounds in the vitamin K series, e.g. **XVI** (scheme *ix*; Dötz *et al.*, 1982).

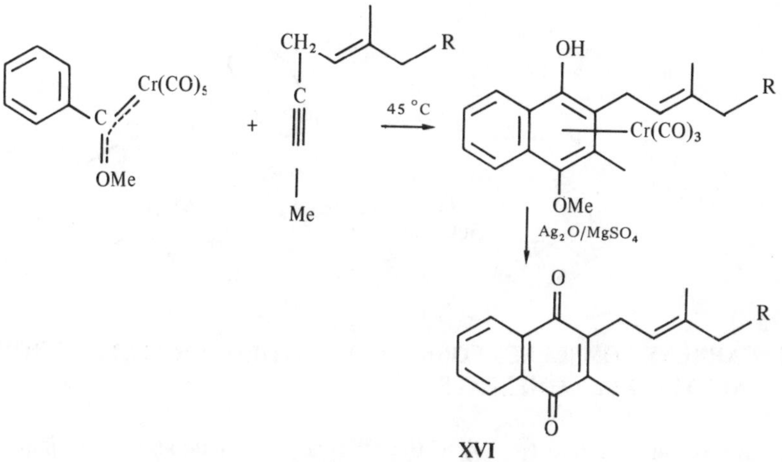

Scheme *ix*

7.4 CONVERSION OF HETEROCARBENE COMPLEXES TO CARBYNE COMPLEXES

In 1973 Fischer and co-workers reported that the reaction of methoxycarbene complexes of Group VI metals with boron halides leads to carbyne complexes which contain a metal–carbon triple bond (equation (7.8); Fischer and Schubert, 1975).

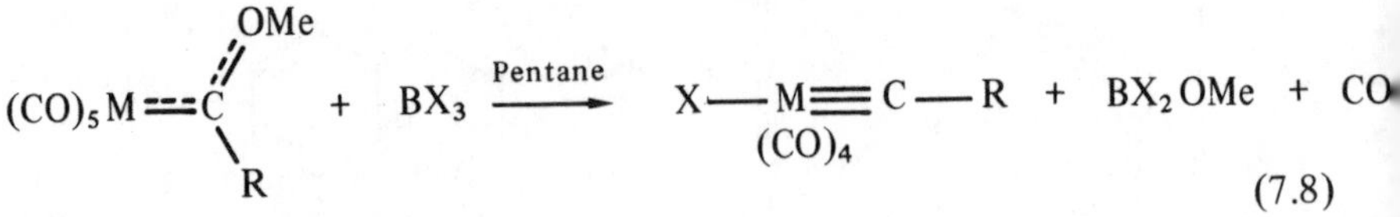

$$(CO)_5 M = C \overset{OMe}{\underset{R}{\Big\langle}} \quad + \quad BX_3 \quad \overset{Pentane}{\longrightarrow} \quad X - M \equiv C - R \quad + \quad BX_2 OMe \quad + \quad CO$$
$$(CO)_4$$

$$(7.8)$$

The halide ion enters the coordination sphere of the metal *trans* to the carbyne ligand. An interesting variation on the reaction is to use a carbene complex in which the *trans* position is blocked by a η^5-cyclopentadienyl group. In this case the reaction with boron trichloride produces a cationic carbyne complex which is susceptible to nucleophilic attack. A dimethyl carbene complex of manganese, **XVII**, has been prepared in this way (scheme *x*; Fischer *et al.*, 1976).

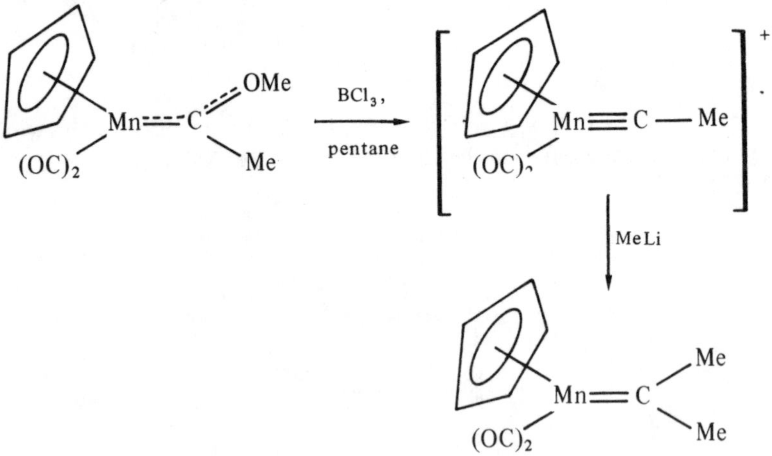

Scheme *x*

XVII

7.5 CARBENE COMPLEXES FORMED BY α-HYDROGEN ABSTRACTION (ALKYLIDENE COMPLEXES)

We have already referred (p. 19) to the tendency for metal alkyls with β-hydrogen atoms to form metal hydrides and eliminate alkenes (equation 7.9).

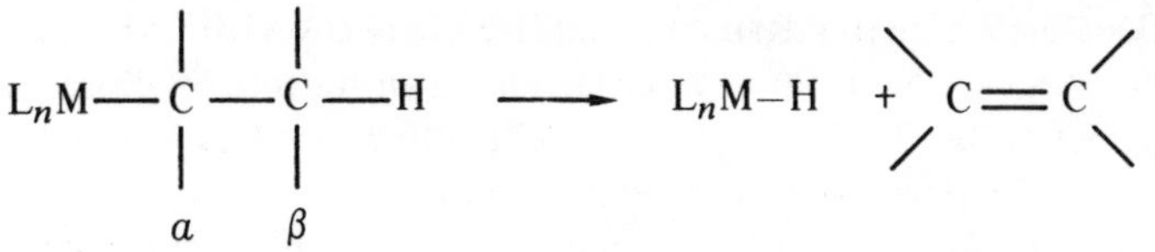

$$L_nM \overset{\displaystyle |}{\underset{\displaystyle |}{C}} \overset{\displaystyle |}{\underset{\displaystyle |}{C}} H \longrightarrow L_nM{-}H \ + \ \overset{\diagdown}{\underset{\diagup}{C}}{=}\overset{\diagup}{\underset{\diagdown}{C}} \qquad (7.9)$$

$$\quad\ \ \ a \quad \ \beta$$

However, there is another effect concerned with the α-hydrogen atoms which is important when the metal is in a high oxidation state. Alkyls of metals in high oxidation states have relatively acidic hydrogen atoms on the α-carbon atom. As a result of this, deprotonation occurs on treatment with strong base, or the deprotonation may take place spontaneously by an intramolecular route. Tantalum(V) provides a good example of intramolecular α-hydrogen abstraction (scheme *xi*). Tantalum pentachloride can be alkylated using dineopentylzinc to give **XVIII**, which is reasonably stable because with neopentyl ligands the β-elimination pathway is blocked. Further alkylation of **XVIII** can be carried out using neopentyllithium, but the reaction does not give the pentaalkyl, instead the alkylidene complex **II** is formed. The reaction is thought to proceed through the tetraalkyl chloride, **XIX**, one neopentyl group of which abstracts a proton from a second neopentyl group to give tetramethylmethane (neopentane) and the alkylidene chloride, **XX**. Further alkylation of **XX** gives the final product **II** (Schrock, 1979).

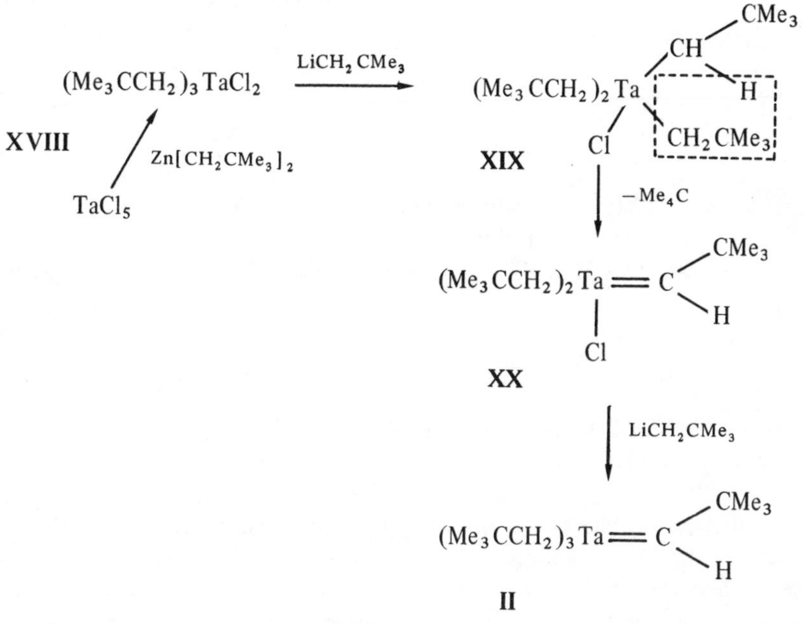

Scheme *xi*

In some reactions the neopentyl ligand may undergo two stages of deprotonation. This occurs in the reaction of tungsten hexachloride with neopentyllithium. The resultant product is the alkylidyne complex **XXI** with a metal–carbon triple bond (equation (7.10); Clark and Schrock, 1978).

$$WCl_6 + 6LiCH_2CMe_3 \longrightarrow Me_3C-C\equiv W-(CH_2CMe_3)_3 + 2Me_4C + 6LiCl$$

$$\textbf{XXI} \qquad (7.10)$$

On reaction with tertiary phosphines **XXI** undergoes further loss of tetramethylmethane to give **XXII** (equation 7.11).

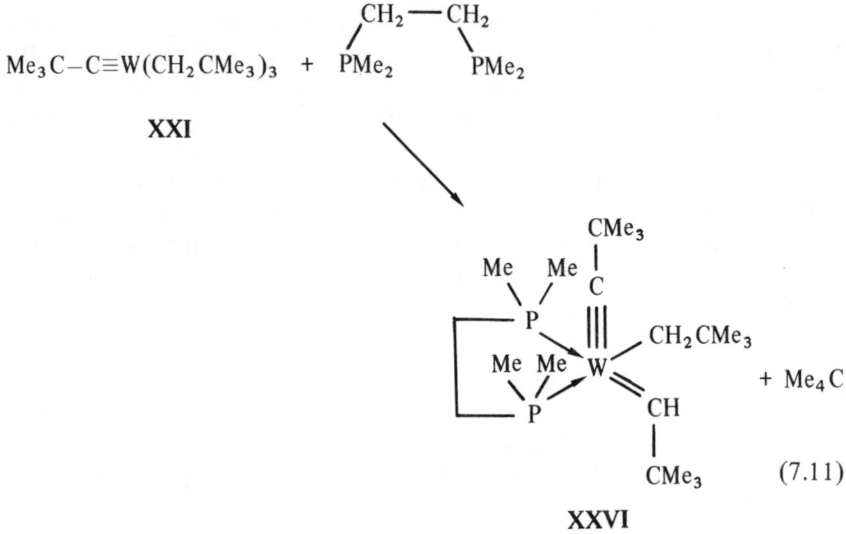

$$Me_3C-C\equiv W(CH_2CMe_3)_3 \quad + \quad \overset{\displaystyle CH_2-CH_2}{\underset{\displaystyle PMe_2 \qquad PMe_2}{\diagup \qquad \diagdown}}$$

XXI

$$+ Me_4C$$

$$(7.11)$$

XXVI

The structure of **XXII** has been investigated by x-ray crystallography; it contains an alkyl, an alkylidene and an alkylidyne ligand bonded to tungsten. The metal–carbon distances are 2.26, 1.94 and 1.79 Å respectively (Churchill and Youngs, 1979).

7.6 REACTIONS OF HIGH OXIDATION STATE ALKYLIDENE AND ALKYLIDYNE COMPLEXES

The reactions of the high oxidation state alkylidene complexes are very different to those of the low oxidation state complexes described in section 7.3. Complexes such as **II** or **XXIII** behave as ylides with a $Ta^+-CH_2^-$ unit. Thus **II** reacts with acetone in the same way as a phosphorus ylide to give the alkene **XXIV** (equation 7.12).

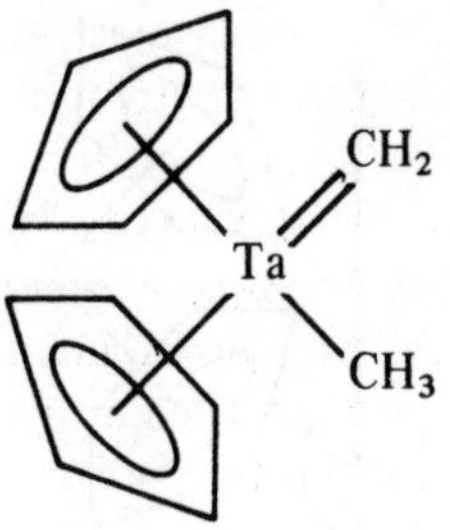

XXIII

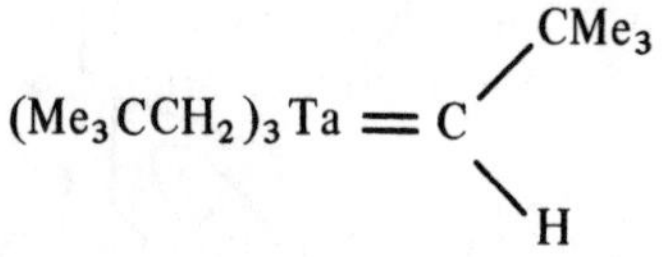

$$(Me_3CCH_2)_3Ta = C \overset{CMe_3}{\underset{H}{}}$$

II

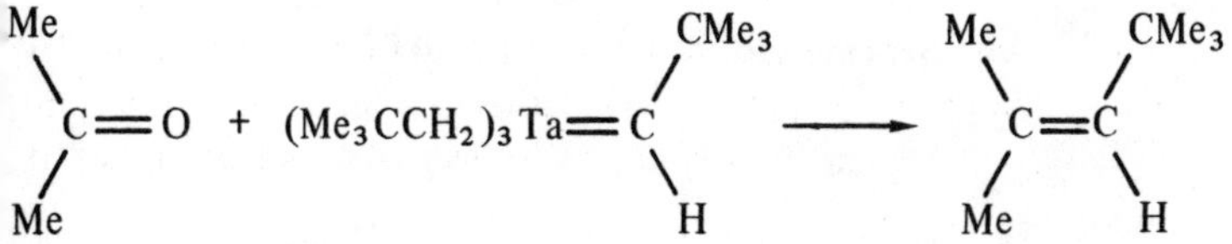

$$\underset{Me}{\overset{Me}{}}C=O \ + \ (Me_3CCH_2)_3Ta = C \overset{CMe_3}{\underset{H}{}} \longrightarrow \underset{Me}{\overset{Me}{}}C=C \overset{CMe_3}{\underset{H}{}} \qquad (7.12)$$

II   **XXIV**

II also reacts with esters to give enol ethers (equation 7.13). This reaction cannot be carried out using phosphorus ylides, although bimetallic $TiCH_2Al$ reagents are an alternative (see p. 189; Schrock, 1976).

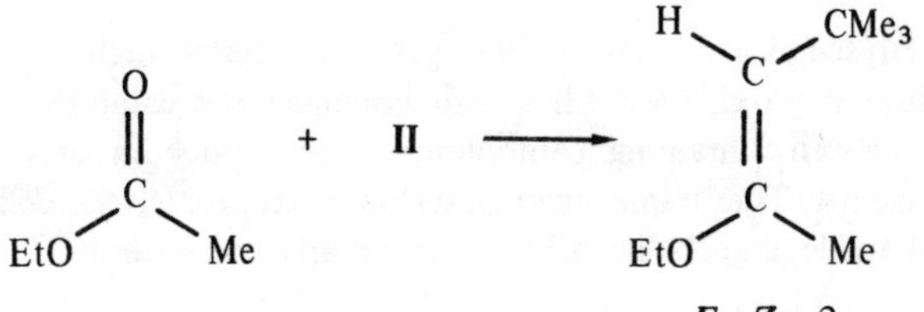

$$\underset{EtO}{\overset{O}{\underset{\|}{C}}}Me \ + \ II \longrightarrow \underset{EtO}{\overset{H \diagdown C \diagup CMe_3}{\underset{\|}{C}}}Me \qquad (7.13)$$

$$E:Z = 2$$

II does not react with methyl iodide, but **XXIII** displaces iodide to give an ethyl compound, **XXV**, which undergoes β-elimination to give an ethene π-complex, **XXVI**. This is shown for the reaction with CD_3I in scheme *xii*.

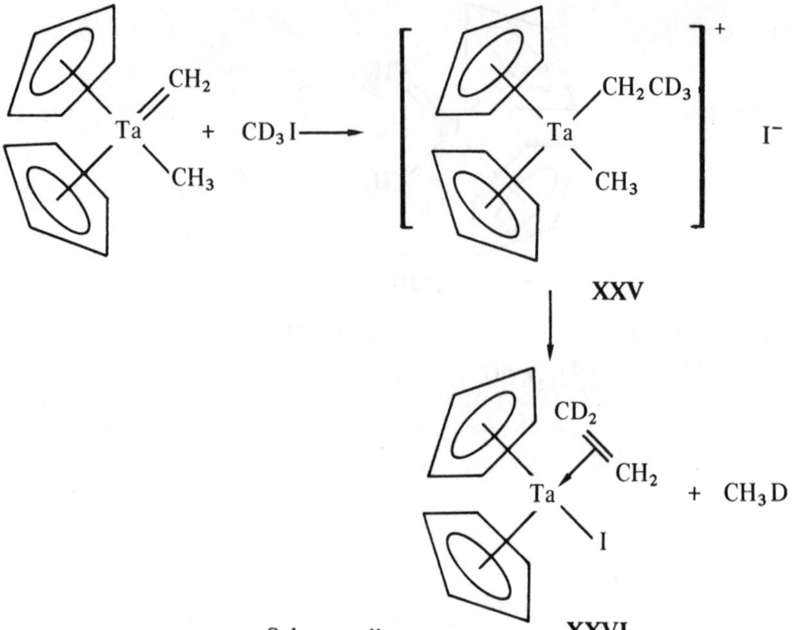

Scheme *xii*   **XXVI**

7.7 CARBENE COMPLEXES AS INTERMEDIATES IN OLEFIN METATHESIS

The olefin metathesis reaction is the exchange of alkylidene groups between olefins (equation 7.14).

$$R^1-CH=CH-R^1 + R^2-CH=CH-R^2 \; \overset{\text{Catalyst}}{\rightleftharpoons} \; 2R^1-CH=CH-R^2$$

$$(7.14)$$

The reaction is catalysed by a number of heterogeneous systems, such as tungsten trioxide or rhenium heptoxide on silica, and homogeneous catalytic systems often based on halogen-containing compounds of molybdenum or tungsten with alkylating agents. The homogeneous systems $WCl_6/EtOH/EtAlCl_2$ and $(Ph_3P)_2Mo(NO)_2Cl_2/Me_3Al_2Cl_3$ have been particularly widely used (Calderon *et al.*, 1979).

Numerous mechanisms have been proposed for olefin metathesis, but the most widely accepted is that of Hérisson and Chauvin (1970). This involves addition of the olefin to a carbene complex to give a four-member metallacycle (**XXVII**). The metallacycle can regenerate a different olefin and carbene complex by scission of the original olefin (scheme *xiii*).

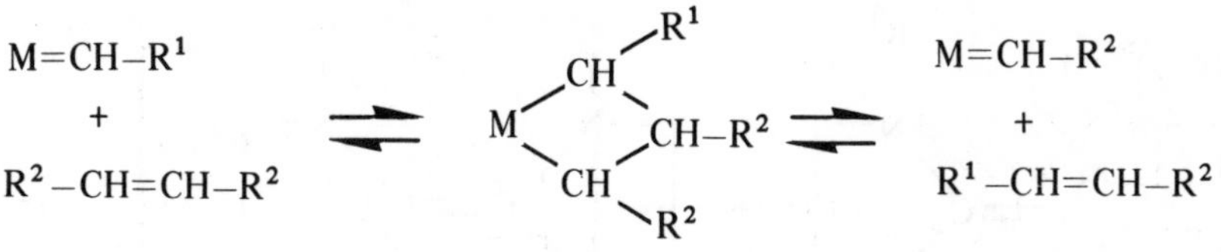

$$M=CH-R^1 + R^2-CH=CH-R^2 \rightleftharpoons \begin{array}{c} CH^{\diagup R^1} \\ M \diagdown \begin{array}{c} \diagup CH-R^2 \\ CH \diagdown R^2 \end{array} \end{array} \rightleftharpoons M=CH-R^2 + R^1-CH=CH-R^2$$

XXVII

Scheme *xiii*

The reactions of carbene complexes with olefins give much support to the idea that they are involved in olefin metathesis. Thus the reaction of the ditolyl carbene complex, **XXVIII**, with isobutene gives three hydrocarbon products, the main one involving an exchange of an alkylidene group of the olefin (equation (7.15), Tol = *p*-tolyl; Casey *et al.*, 1976).

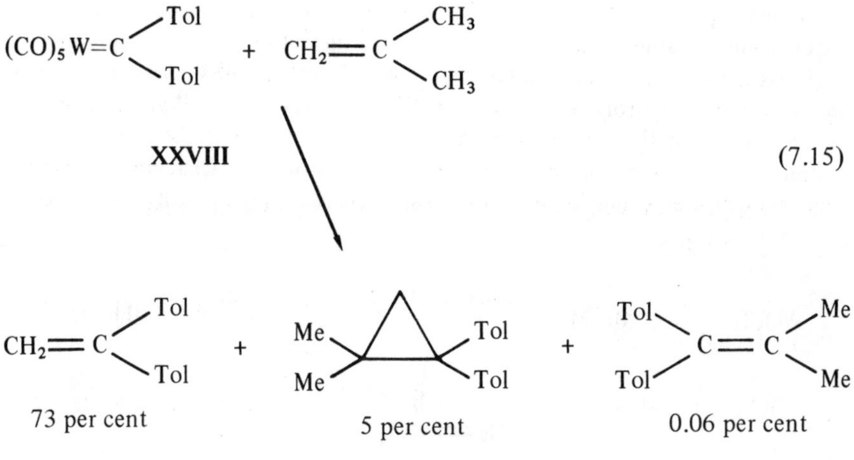

$$(CO)_5W=C\diagup{Tol}\diagdown{Tol} \quad + \quad CH_2=C\diagup{CH_3}\diagdown{CH_3}$$

XXVIII   (7.15)

$$CH_2=C\diagup{Tol}\diagdown{Tol} \quad + \quad \triangle \quad + \quad \diagup{Tol}\diagdown{Tol}C=C\diagup{Me}\diagdown{Me}$$

73 per cent   5 per cent   0.06 per cent

This reaction is very similar to that mentioned earlier for heterocarbene complexes (scheme *vi*).

Experiments with electron-rich olefins also suggest the intermediacy of carbene complexes. Thus the reaction of two different but symmetrical tetra-aminoalkenes with a catalytic amount of $(Ph_3P)_3RhCl$ led to the exchange of alkylidene groups (that is, metathesis) and formation of a rhodium carbene complex. (The metathesis of olefins is not generally catalysed by rhodium complexes, but the tetraaminoalkenes are a special case (see p. 141; equation (7.16); Lappert, 1975).

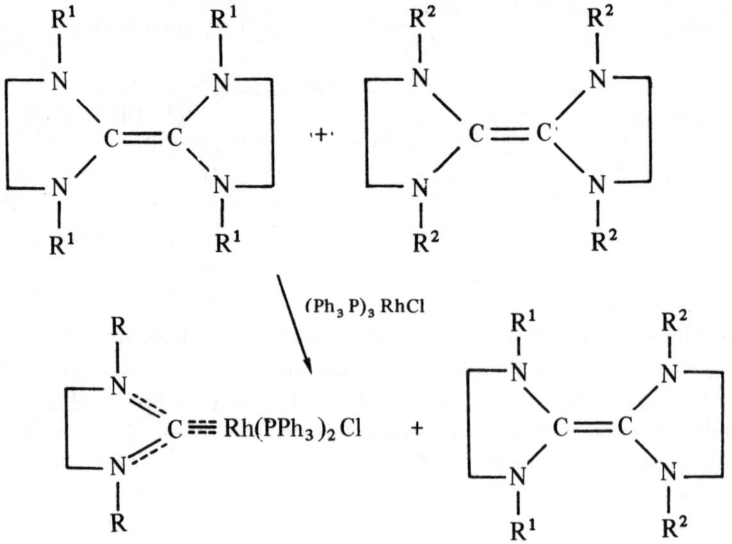

$$(7.16)$$

Although equations (7.15) and (7.16) provide good examples of the role of carbenes in metathesis, they are atypical, since the active systems usually involve high oxidation state compounds. The WCl_6/EtOH/EtAlCl$_2$ system probably involves initial hydrolysis of WCl_6 to $WOCl_4$ followed by alkylation. Recent work has shown that an α-elimination reaction is the most likely route to the initial carbene complex required for the Hérisson and Chauvin mechanism (scheme *xiv*; Band and Muetterties, 1980). The exchange of alkylidene groups

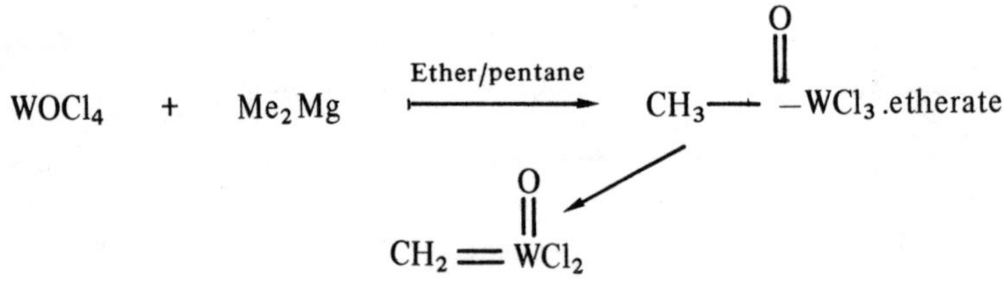

Scheme *xiv*

between tungsten and olefins has been observed directly in stoicheiometric reactions (Schrock, 1983) and spectroscopically under catalytic conditions (Kress *et al.*, 1982).

7.8 THE FISCHER-TROPSCH SYNTHESIS

The Fischer-Tropsch synthesis was discovered in Germany in the 1920s, and was used during the Second World War to produce liquid hydrocarbon fuel from coal. The basic reaction is given as equation (7.17).

$$nCO + 2nH_2 \xrightarrow{\text{Iron catalyst}} -(CH_2)_n- + nH_2O$$

$$(7.17)$$

$$\Delta H = -165 \text{ kJ mol}^{-1} \text{ at } 500 \text{ K}$$

This equation is a gross simplification as oxygen-containing products such as alcohols, esters and acids are formed, in addition to hydrocarbons. The original process used pressures of carbon monoxide and hydrogen of about 100 atm and a temperature of about 400 °C over an alkaline-doped iron heterogeneous catalyst. The process has undergone considerable refinement, and is currently operated in South Africa (Masters, 1979).

The mechanism of the Fischer–Tropsch synthesis has become a widely studied topic in organometallic chemistry. The overall picture is complicated. Work has been carried out on heterogeneous and homogeneous systems, using mono- and poly-nuclear complexes. Studies of the interaction of carbon monoxide with metal surfaces have established the formation of carbides. These carbides could, in principle, be subsequently hydrogenated to carbenes, i.e. methylene; although there is no direct evidence for this. So far as the formation of hydrocarbons is concerned (as distinct from oxygen-containing products), there is good evidence that these are derived from the coupling of methylene groups. This was demonstrated by Brady and Pettit (1980) who showed that diazomethane in the presence of hydrogen decomposed over a Fischer–Tropsch catalyst to give the same hydrocarbon product distribution as the Fischer–Tropsch synthesis itself. Thus the following mechanism can be proposed (scheme xv). This very simplified mechanism is one of several which have been suggested.

Scheme xv

The mechanism does make the point that the insertion of CO in an M–H bond is very unlikely (see p. 129). On the other hand, the formation of some M–CH$_3$ species and insertion of CO to give

$$\underset{\underset{O}{\|}}{M-C-CH_3}$$

is quite likely to be a pathway in addition to that shown in scheme xv (Herrmann, 1982b).

7.9 MERCURY DERIVATIVES AS A SOURCE OF CARBENES

We have seen that many transition metals form isolable carbene complexes which can be used in organic synthesis, although the reactions do not involve free carbenes. The position with mercury is reversed; the carbene complexes of mercury are not stable, but organomercury compounds can be used as sources of free carbenes. As we discuss on p. 168, α-halomethylmercury(II) derivatives act as sources of carbenes. The phenyl derivatives Ph–Hg–CXYHal are often used, and X and Y are frequently halogen atoms. When heated in a variety of solvents in the presence of an alkene, a cyclopropane, **XXIX**, is formed (equation (7.18); Seyferth, 1972).

$$\text{Ph–HgCXYHal} \; + \; \underset{C}{\overset{C}{\|}} \quad \xrightarrow[\text{solvent}]{\text{Heat in}} \quad \triangle\!\!<^{X}_{Y} \; + \; \text{PhHgHal} \qquad (7.18)$$

$$\textbf{XXIX}$$

These thermal reactions are generally considered to involve the free carbene as intermediate, but the reaction of *bis*(bromomethyl)mercury, **XXX**, with alkenes is an exception and thought to proceed by bimolecular reaction between **XXX** and the alkene (equation (7.19); Seyferth *et al.*, 1969).

$$(\text{BrCH}_2)_2\text{Hg} \; + \; \bigcirc \quad \longrightarrow \quad \bigcirc\!\!\triangleright \; + \; \text{BrCH}_2\text{HgBr} \qquad (7.19)$$

$$\textbf{XXX}$$

An interesting application of mercury derivatives to organometallic compounds is the synthesis of a dichlorocarbene complex, **XXXI**, using *bis*(trichloromethyl)-mercury (equation 7.20) (Clark *et al.*, 1980). *al.*, 1980).

$$(\text{Ph}_3\text{P})_3\text{OsCl(CO)H} \; + \; (\text{Cl}_3\text{C})_2\text{Hg} \quad \xrightarrow[\text{toluene}]{\text{Reflux}} \quad \overset{\text{PPh}_3}{\underset{\text{PPh}_3}{\overset{\displaystyle|}{\underset{\displaystyle|}{\text{Cl}\cdots\overset{\text{Cl}}{\underset{\text{Cl}}{\text{Os}}}}}}\!\!=\!C\overset{\text{Cl}}{<}_{\text{CO}} \qquad (7.20)$$

$$\textbf{XXXI}$$

REFERENCES

Badley, E. M., Chatt, J., and Richards, R. L. (1971). *J. chem. Soc. (A)* 21

Band, E., and Muetterties, E. L. (1980). *J. Am. chem. Soc.* **102**, 6572

Brady, R. C., III and Pettit, R. (1980). *J. Am. chem. Soc.* **102**, 6181

Calderon, N., Lawrence, J. P., and Ofstead, E. A. (1979). *Adv. organomet. Chem.* **17**, 449

Casey, C. P., and Anderson, R. L. (1975). *J. chem. Soc. chem. Commun.* 895

Casey, C. P., Burkhardt, T. J., Bunnell, C. A., and Calabrese, J. C. (1977). *J. Am. chem. Soc.* **99**, 2127

Casey, C. P., Tuinstra, H. E., and Saeman, M. C. (1976). *J. Am. chem. Soc.* **98**, 608

Chisholm, M. H., Clark, H. C., Johns, W. S., Ward, J. E. H., and Yasafuka, K. (1975). *Inorg. Chem.* **14**, 900

Churchill, M. R., and Youngs, W. J. (1979). *Inorg. Chem.* **18**, 2454

Clark, D. N., and Schrock, R. R. (1978). *J. Am. chem. Soc.* **100**, 6774

Clark, G. R., Marsden, K., Roper, W. R., and Wright, L. J. (1980). *J. Am. chem. Soc.* **102**, 1206

Connor, J. A., and Fischer, E. O. (1969). *J. chem. Soc. (A)*, 578

Cooke, M. D., and Fischer, E. O. (1973). *J. organomet. Chem.* **56**, 279

Dötz, K. H., Praskil, I., and Mühlemeier, J. (1982). *Chem. Ber.* **115**, 1278

Dötz, K. H., Mühlemeier, J., Schubert, U. and Orama, O. (1983). *J. organomet. Chem.* **247**, 187

Fischer, E. O. (1976). *Adv. organomet. Chem.* **14**, 1

Fischer, E. O., and Dötz, K. H. (1972). *Chem. Ber.* **105**, 3966

Fischer, E. O., and Maasböl, A. (1964). *Angew. Chem. int. Edn Engl.* **3**, 580

Fischer, E. O., and Plabst, D. (1974). *Chem. Ber.* **107**, 3326

Fischer, E. O., and Schubert, U. (1975). *J. organomet. Chem.* **100**, 59

Fischer, E. O., Clough, R. L., Besl, G., and Kreissl, F. R. (1976). *Angew. Chem. int. Edn Engl.* **15**, 543

Fischer, E. O., Held, W., Kreissl, F. R., Frank, A. and Huttner, G. (1977). *Chem. Ber.* **110**, 656

Fischer, E. O., Kreissl, F. R., Winkler, E., and Kreiter, C. G. (1972). *Chem. Ber.* **105**, 588

Fischer, E. O., Schneider, J., and Neugebauer, D. (1983). *Monatsh. Chem.* **114**, 851

Goddard, R. J., Hoffmann, R., and Jemmes, E. D. (1980). *J. Am. chem. Soc.* **102**, 7667

Hérisson, J. L., and Chauvin, Y. (1970). *Makromol. Chem.* **141**, 161

Herrmann, W. A. (1982a). *Adv. organomet. Chem.* **20**, 160

Herrmann, W. A. (1982b). *Angew. Chem. int. Edn Engl.* **21**, 117

Hoffmann, R. W. (1968). *Angew. Chem. int. Edn Engl.* **7**, 754

Holton, J., Lappert, M. F., Pearce, R., and Yarrow, P. I. W. (1983). *Chem. Rev.* **83**, 135

Jensen, C. M., and Kaesz, H. D. (1983). *J. Am. chem. Soc.* **105**, 6969

Kirmse, W. (1971). *Carbene Chemistry*, 2nd edn. Academic Press, New York

Kress, J., Wesolek, M., and Osborn, J. A. (1982). *J. chem. Soc. chem. Commun.* 514

Lappert, M. F. (1975). *J. organomet. Chem.* **100**, 139

Masters, C. (1979). *Adv. organomet. Chem.* **17**, 61

Miki, K., Taniguchi, H., Kai, Y., Kasai, N., Nishiwaki, K., and Wada, M. (1982). *J. chem. Soc. chem. Commun.* 1178

Nugent, W. A., and Haymore, B. L. (1980). *Co-ord. Chem. Rev.* **31**, 123

Schrock, R. R. (1974). *J. Am. chem. Soc.* **96**, 6796

Schrock, R. R. (1976). *J. Am. chem. Soc.* **98**, 5399

Schrock, R. R. (1979). *Acc. chem. Res.* **12**, 98

Schrock, R. R. (1983). *Science* **219**, 13

Semmelhack, M. F., and Tamura, R. (1983). *J. Am. chem. Soc.* **105**, 4099

Seyferth, D. (1972). *Acc. chem. Res.* **5**, 65

Seyferth, D., Turkel, R. M., Eistert, M. A. and Todd, L. J. (1969). *J. Am. chem. Soc.* **91**, 5027

GENERAL READING

Brown, F. J. (1980). Stoichiometric reactions of transition metal carbene complexes. *Prog. inorg. Chem.* **27**, 1

Fischer, H. (1982). Synthesis of transition metal carbene complexes. In *The Chemistry of the Metal–Carbon Bond*, Vol. 1 (F. R. Hartley and S. Patai, eds). Wiley, Chichester, p. 181

Fischer, H., Kreissl, F. R., Schubert, U., Hoffman, P., Dötz, K.-H., and Weiss, K. *Transition Metal Carbene Complexes*, Festschrift für E. O. Fischer. Verlag Chemie, Weinheim, 1983

8

STOICHEIOMETRIC APPLICATIONS OF ORGANOMETALLIC COMPOUNDS TO ORGANIC CHEMISTRY

The largest and most rapidly developing area of organometallic chemistry, with numerous research papers, reviews, books and conferences devoted to it, is that of application to organic chemistry. Catalytic uses of organometallic compounds are dealt with in chapter 9, but here we discuss the use of these compounds as reagents and intermediates.

We have already seen that, if an organic group is linked to a metal, its reactivity is modified. Groups attached by σ-bonds are activated to participation in nucleophilic and other reactions. Although the difference between reagents and substrates is largely semantic, we generally think of this class of organometallics as reagents. Unsaturated compounds, π-bonded to transition metals, are more likely to be substrates with the unsaturated residue electronically activated as well as regio- and stereo-chemically disposed for attack by a reagent. We shall also encounter examples where an unsaturated centre is deactivated, or protected, by metal complex formation.

In this chapter, ruthless selection has been necessary, particularly in the burgeoning transition metal field.

8.1 MAIN GROUP ELEMENT COMPOUNDS

Despite spectacular developments elsewhere, the organic derivatives of main group metals remain of central importance to organic synthesis. A relatively small number of metals is involved but, nevertheless, some selection has been unavoidable. Three transition metals, whose organic derivatives resemble those of the typical elements, present classification dilemmas. These have been arbitrarily resolved by including mercury in this section, but treating copper and titanium under section 8.2.

8.1.1 Organolithium Compounds

Organic derivatives of lithium are common laboratory reagents for a number of reasons. These include ease of synthesis, commercial availability of the methyl, butyl and phenyl compounds, high reactivity, and clean work-up of reaction mixtures. They are particularly useful as sources of carbanions (see chapter 4).

The classical synthesis of alcohols by addition of an organometallic compound to a carbonyl species is a widely used reaction (scheme i). Organolithium

$$RLi + R^1COR^2 \longrightarrow RR^1R^2COLi \xrightarrow{H_2O} RR^1R^2COH + LiOH$$

Scheme i

compounds show higher reactivities than Grignard reagents in these reactions and fewer by-products are formed. This is particularly important with sterically hindered ketones which may be reduced by Grignard reagents to the corresponding alcohols instead of giving the addition product. An example is the reaction of diisopropyl ketone in scheme ii (Young and Roberts, 1944).

$$iPr_2CHOH \xleftarrow[2. H_2O]{1. iPrMgBr} iPrCOiPr \xrightarrow[2. H_2O]{1. iPrLi} iPr_3COH$$

(no addition)   19 per cent
(no reduction)

Scheme ii

Another important difference between lithium and magnesium compounds is shown in their behaviour with α,β-unsaturated carbonyl compounds. Generally, lithium derivatives give predominantly 1,2-addition whereas, the organic group of Grignard reagents is introduced mainly in the β-position via 1,4-addition (scheme *iii*). If the two reactants contain bulky substituents, the situation

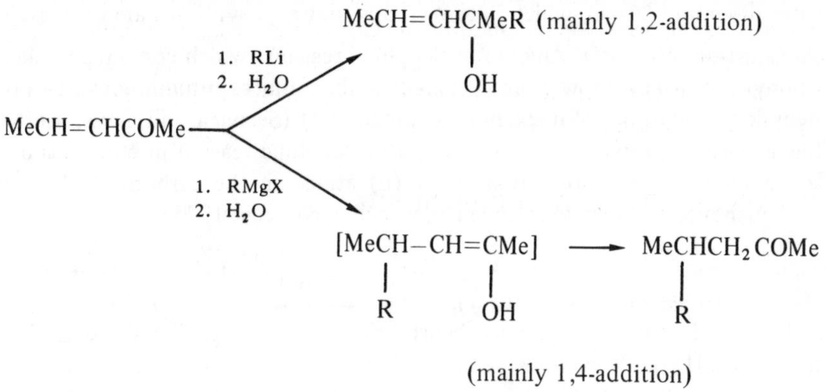

$$\text{MeCH=CHCMeR (mainly 1,2-addition)}$$
$$\quad\quad\quad\quad\;\;|$$
$$\quad\quad\quad\quad\;\;\text{OH}$$

MeCH=CHCOMe
— 1. RLi / 2. H$_2$O (upper)
— 1. RMgX / 2. H$_2$O (lower)

$$[\text{MeCH–CH=CMe}] \longrightarrow \text{MeCHCH}_2\text{COMe}$$
$$\quad\;|\quad\quad\;\;|\quad\quad\quad\quad\quad\quad\;|$$
$$\quad\;\text{R}\quad\quad\text{OH}\quad\quad\quad\quad\quad\;\text{R}$$

(mainly 1,4-addition)

Scheme *iii*

becomes more complicated as in the reaction between *t*-butyl organometallic compounds and 2,2,6,6-tetramethylhepta-4-en-3-one (scheme *iv*; Ashby and Weisemann, 1978). Here the lithium compound shows substantial 1,4-addition, though the 1,2-mode still predominates, but the Grignard reagent now gives mainly reduction product.

tBuCH=CHCOtBu → 1. tBuM / 2. H$_2$O

	Percentage yield	
	M = Li	M = MgCl
tBuCH=CHCH(OH)tBu (1,2-reduction)	1.9	50.2
tBu$_2$CHCH$_2$COtBu (1,4-addition)	46.9	15.9
tBuCH=CHC(OH)tBu$_2$ (1,2-addition)	51.2	33.9

Scheme *iv*

Organolithium compounds add readily to carbon–carbon double bonds (equation 8.1).

$$\text{RLi} + \;>\!\text{C=C}\!< \longrightarrow \text{R}-\underset{|}{\overset{|}{\text{C}}}-\underset{|}{\overset{|}{\text{C}}}-\text{Li} \quad\quad\quad (8.1)$$

This reaction is of particular interest because the product will add to further molecule of alkene resulting, eventually, in polymerisation (see p. 232).

More recently, the value of organolithium reagents carrying heteroatom substituents has been widely exploited. One motivation here has been to overcome the synthetic problems caused by the polarity of the carbonyl group being in the direction $\overset{\delta+}{\underset{}{>}}\overset{\delta-}{C=O}$, which in turn means that acyl synthons are only available as cations $R-\overset{+}{C}=O$. Anionic nucleophilic reagents which contain a masked carbonyl group have now been prepared in the form of lithium derivatives of aldehyde dithioacetals (for example compound **I**) (Seebach and Geiss, 1976). Thus a masked propionyl group acts as a nucleophilic reagent in either (a) displacing iodine from 2-iodopropane, or (b) attacking the carbonyl carbon in benzophenone, as shown in scheme ν (Seebach and Corey, 1975).

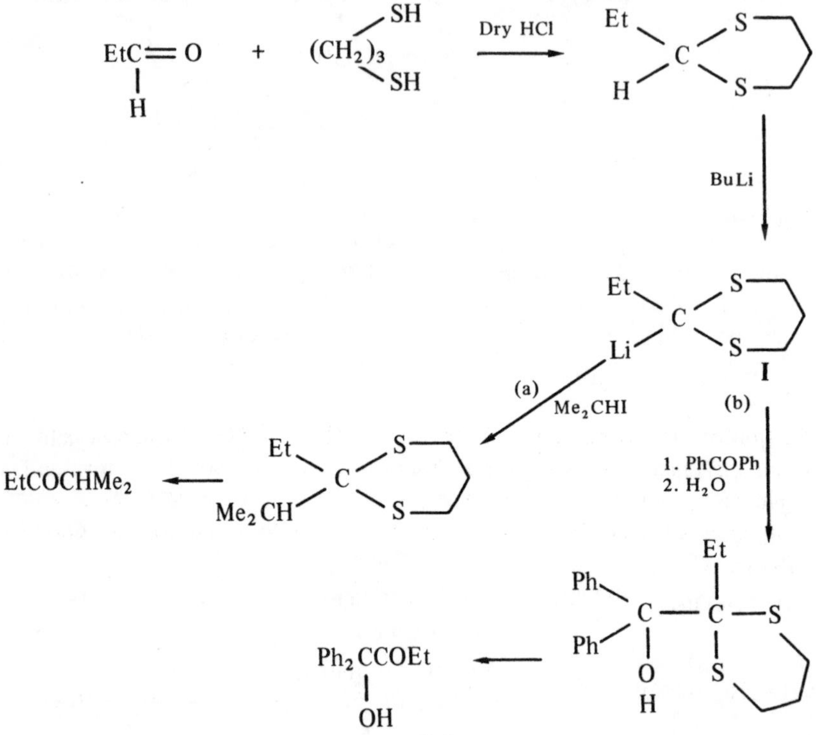

Scheme ν

Substituted allyllithium species of the type **II** add to carbonyl compounds with either the α- or γ-position becoming attached to carbonyl carbon. Relatively

II

small changes in the structure of **II** make considerable differences to the composition of the products formed, as indicated in equation (8.2).

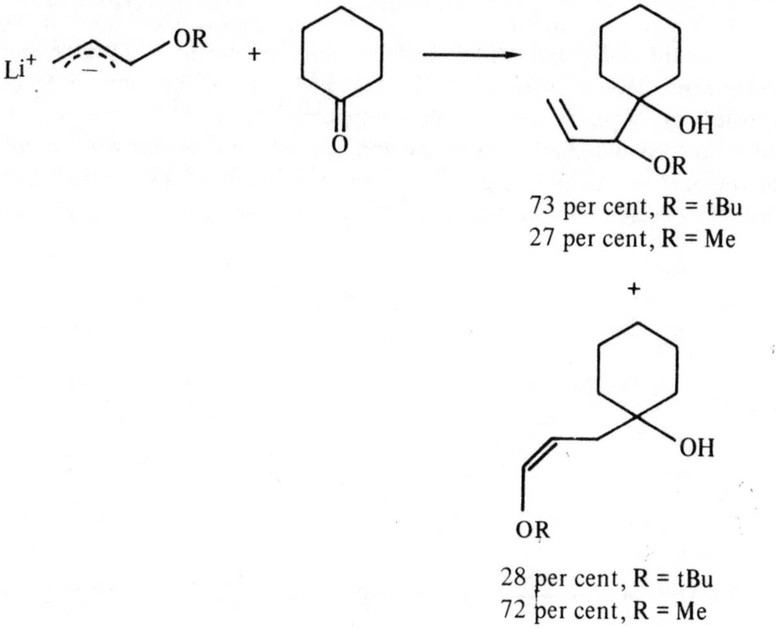

73 per cent, R = tBu
27 per cent, R = Me

+   (8.2)

28 per cent, R = tBu
72 per cent, R = Me

Lithiation of diethyl but-2-enylphosphonate (**III**) gives **IV**, which, on treatment with an alkyl halide, gives exclusively the α-alkylated product **V**. Removal of the phosphonate group by hydrogenolysis gives the *trans*-alkene (**VI**) even when the starting material (**III**) is a mixture of *cis* and *trans* isomers (scheme *vi*; Kondo *et al.*, 1974).

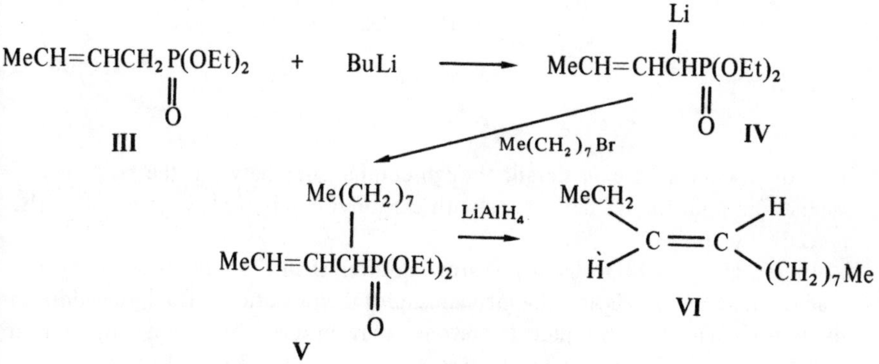

Scheme *vi*

8.1.2 Organomagnesium Compounds

The value of organomagnesium reagents to the organic chemist was recognised by a Nobel Prize award in 1912 and, despite many alternative reagents now available, Grignard reagents are still of prime importance. They are readily prepared from the metal and organic halide in an ether medium and are easily handled (in many cases an inert atmosphere is unnecessary). The common preparative reactions of organomagnesium compounds are well documented in standard textbooks of organic chemistry and a more comprehensive survey of classical Grignard reactions is available in the excellent book by Kharash and Reinmuth (1954). Some important reactions are summarised in scheme *vii*, but

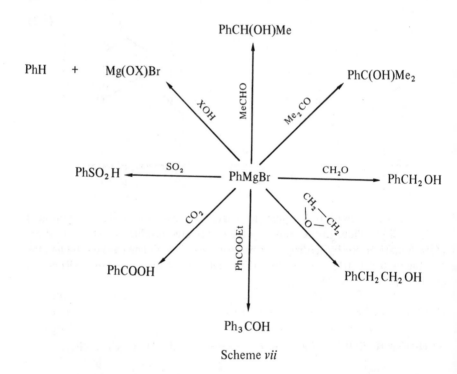

Scheme *vii*

are not discussed here in detail, though comparisons between the reactions of magnesium and lithium derivatives with carbonyl compounds were noted in the preceding section.

In recent years there has been growing interest in the structures of organomagnesium compounds and the mechanisms of their reactions. Thus, the addition of methylmagnesium bromide to ketones may proceed by a polar mechanism (electron pair displacements) or by a single electron transfer mechanism (scheme *viii*; Ashby, 1980). It is seen that the latter mechanism can also give the pinacol

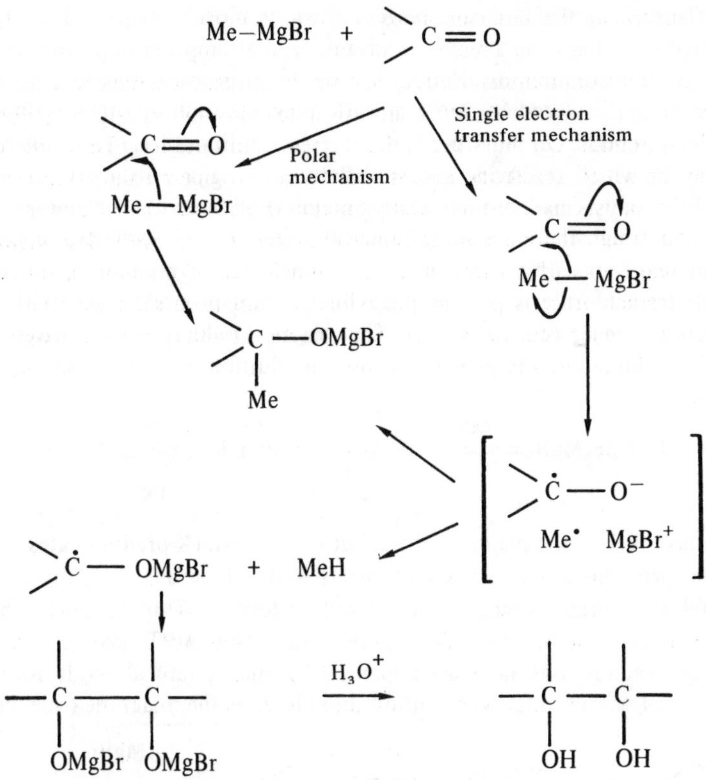

$$Me-MgBr \quad + \quad \overset{\diagdown}{\diagup}C = O$$

Single electron
transfer mechanism

Polar
mechanism

$$\overset{\diagdown}{\diagup}C = O$$
$$Me-MgBr$$

$$\overset{\diagdown}{\diagup}C = O$$
$$Me-MgBr$$

$$\overset{\diagdown}{\diagup}C-OMgBr$$
$$\underset{Me}{\big|}$$

$$\left[\begin{array}{c} \overset{\diagdown}{\diagup}\dot{C}-O^- \\ Me^{\bullet} \quad MgBr^+ \end{array} \right]$$

$$\overset{\diagdown}{\diagup}\dot{C}-OMgBr \quad + \quad MeH$$

$$\underset{OMgBr \;\; OMgBr}{-\overset{|}{\underset{|}{C}}-\overset{|}{\underset{|}{C}}-} \quad \xrightarrow{H_3O^+} \quad \underset{OH \quad OH}{-\overset{|}{\underset{|}{C}}-\overset{|}{\underset{|}{C}}-}$$

Scheme *viii*   **VII**

(**VII**) by bimolecular reduction. The actual mechanism and product mix will depend upon such factors as solvent, concentration, nature of the ketone and purity of the magnesium.

Monomolecular reduction of carbonyl groups has already been referred to. This reaction is shown by Grignard reagents which possess β-hydrogen atoms; reduction occurs by addition of a magnesium hydride moiety and the Grignard reagent is converted to the alkene (scheme *ix*).

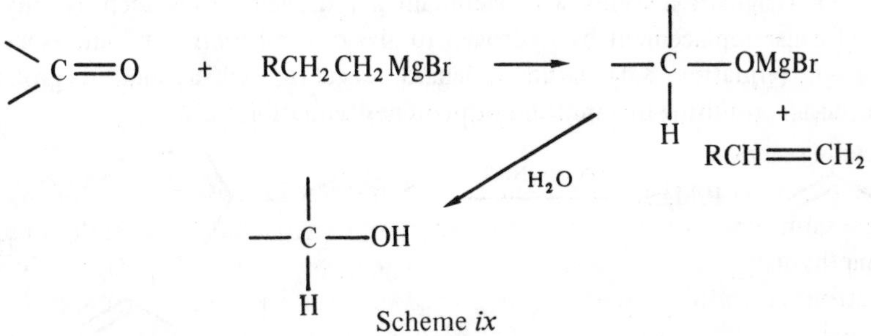

$$\overset{\diagdown}{\diagup}C = O \quad + \quad RCH_2CH_2MgBr \quad \longrightarrow \quad -\overset{|}{\underset{H}{C}}-OMgBr$$
$$+$$
$$RCH = CH_2$$

$$-\overset{|}{\underset{H}{C}}-OH$$

$$\xrightarrow{H_2O}$$

Scheme *ix*

The lability of the Grignard reagent towards most unsaturated groups and active hydrogen has been a major disadvantage in attempting to prepare complex polyfunctional compounds. Hence, for some purposes, Grignard reagents are tending to be displaced by more specific reagents such as organo-copper and -boron compounds. On the other hand, interest continues to grow in the remarkable way in which reactivity and specificity of Grignard reagents can be controlled by catalysts made from transition metals (Felkin and Swierczewski, 1975).

Thus, although there are some exceptions (see p. 166), Grignard reagents are normally inert to carbon–carbon double bonds, but if a catalytic quantity of titanium tetrachloride is present propylmagnesium bromide reacts with monosubstituted ethenes (equation 8.3). This reaction, which can be driven to the right by distilling out the propene, is not an addition but a reversible exchange reaction.

$$2RCH{=}CH_2 + 2PrMgBr \xrightarrow{\text{TiCl}_4} RCH_2CH_2MgBr + RCH(MgBr)Me + 2MeCH{=}CH$$

$$\textbf{VIII} \qquad\qquad \textbf{IX} \qquad\qquad (8.3)$$

If R = alkyl, **VIII** is the major product, but for R = Ph, **IX** predominates.

When propylmagnesium bromide reacts with allylic alcohols, the nature of the catalyst determines what products will be formed. Thus nickel(II) chloride, in the presence of hexamethylphosphoric triamide (HMPT) promotes a similar exchange reaction but now accompanied by magnesium alkoxide formation; treatment of the product with carbon dioxide gives the γ-lactone (scheme x). If

Scheme x ~30 per cent

the catalyst is *bis*(triphenylphosphine)nickel(II) chloride, the hydroxyl group is replaced. Grignard reagents which contain β-hydrogen atoms such as ethyl or propyl cause replacement by hydrogen to give *cis*- and *trans*-but-2-ene as well as but-1-ene (equation 8.4). Other Grignard reagents, such as methyl, give new C–C bonds, resulting in a mixture of pentenes (equation 8.5).

$$(8.4)$$

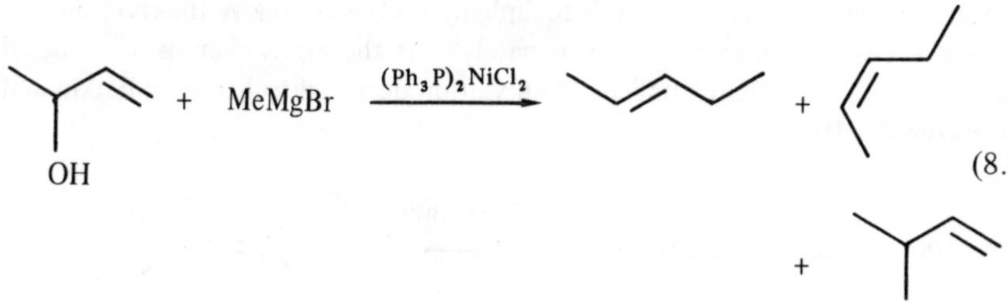

$$(8.5)$$

Cross-coupling reactions of the type shown in equation (8.6) occur in the presence of suitable catalysts and are preparatively useful.

$$RMgX + R'X \longrightarrow R\text{–}R' + MgX_2 \qquad (8.6)$$

They have the particular advantage that the normally unreactive phenyl and vinyl halides can be used (equations (8.7) and (8.8); Kumada, 1980).

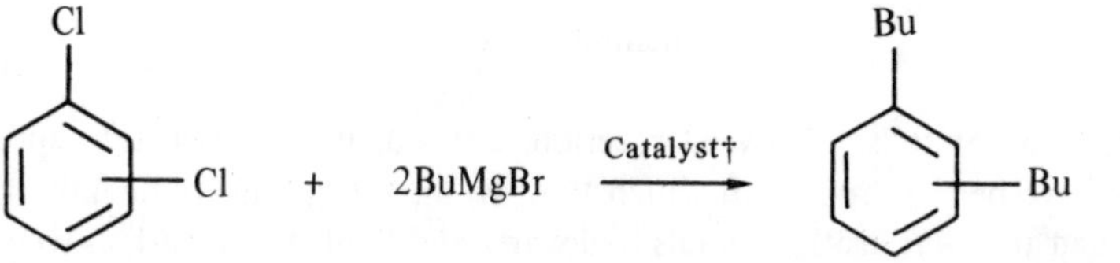

$$(8.7)$$

ortho, 79–83 per cent
meta, 94 per cent
para, 95 per cent

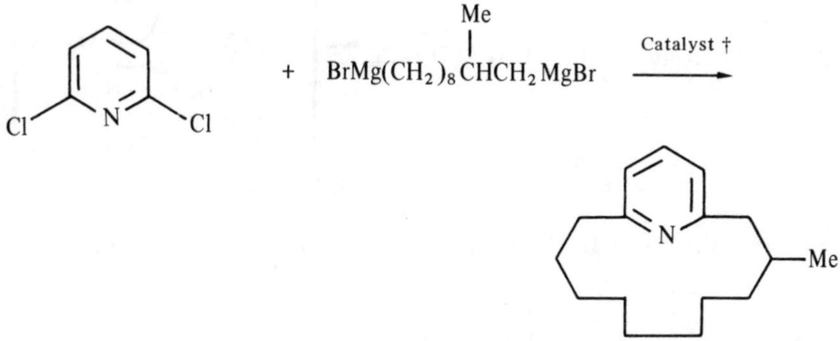

Muscopyridine (20 per cent)

† This catalyst is $(CH_2)_3 \underset{P}{\overset{P}{<}} \; NiCl_2$ with Ph_2

$$(8.8)$$

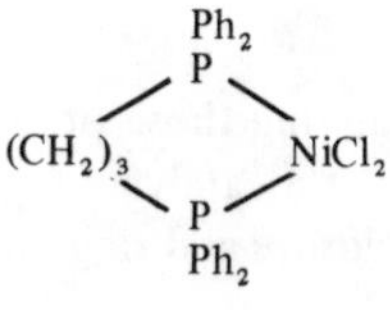

Methylmagnesium bromide adds to diphenylacetylene to give the *trans* adduct when a rhodium complex is used as catalyst or the *cis* isomer using a nickel complex; with palladium catalysts, dialkylation occurred (scheme *xi*; Felkin and Swierczewski, 1975).

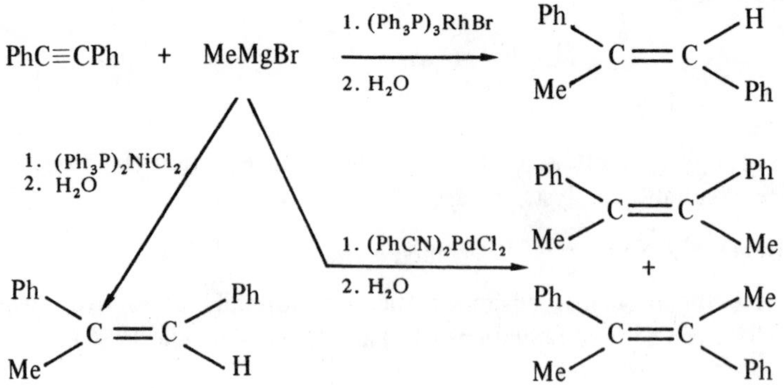

Scheme *xi*

The mechanisms of these catalysed reactions are still under discussion and intermediates have been proposed in which the two organic groups to be united become attached to the transition metals (Schwartz and Labinger, 1980), as well as others containing transition metal–magnesium bonds.

Another recent development has been the discovery that allylmagnesium compounds will add to unconjugated carbon–carbon double bonds in the absence of catalysts; a six-member transition state has been proposed (scheme *xii*; Lehmkuhl, 1981).

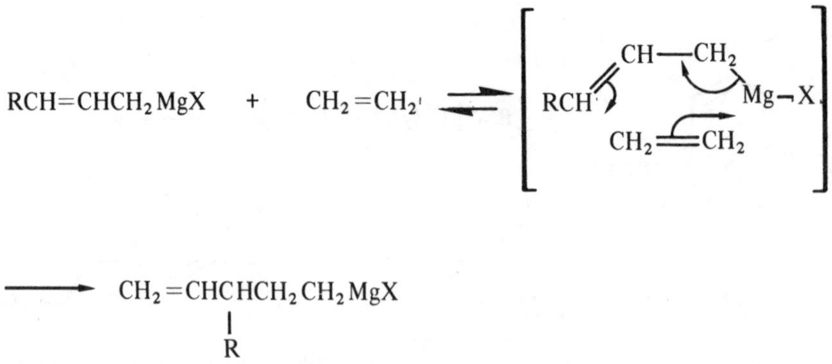

Scheme *xii*

Suitable alkenylmagnesium compounds (including those produced in scheme *xii*) will undergo intramolecular addition without a catalyst. A general formulation for this process, which involves both ring closure and ring opening, is shown

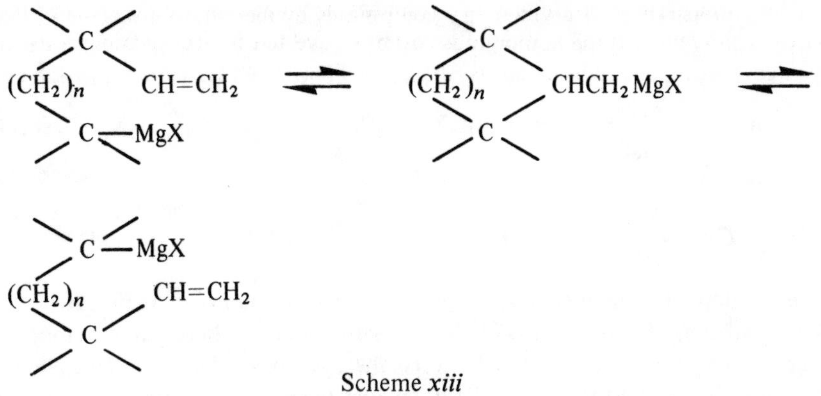

Scheme *xiii*

in scheme *xiii* (Hill, 1977). If the ring is reasonably strain-free, for example when $n = 2$, then the ring-opening reaction becomes unimportant and a stable cyclic product is formed (equation 8.9).

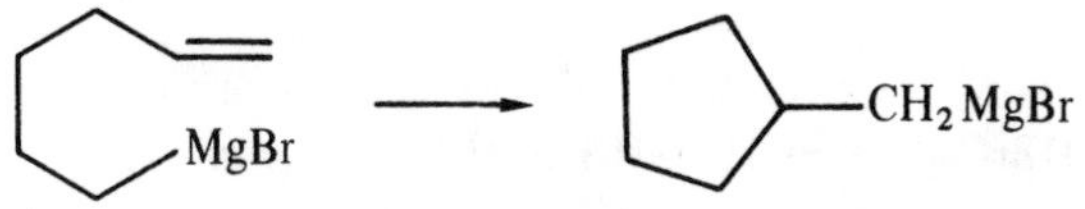

$$(8.9)$$

A similar, but more rapid, intramolecular addition occurs with alkynes (equation (8.10); Hill, 1975).

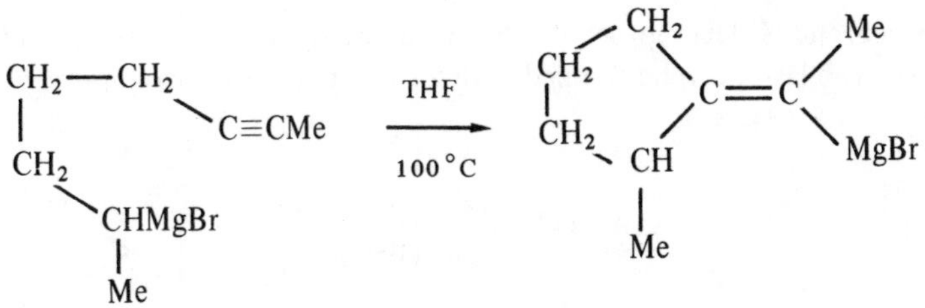

$$(8.10)$$

8.1.3 Organomercury Compounds

Anyone contemplating use of an organomercury reagent should be fully aware of its toxicity; this will usually be high, and special handling techniques may be necessary (Barnes and Magos, 1968). The most striking chemical characteristic of these compounds is their high thermal and chemical stability which can be advantageous in the synthesis of compounds containing functional groups (Larock, 1976). Three areas have been selected for fuller discussion, but we note that the preparation of arylmercury compounds by mercuration was mentioned earlier (p. 27) and these compounds can be converted to aryl halides (equation 8.11).

$$\text{ArHgCl} + X_2 \xrightarrow{\hspace{2cm}} \text{ArX} + \text{HgClX} \qquad (8.11)$$
$$(X = \text{Cl, Br, I})$$

Divalent Carbon Transfer Reactions

The trihalomethylmercury derivatives are particularly convenient precursors for dihalocarbenes (Seyferth, 1972). Most useful is bromodichloromethylphenyl-mercury (**X**) made by reaction between the appropriate haloform and phenyl-mercury chloride in the presence of potassium *t*-butoxide. The reaction is shown in equation (8.12) and the mechanism in scheme *xiv*, where we see that the product (**X**) is formed by a nucleophilic substitution at mercury by the bromo-dichloromethyl carbanion.

$$\text{PhHgCl} + \text{CHBrCl}_2 + \text{tBuOK} \xrightarrow{\hspace{1cm}} \text{PhHgCCl}_2\text{Br} + \text{tBuOH} + \text{KCl} \qquad (8.12)$$
$$\mathbf{X}$$

$$\text{tBuO}^- + \text{CHBrCl}_2 \xrightarrow{\hspace{1cm}} \text{tBuOH} + \text{CBrCl}_2^-$$

$$\text{CBrCl}_2^- + \text{PhHgCl} \xrightarrow{\hspace{1cm}} \text{PhHgCCl}_2\text{Br} + \text{Cl}^-$$
$$\mathbf{X}$$

Scheme *xiv*

On heating in benzene **X** decomposes smoothly to phenylmercury bromide and dichlorocarbene. The latter can be trapped with an alkene to give a cyclopropane derivative (equation 8.13).

$$\text{PhHgCCl}_2\text{Br} \quad + \quad \bighexagon \xrightarrow[\text{2 h}]{\text{C}_6\text{H}_6,\ 80\,^\circ\text{C}} \quad \text{PhHgBr} \quad + \quad \text{(bicyclic } CCl_2\text{ adduct)}$$
$$\mathbf{X}$$

$$(8.13)$$

There is good evidence that dichlorocarbene is an intermediate in these reactions. The advantages of this method of carbene generation are neutral conditions, moderate temperatures, absence of side reactions, and easy product isolation.

This reaction was adapted to give ring expansion of cyclic ketones via addition of dichlorocarbene to the enol acetate (scheme *xv*; Stork *et al*., 1966). Reduction of the adduct with lithium aluminium hydride causes ring expansion as well as deacetylation and replacement of one chlorine by hydrogen. The second chlorine was removed by sodium in liquid ammonia and the product oxidised to give

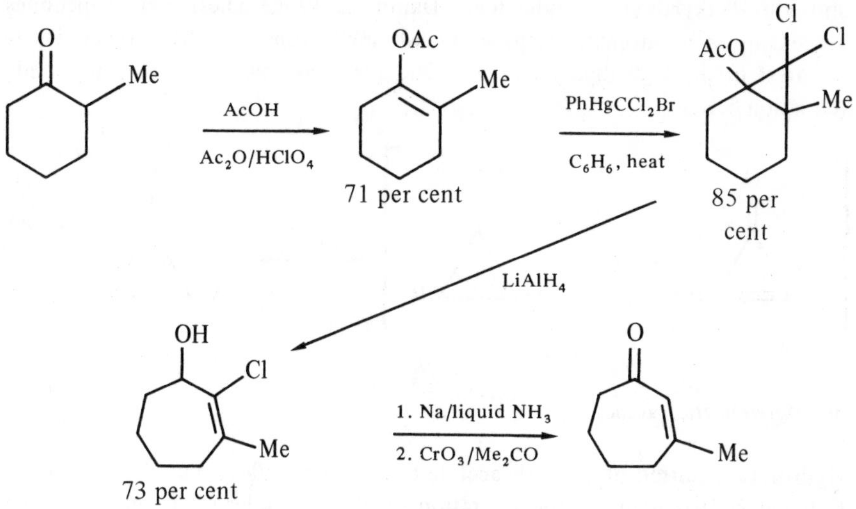

Scheme *xv*

3-methylcyclohept-2-enone. (An alternative procedure for effecting this ring expansion given on p. 177).

Dichlorocarbene will insert into reactive carbon–hydrogen bonds (equation 8.14) and, occasionally, inadvertent insertion may complicate an addition reaction (equation 8.15).

The carbene will add to diaryl- or arylalkyl-alkynes to give, after hydrolysis, the otherwise inaccessible cyclopropenones (scheme *xvi*); the reaction is unsucces-

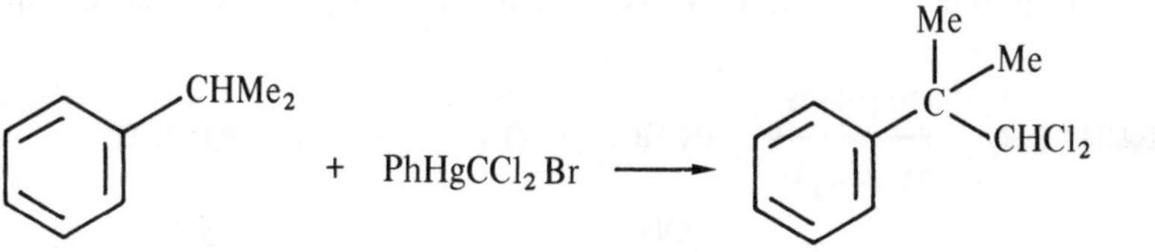

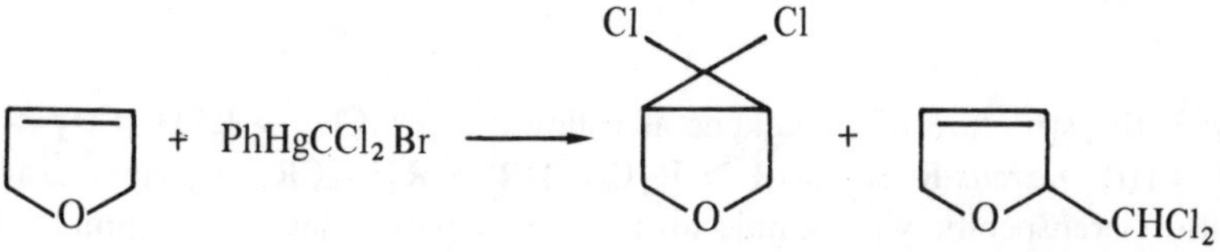

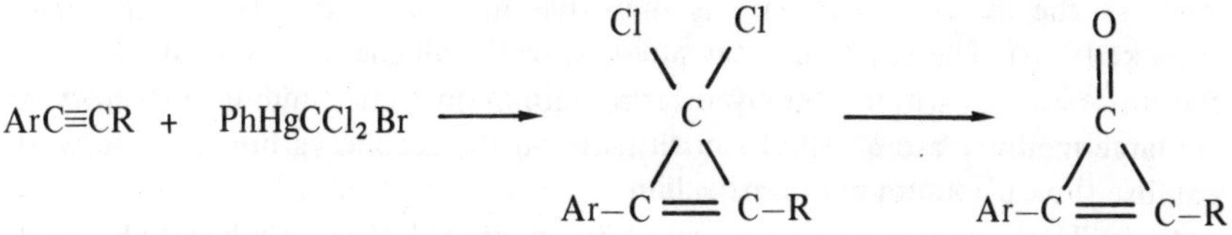

Scheme *xvi*

sful with dialkylalkynes (Seyferth and Damrauer, 1966). These cyclopropenones are examples of aromatic rings with two π-electrons and this aromaticity is manifest in the high dipole moments and high basicities of these compounds (equation 8.16).

$$\left[\begin{array}{c} \underset{Ar}{\overset{O}{\triangle}} R \end{array} \longleftrightarrow \begin{array}{c} \underset{Ar}{\overset{O^-}{\triangle}} R \end{array} \right] \xrightarrow{HClO_4} \begin{array}{c} \underset{Ar}{\overset{OH}{\triangle}} R \ \ ClO_4{}^- \end{array} \qquad (8.16)$$

Solvomercuration Reactions

Hydrolytic addition of mercuric acetate to an alkene followed by demercuration (scheme *xvii*) is used for the hydration of alkenes under mild conditions; the reaction is regiospecific and, overall, Markownikoff's rule is obeyed. Hence this is a useful alternative to the hydroboration/oxidation sequence which leads, overall, to 'anti-Markownikoff' hydration (p. 172). The reaction was initially called 'oxymercuration', but when it was extended to cover the addition of other HX species, including cases where X does not contain oxygen, the more general term of solvomercuration came into use.

Alkenes are hydrated by treatment with mercury diacetate in a mixture of THF and water to give the hydroxyalkylmercury acetate which is demercurated with alkaline sodium borohydride (scheme *xvii*). The rate of oxymercuration

$$RCH{=}CH_2 \xrightarrow[\text{THF/H}_2\text{O}]{Hg(OAc)_2} \underset{\overset{|}{OH}}{RCHCH_2\,HgOAc} \xrightarrow[\text{NaBH}_4]{NaOH} \underset{\overset{|}{OH}}{RCHCH_3}$$

Scheme *xvii*

varies with the structure of the alkene as follows: $R_2C{=}CH_2 > RCH{=}CH_2 >$ *cis*-RCH=CHR > *trans*-RCH=CHR > $R_2C{=}CHR > R_2C{=}CR_2$. Oxymercuration is also stereospecific with simple unstrained alkenes giving *cis* addition but strained monocylic olefines give the *trans* adduct. A mechanism which accommodates the experimental data is indicated in scheme *xviii* (Kitching, 1972; Larock, 1976). The reaction is favoured when the alkene has no substituents on one unsaturated carbon, thereby offering minimum steric hindrance to entry of the large mercury atom. Alkyl substituents on the second carbon atom serve to stabilise the carbonium ion intermediate.

If the THF–water mixture is replaced by an alcohol, then an ether is obtained. In this case mercury *bis*(trifluoroacetate) gives better results, particularly when

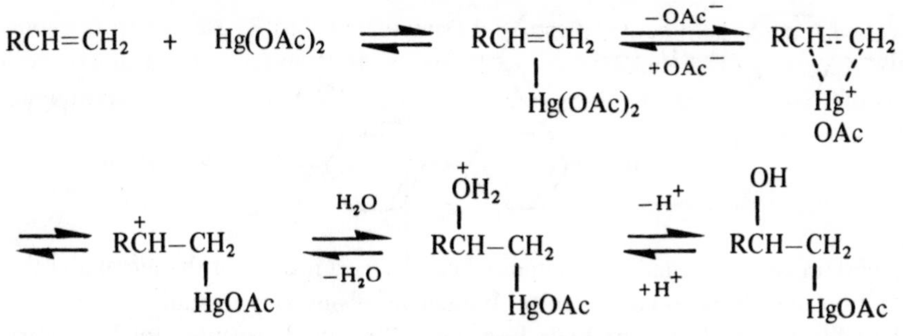

$$RCH=CH_2 \; + \; Hg(OAc)_2 \; \rightleftharpoons \; \underset{\underset{Hg(OAc)_2}{|}}{RCH=CH_2} \; \underset{+OAc^-}{\overset{-OAc^-}{\rightleftharpoons}} \; \underset{\underset{OAc}{Hg^+}}{RCH\dot{=}CH_2}$$

$$\rightleftharpoons \; \underset{\underset{HgOAc}{|}}{\overset{+}{RCH}-CH_2} \; \underset{-H_2O}{\overset{H_2O}{\rightleftharpoons}} \; \underset{\underset{HgOAc}{|}}{\overset{\overset{+}{OH_2}}{RCH-CH_2}} \; \underset{+H^+}{\overset{-H^+}{\rightleftharpoons}} \; \underset{\underset{HgOAc}{|}}{\overset{OH}{RCH-CH_2}}$$

Scheme *xviii*

using secondary or tertiary alcohols (scheme *xix*). In schemes *xx* and *xxi* we see how the reaction has been extended to give, respectively, the secondary alkyl peroxide (mixed with some epoxide) and the amide.

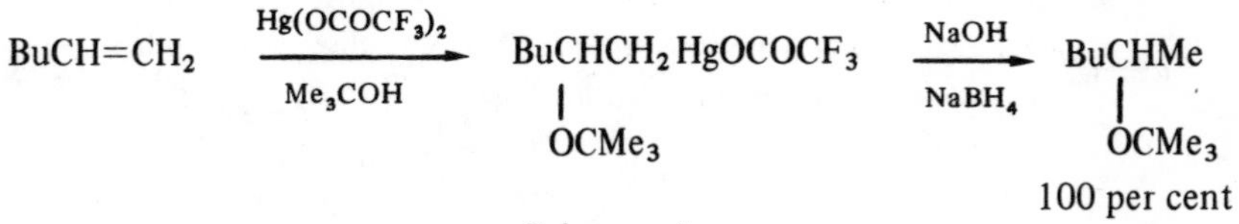

$$BuCH=CH_2 \; \underset{Me_3COH}{\overset{Hg(OCOCF_3)_2}{\longrightarrow}} \; \underset{\underset{OCMe_3}{|}}{BuCHCH_2HgOCOCF_3} \; \underset{NaBH_4}{\overset{NaOH}{\longrightarrow}} \; \underset{\underset{OCMe_3}{|}}{BuCHMe}$$

100 per cent

Scheme *xix*

$$RCH=CH_2 \; \underset{Hg(OCOCF_3)_2}{\overset{Me_3COOH}{\longrightarrow}} \; \underset{\underset{OOCMe_3}{|}}{RCHCH_2HgOCOCF_3} \; \underset{NaBH_4}{\overset{NaOH}{\longrightarrow}} \; \underset{\underset{OOCMe_3}{|}}{RCHMe} \; + \; \underset{O}{RCH-CH_2}$$

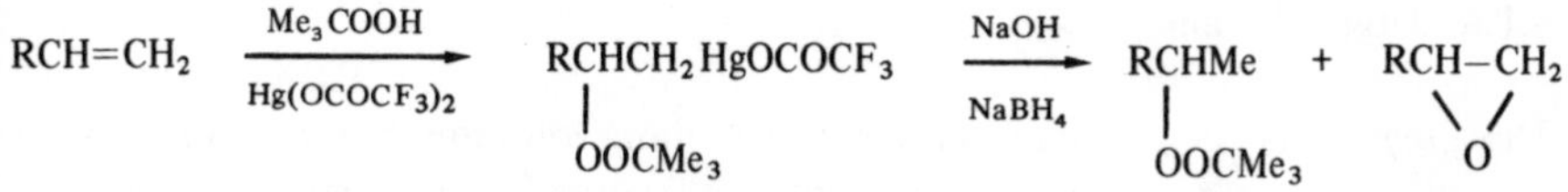

Scheme *xx*

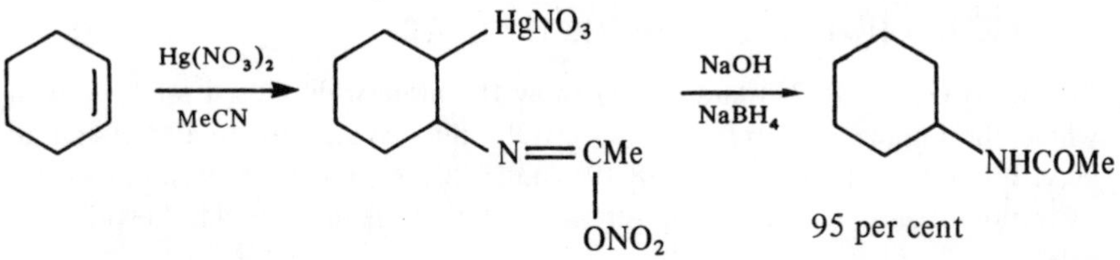

Scheme *xxi*

Dimerisation Reactions

When diarylmercury compounds are heated with a powdered transition metal (such as copper, nickel, palladium, platinum, silver or gold) mercury is eliminated and the biaryl is formed (equation 8.17).

$$\text{Me} - \underset{}{\bigcirc} - \text{Hg} - \underset{}{\bigcirc} - \text{Me} \xrightarrow[\text{heat}]{\text{Pd}}$$

$$\text{Me} - \underset{}{\bigcirc} - \underset{}{\bigcirc} - \text{Me} \quad + \quad \text{Hg}$$

$$(8.17)$$

In certain cases a palladium compound can be used instead of the metal and this reaction has been used to form butadienes from vinylmercury compounds (Larock, 1975). These methods have been little used because stoicheiometric amounts of an expensive metal or compound were required. Recently, however, vinyl groups from vinylmercury chlorides have been coupled using catalytic quantities of certain rhodium compounds (equation 8.18).

$$\underset{H}{\overset{Bu}{>}}C = C \overset{H}{\underset{HgCl}{<}} \xrightarrow[\text{2 mol LiCl, HMPT; r.t.}]{[\text{ClRh(CO)}_2]_2,\ \text{catalytic amount}} \underset{H}{\overset{Bu}{>}}C = C \overset{H}{\underset{C = C}{<}} \overset{H}{\underset{Bu}{<}} \quad (8.18)$$

95 per cent

8.1.4 Organoboranes

The preparation of organoboranes by hydroboration (equation 8.19) followed by replacement of boron with various atoms or groups has been referred to earlier (p. 29) and is a well-established procedure for alkene transformations (Brown, 1975).

$$6RCH{=}CH_2 + B_2H_6 \longrightarrow 2(RCH_2CH_2)_3B \tag{8.19}$$

Equations (8.20), (8.21) and (8.22) show the more widely used procedures in which the borane is converted, respectively, into an alcohol, an alkane and an alkyl bromide. Taking reactions (8.19) and (8.20) together, the overall process is hydration of an alkene in the opposite sense to that predicted by Markownikoff's rule.

$$(RCH_2CH_2)_3B + 3H_2O_2 + OH^- \longrightarrow 3RCH_2CH_2OH + B(OH)_4^- \tag{8.20}$$

$$RCH_2CH_2B{<} + MeCOOH \longrightarrow RCH_2Me + {>}BOCOMe \tag{8.21}$$

$$(RCH_2CH_2)_3B + 3Br_2 + 4NaOMe \longrightarrow 3RCH_2CH_2Br + Na^+[B(OMe)_4]^- + 3NaBr$$
$$\tag{8.22}$$

Comparing the oxymercuration (p. 170) and hydroboration procedures, we note that the electropositive mercury or boron becomes attached to the carbon carrying most hydrogens but, subsequently, mercury is replaced by hydrogen and boron by hydroxyl leading to the difference in overall regiospecificity of hydration. Equations (8.19) and (8.22) illustrate the overall addition of hydrogen bromide in the anti-Markownikoff manner.

When a terminal alkyne is treated with diborane the initially formed alkenylborane usually reacts further to give the fully saturated borane. This second addition can be prevented by starting with a borane containing one or two bulky alkyl substituents. In the example shown in scheme *xxii* (Zweifel and Brown, 1963), oxidation of the alkenylborane formed gives an aldehyde which could not have been obtained by the usual acid-catalysed hydration of the alkyne since this procedure gives the methyl ketone (equation 8.23).

$$C_6H_{13}C{\equiv}CH + (Me_2CHCHMe)_2BH \longrightarrow C_6H_{13}CH{=}CHB(CHMeCHMe_2)_2$$

$$\diagup \begin{array}{c} H_2O_2 \\ NaOH \end{array}$$

$$C_6H_{13}CH_2CHO$$

Scheme *xxii*

$$C_6H_{13}C{\equiv}CH + H_2O \xrightarrow{H^+} C_6H_{13}COMe \qquad (8.23)$$

Other transformations of hydroboration products (Brown, 1975) include the conversion to an amine (equation 8.24).

$$RB{\Big\langle} \xrightarrow[\text{NH}_3^+ \text{OSO}_3^-)/\text{NaOH}]{\text{NH}_2\text{Cl (or}} RNH_2 \qquad (8.24)$$

Here the reagent is either chloramine or hydroxylamine-*O*-sulphonic acid in the presence of alkali. The reaction involves a progressive transfer of alkyl groups from boron to nitrogen. Scheme *xxiii* shows the essential steps.

$$R_3B \;+\; NH_2Cl \longrightarrow R_2\underset{-}{B}{-}\underset{+}{NH_2}{-}Cl \longrightarrow$$

$$R_2\overset{+}{B}NH_2RCl^- \xrightarrow{NaOH} RNH_2$$

Scheme *xxiii*

Trialkylboranes with their low-polarity boron–carbon bonds and electron-deficient central atom are poor anion sources and weak nucleophiles. They are, of course, organometalloids rather than true organometallics so that lower reactivity and reactions atypical of polar metal–carbon bonds may be expected. They can be used to alkylate certain heavy metal halides, such as mercury compounds (equation (8.25); Honeycutt and Riddle, 1960).

$$2Et_3B + 3HgCl_2 + 12NaOH \longrightarrow 3Et_2Hg + 2B(ONa)_3 + 6NaCl + 6H_2O \qquad (8.25)$$

$$95 \text{ per cent}$$

If a trialkylborane obtained from hydroboration (or elsewhere) reacts with an alkyllithium compound (equation 8.26), the resulting lithium tetraalkylborate has very different properties from either of the starting materials.

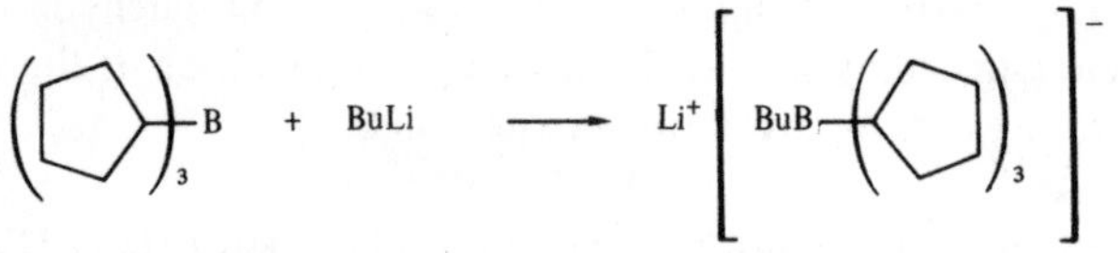

$$(8.26)$$

These borates are not particularly reactive to alkyl halides but, on treatment with acyl halides, give good yields of ketones. Unlike alkyllithium compounds, the tetraalkylborates do not add to carbonyl groups so that no tertiary alcohol is formed. The example given in equation (8.27) illustrates the point that primary alkyl groups are transferred from boron in preference to secondary.

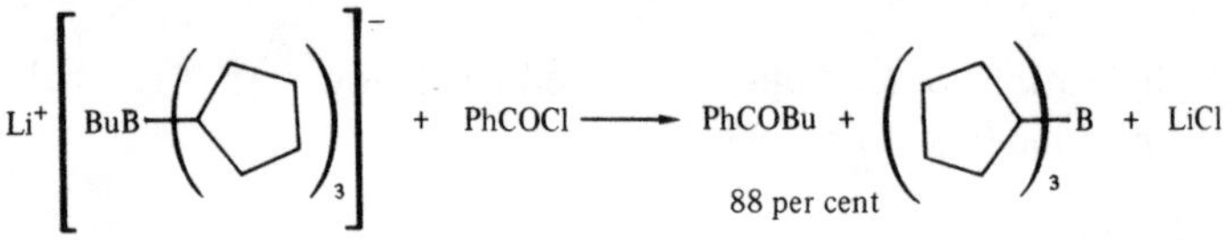

$$(8.27)$$

88 per cent

Some synthetically useful reactions of organoborates are known in which intramolecular transfer of an alkyl group occurs. An example is shown in scheme *xxiv*, where the trialkylalkynylborate reacts with benzyl bromide by transfer of an

$$(C_8H_{17})_3B \quad + \quad LiC{\equiv}CBu \quad \longrightarrow \quad Li^+[(C_8H_{17})_3BC{\equiv}CBu]^-$$

PhCH$_2$Br

$$(C_8H_{17})_2BC{=}CBuCH_2Ph \quad \xrightarrow[\text{NaOH}]{H_2O_2} \quad C_8H_{17}COCHBuCH_2Ph$$
$$|$$
$$C_8H_{17}$$

XI

Scheme *xxiv*

octyl group from boron to the α-acetylenic carbon with accompanying benzylation of the β-carbon. The resulting dioctylalkenylborane is oxidised by alkaline hydrogen peroxide to give, transiently, an enol which ketonises to give 5-benzyltetradecan-6-one (**XI**).

8.1.5 Organoaluminium Compounds

Trialkylaluminium compounds (or organoalanes) react dramatically with water or moist air, the hydrolytic cleavage of the aluminium–carbon bonds being often sufficiently exothermic to ignite the alkane produced. This property and their ability to cause skin burns are the main reasons why organoaluminium compounds have not found wide use as laboratory reagents. There has, nevertheless, been some recent interest in vinylaluminium compounds which are accessible by the addition of hydrides to alkynes (equation (8.28); Negishi, 1976).

$$R_2AlH + R'C{\equiv}CH \longrightarrow R_2AlCH{=}CHR' \qquad (8.28)$$

Vinylaluminium compounds can also be obtained by reaction between a trialkylalane and a disubstituted alkyne when *cis* addition occurs (equation (8.29); Eisch and Hordis, 1971).

$$R_3^1Al \quad + \quad R^2C{\equiv}CR^2 \longrightarrow \quad R_2^1Al{\diagdown}{}{\diagup}R^1 \atop R^2{\diagup}\overset{\displaystyle C=C}{}{\diagdown}R^2 \qquad (8.29)$$

Monosubstituted alkynes also undergo *cis* addition but, in this case, addition competes with metallation because of the acidic alkyne hydrogen (equation (8.30); Mole and Surtees, 1964).

$$R_3Al \quad + \quad PhC{\equiv}CH \Big\langle {R_2Al{\diagdown}{}{\diagup}R \atop Ph{\diagup}\overset{\displaystyle C=C}{}{\diagdown}H} \atop {R_2AlC{\equiv}CPh \quad + \quad RH} \qquad (8.30)$$

These vinylalanes show similar reactions to the corresponding, but less accessible, vinyl-magnesium and -lithium compounds.

The most important use of organoaluminium compounds is for the industrial scale polymerisation and oligomerisation of alkenes (chapter 9).

8.1.6 Organosilicon compounds

If we had to choose the main group element whose organic derivatives were currently having most impact on synthetic organic chemistry, the choice would probably be silicon. The ever-extending range of application of the surprisingly versatile organosilicon reagents is quite remarkable. The large number of review articles in this area (e.g. Hudrlik, 1976; Fleming, 1979, 1981) is now being followed by some books (Colvin, 1981; Weber, 1982).

Compounds containing Si–O bonds

Alcohols are readily converted to trimethylsilyl ethers by treatment with, for example, chlorotrimethylsilane and base, the alcohol being easily regenerated by exposure to water (scheme *xxv*). The first interest in this reaction was its use for

$$ROH + Me_3SiCl + pyridine \longrightarrow ROSiMe_3 + pyridine \cdot HCl$$

$$\swarrow H_2O$$

$$Me_3SiOH + ROH$$

Scheme *xxv*

eliminating the hydrogen-bonded association of carbohydrates by masking the hydroxyl groups. This led to increased volatility and made gas chromatography of carbohydrates possible. Silylation also occurs, in the absence of base, with a 1:1 mixture of hexamethyldisilazane, $(Me_3Si)_2NH$, and chlorotrimethylsilane. As the process began to be used more widely for masking hydroxyl groups in a variety of organic compounds, with subsequent reaction at another site in the molecule, the need arose for derivatives which were more stable to water and other nucleophiles. It was demonstrated that the rate of hydrolysis of silyl ethers was related to the size of the alkyl groups on silicon and soon the *t*-butyldimethylsilyl group emerged as a particularly useful hydroxyl-protecting moiety (Stork and Hudrlik, 1968). This group, besides being more stable to water than its trimethyl analogue, resists attack by many other reagents and can be removed smoothly and specifically by fluoride ion using tetrabutylammonium fluoride (scheme *xxvi*).

$$ROH + tBuMe_2SiCl \xrightarrow[DMF]{Imidazole} ROSiMe_2tBu$$

$$\swarrow Bu_4N^+F^-/THF$$

$$tBuMe_2SiF + ROH$$

Scheme *xxvi*

Besides protection, silyl groups are valuable for the stabilisation of enols. For simple aldehydes and ketones the keto–enol equilibrium is overwhelmingly in favour of the keto form (equation (8.31); Bell and Smith, 1966).

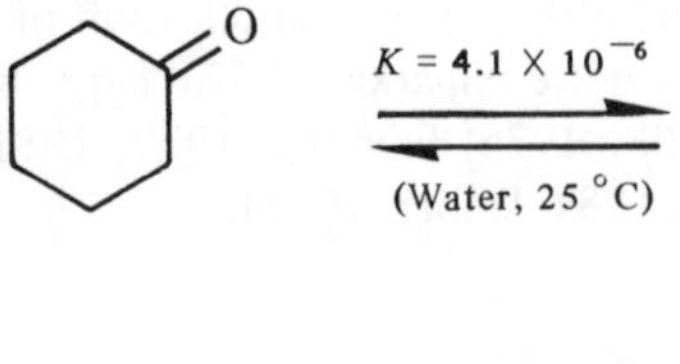

$$K = 4.1 \times 10^{-6}$$

(Water, 25 °C)

(8.31)

Clearly, an α-proton must be removed to form the enol and this will be a very slow process because of the low acidity of the α-CH_2 group, but equilibrium is established rapidly in the presence of a catalytic amount of acid or base. Addition of a stoicheiometric amount of lithium diisopropylamide effectively converts the ketone entirely to the lithium salt of the enol (equation 8.32).

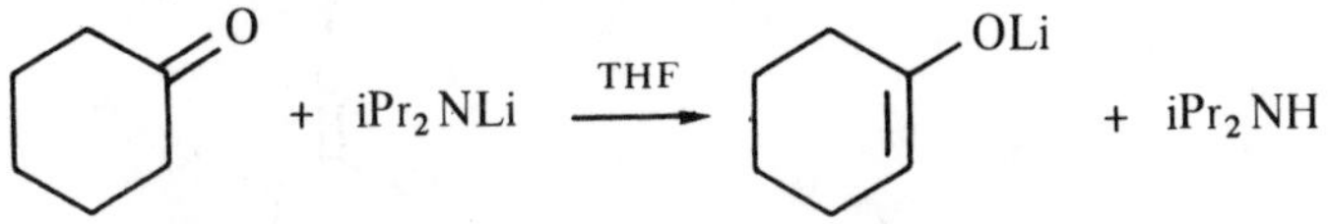

$$+ \ iPr_2NLi \xrightarrow{THF} + \ iPr_2NH \qquad (8.32)$$

Enols would be very useful synthetic intermediates but, at best, they have a very transient existence and acidification of the lithium enolate formed in reaction (8.32) gives the keto form. Addition of chlorotrimethylsilane, however, gives the enol ether which, in many ways, behaves as the free enol. In the example given in scheme *xxvii*, methyl vinyl ketone is converted to its silylated enol which, as a 1,3-diene, undergoes a Diels–Alder reaction with another molecule of methyl vinyl ketone; desilylation of the product gives 4-acetylcyclohexanone (Jung and McCombs, 1976).

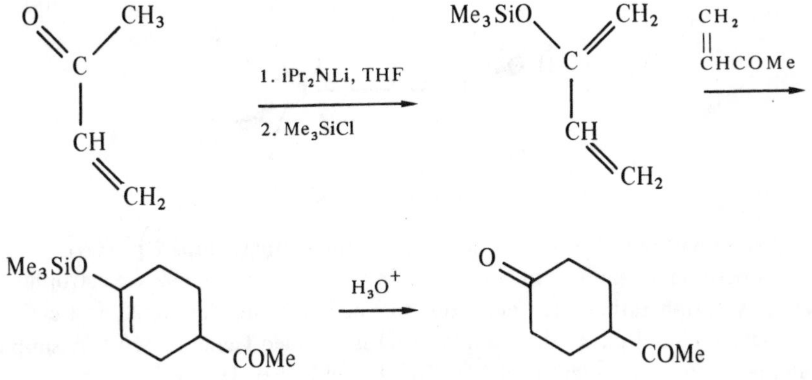

Scheme *xxvii*

Carbonyl compounds can also be converted directly to the silylated enols as illustrated in scheme *xxviii*. Here, dichlorocarbene is added to the enol double bond and the resulting bicyclic system then undergoes spontaneous rearrangement to give the chloromethylcycloheptenone (Hudrlik, 1976). A similar reaction, in which the enol was protected by acetylation and the dichlorocarbene was generated by an alternative method, is mentioned on p. 168.

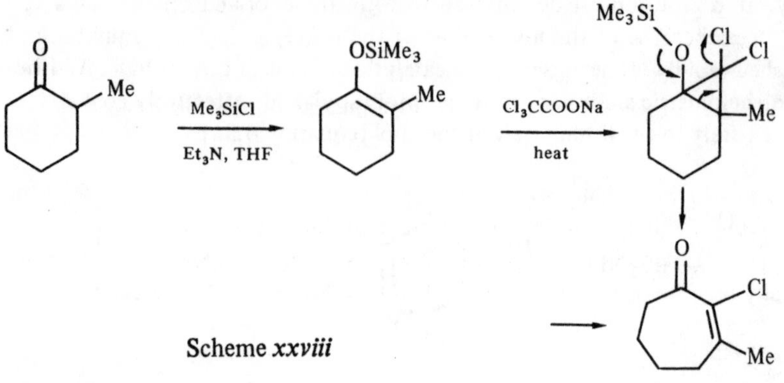

Scheme *xxviii*

Vinylsilanes and arylsilanes

Two properties of organosilicon compounds have special significance in preparative organic chemistry. Firstly, the trimethylsilyl cation, Me_3Si^+, departs from a carbon atom more readily than does a proton. Secondly, the silicon atom stabilises a β-positive charge, that is to say that the ion $Me_3SiCH_2CH_2^+$ is more stable than either $CH_3CH_2CH_2^+$ or $CH_3CH_2^+$. The source of this stabilisation lies in an interaction between an empty p-orbital at the β-carbon atom and electrons at silicon which occurs in the conformation shown in **XII** (Fleming, 1976, 1981).

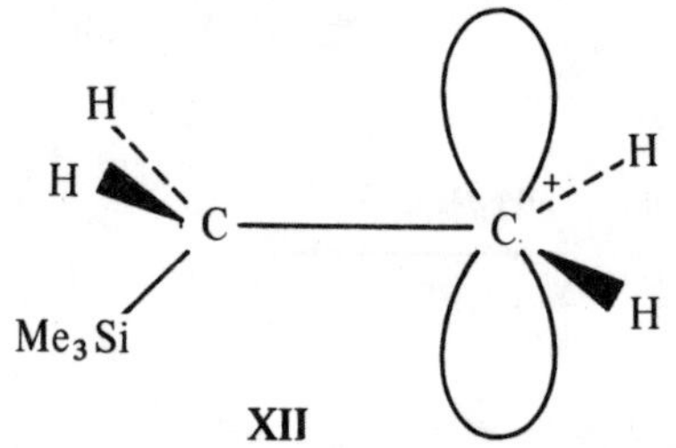

XII

The precise nature of this interaction remains uncertain (compare p. 100). These two properties are manifest in reactions of vinylsilanes. Thus, many electrophiles give substitution rather than addition, the silicon being eliminated as a cation and the configuration at the double bond is retained (scheme *xxix*). A simple example is shown in equation (8.33) (Koenig and Weber, 1973).

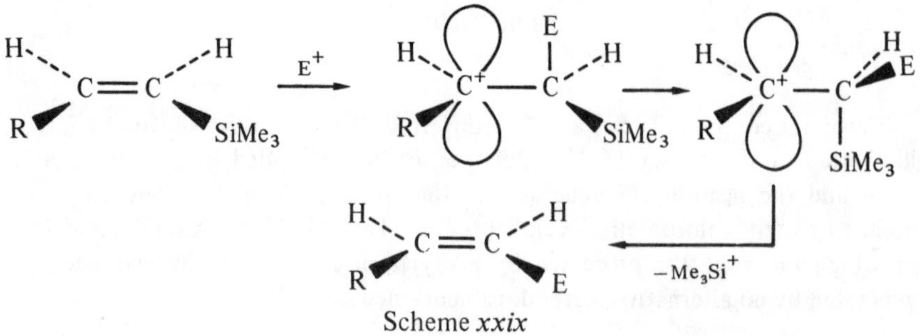

Scheme *xxix*

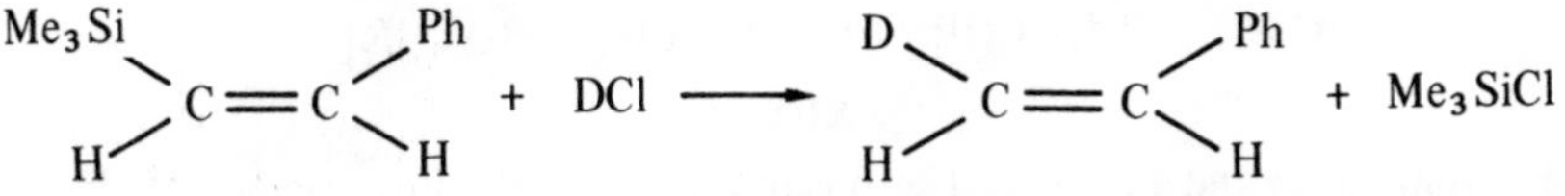

$$\underset{H}{\overset{Me_3Si}{>}}C=C\underset{H}{\overset{Ph}{<}} + DCl \longrightarrow \underset{H}{\overset{D}{>}}C=C\underset{H}{\overset{Ph}{<}} + Me_3SiCl$$

(100 per cent retention
of configuration)   (8.33)

Additions to vinylsilanes can also occur and then the electrophilic part of the addend tends to link to the carbon bearing the silicon. In the case of hydride additions, steric factors may favour the opposite mode and such reactions are not totally regiospecific (equation (8.34); Musker and Larson, 1968).

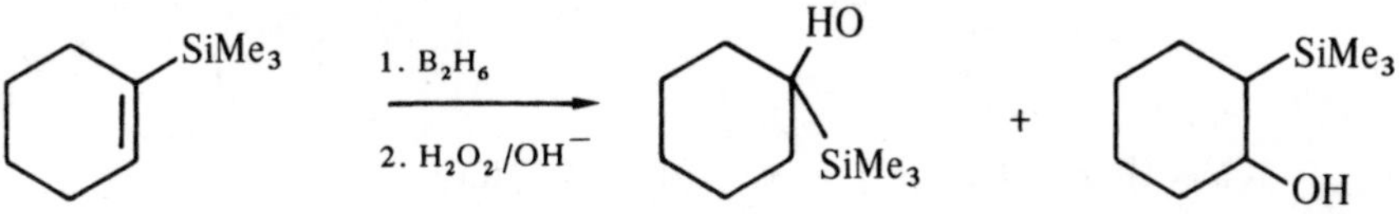

90 per cent   10 per cent

(8.34)

This stabilisation of a positive charge in the β-position, together with ready loss of the trimethylsilyl cation, is the reason why arylsilanes often undergo *ipso*-substitution† when treated with electrophiles (compare equation 4.15). This propensity can be utilised to prepare inaccessible polysubstituted aromatic compounds such as 1,2,4,5-tetraiodobenzene (scheme *xxx*; Félix *et al.*, 1977).

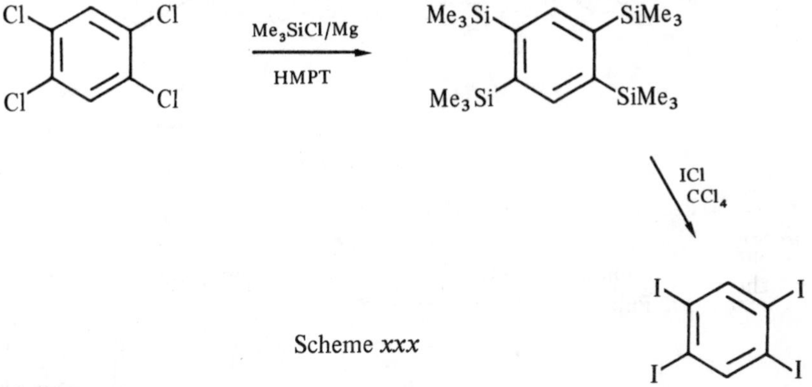

Scheme *xxx*

Allylsilanes

Most allylmetal compounds are best represented by the ionic structure **XIII** in which the charge on the anion is delocalised and reaction may occur at the 1 and 3 positions. This point has been briefly referred to in connection with allyllithium compounds (p. 160) and is further illustrated by the reaction between butenyl-

---

†Attack by an electrophile at a substituted benzene ring usually leads to the electrophile entering the ring by displacing a hydrogen atom. If, instead, it enters by displacing the substituent, this is an *ipso*-substitution.

$$M^+[\bar{C}H_2-CH=CHR \quad \longleftrightarrow \quad CH_2=CH-\bar{C}HR]$$

XIII

magnesium halide and diisopropyl ketone (equation (8.35); Courtois and Miginiac, 1974).

$$[Me\bar{C}HCH=CH_2 \quad \longleftrightarrow \quad MeCH=CH\bar{C}H_2]M\overset{+}{g}X \quad + \quad iPrCOiPr$$

1. React
2. H_2O/H^+

(8.35)

$$\underset{\underset{\text{34 per cent}}{OH}}{MeCH=CHCH_2\,C(iPr)_2} \quad + \quad \underset{\underset{\text{66 per cent}}{OH}}{\overset{\overset{Me}{|}}{CH_2=CHCHC(iPr)_2}}$$

However, the covalent nature of the silicon–carbon bond and the preference for a charge in the β-position in the cationic intermediate ensure that allylsilanes react regiospecifically and do not undergo rearrangement at ordinary temperatures (scheme *xxxi*). A specific example of this reaction is the preparation of

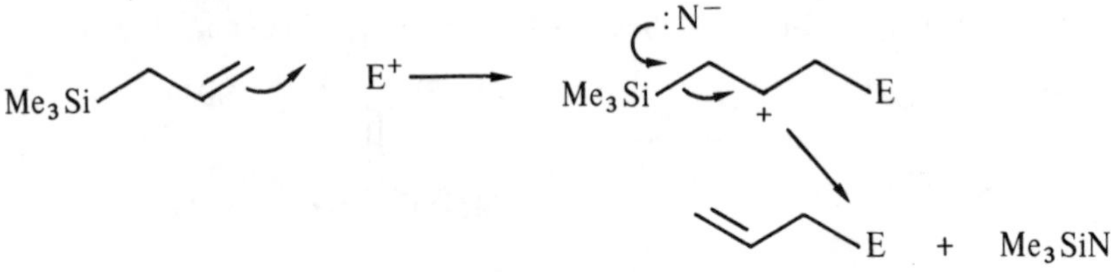

Scheme *xxxi*

artemesia ketone (2,5,5-trimethylhepa-2,6-dien-4-one; **XIV**) where an appropriately substituted allylsilane reacts regiospecifically with an acyl halide, in the presence of aluminium chloride, to give the required product in high yield (scheme *xxxii*; Pillot *et al.*, 1976).

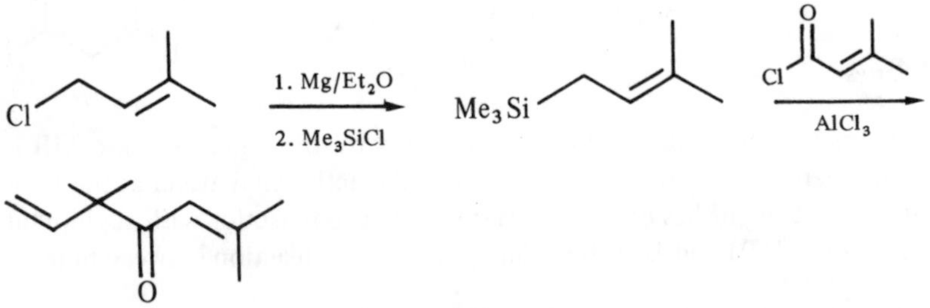

XIV   90 per cent   Scheme *xxxii*

Other Classes of Organosilicon Compounds

On p. 160 we referred to the value of organolithium reagents which carry hetero-atom substituents. In many cases the heteroatom of choice is silicon. Thus an alkene synthesis which, in terms of the substituents which can be introduced on to the alkene, is much more flexible than the Wittig reaction, utilises attack on a carbonyl compound by an α-trimethylsilylorganolithium reagent. The general form of the synthesis is shown in scheme *xxxiii* and a specific example, which indicates how the reagent is prepared, is given in scheme *xxxiv* (Magnus, 1980).

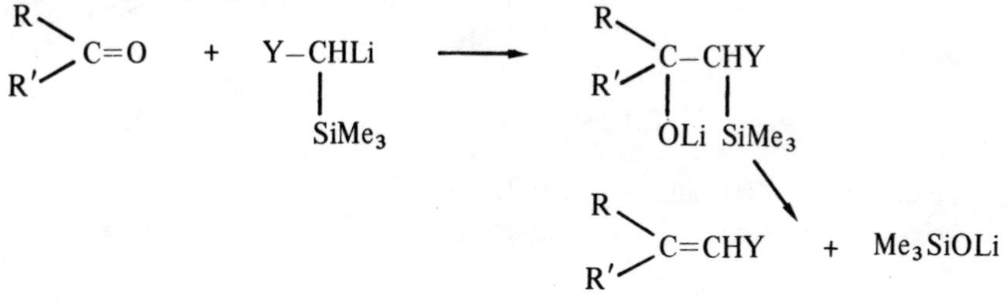

$$\begin{array}{c}
R \\
{>}C{=}O \quad + \quad Y-CHLi \quad \longrightarrow \\
R'
\end{array}
\qquad
\begin{array}{c}
R \\
{>}C-CHY \\
R' \quad | \quad | \\
\quad OLi \; SiMe_3
\end{array}$$

$$\underset{SiMe_3}{\overset{|}{}}$$

$$\begin{array}{c}
R \\
{>}C{=}CHY \quad + \quad Me_3SiOLi \\
R'
\end{array}$$

Scheme *xxxiii*

$$Ph_2P(S)CH_2SiMe_3 \quad + \quad BuLi \quad \longrightarrow \quad Ph_2P(S)\overset{\overset{\displaystyle SiMe_3}{|}}{C}HLi \quad + \quad BuH$$

$$\begin{array}{c}
Ph \\
{>}C{=}CHP(S)Ph_2 \quad \xleftarrow{\quad Ph_2CO \quad} \\
Ph
\end{array}$$

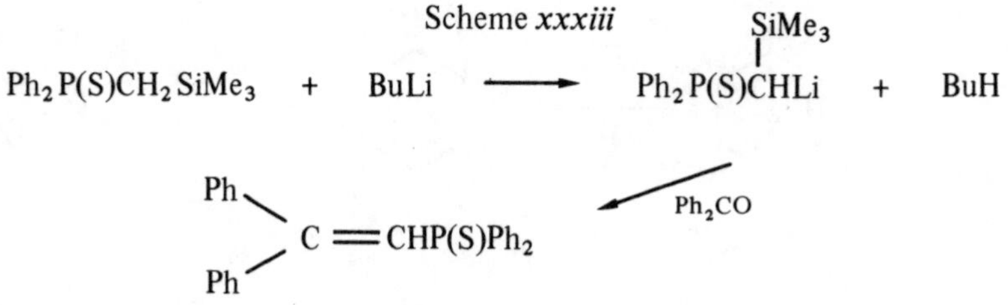

Scheme *xxxiv*

Allylsilanes

Iodotrimethylsilane can often be used for ether cleavage, sometimes under surprisingly mild conditions (equation (8.36); Jung and Lyster, 1977).

$$PhOMe + Me_3SiI \xrightarrow[48\,h]{25\,^\circ C} PhOSiMe_3 + MeI \qquad (8.36)$$
$$\text{100 per cent}$$

The same reagent can be used for the conversion of ketals to ketones in a neutral, non-aqueous reaction, invaluable when the standard aqueous acidic conditions have to be avoided (equation (8.37); Jung *et al.*, 1977).

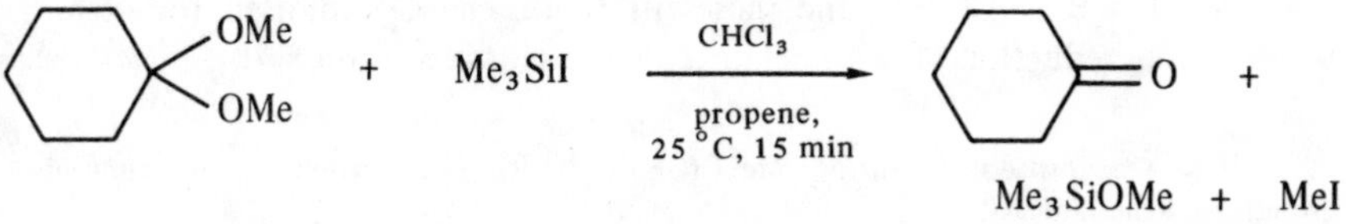

$$Me_3SiOMe \quad + \quad MeI$$
$$(8.37)$$

The solvent in this reaction is saturated with propene, which absorbs hydrogen iodide and so minimises side reactions.

Finally, benzene and its alkyl derivatives can be reduced to 1,4-cyclohexadienes by treatment with chlorotrimethylsilane and lithium in THF (scheme *xxxv*; Dunoguès, 1982). The ready separation, by distillation, of the intermediate

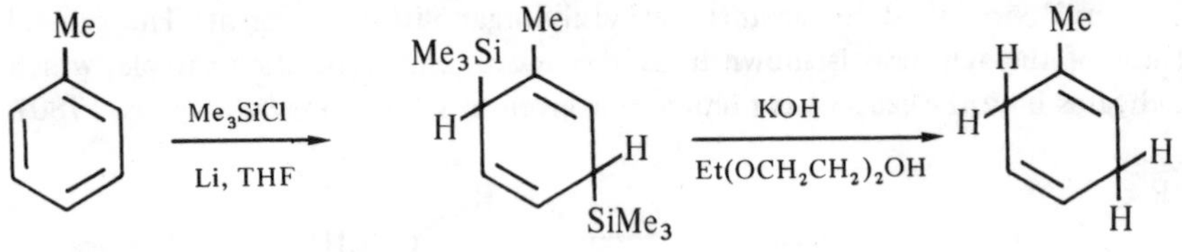

Scheme *xxxv*

disilyl derivative often allows the isolation of a purer product than can be obtained from the Birch reduction (equation 8.38).

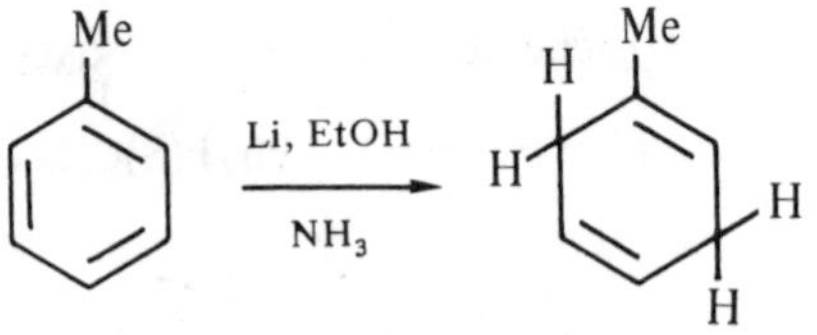

$$(8.38)$$

8.1.7 Organotin Compounds

Organotin hydrides are widely used in preparative organic chemistry for the specific reduction of organic halides giving high yields and clean separation of products (equation 8.39).

$$R_3SnH + R'X \longrightarrow R_3SnX + R'H \qquad (8.39)$$

(R, R' = alkyl or aryl; X = F, Cl, Br, I)

The reactions occur by a radical mechanism and are accelerated by radical sources such as AIBN† or by ultra-violet irradition (Kuivila, 1968). The rate of reduction depends upon which halogen atom, X, is being replaced in R'X, in the sense X = I > Br > Cl > F, and these differences can be exploited, for example in the partial reduction of bromochloro compounds (equation 8.40).

---

†AIBN = azo*bis*isobutyronitrile, $Me_2C(CN)N=NC(CN)Me_2$, which, on warming gives isobutyronitrile radicals:

$$Me_2C(CN)N=NC(CN)Me_2 \xrightarrow{\Delta} 2Me_2\dot{C}(CN) + N_2$$

$$\text{(Cl, Br benzene)} + Ph_3SnH \xrightarrow{154\,^\circ C} \text{(Cl benzene)} + Ph_3SnBr \qquad (8.40)$$

97 per cent

Organotin hydrides can also be used for reducing carbonyl compounds to alcohols, but the reaction does not offer any particular advantages over traditional reducing agents.

The hydroxyl groups of primary alcohols can be replaced by hydrogen following an S_N2 substitution by bromide of the corresponding p-toluenesulphonyl ester (scheme *xxxvi*). In contrast, secondary alcohols, where S_N2 processes are less

$$RCH_2OH \xrightarrow{p\text{-MeC}_6H_4SO_2Cl} RCH_2OSO_2C_6H_4Me\text{-}p \xrightarrow{Br^-} RCH_2Br$$

$$\xrightarrow{Ph_3SnH} RCH_3$$

Scheme *xxxvi*

favoured, are more difficult to deoxygenate. The problem is solved by converting the secondary alcohol to its thiobenzoate which can be directly reduced to the hydrocarbon (scheme *xxxvii*; Barton and McCombie, 1975).

$$ROH + ClC\overset{+}{=}NMe_2\,Cl^- \xrightarrow{C_5H_5N} RO-C\overset{+}{=}NMe_2\,Cl^- \xrightarrow{H_2S} ROCPh$$
$$\qquad\quad |\qquad\qquad\qquad\qquad\quad |\qquad\qquad\qquad\qquad \|$$
$$\qquad\quad Ph\qquad\qquad\qquad\qquad\quad Ph\qquad\qquad\qquad\qquad S$$

$$RH \xleftarrow{Bu_3SnH}$$

Scheme *xxxvii*

Although there are other applications of organotin reagents in organic chemistry, they are mostly rather specialised, but two areas are worthy of mention. One of these procedures allows regiospecific alkylation of a *cis* diol, a reaction of value in carbohydrate chemistry. Thus, conventional benzylation of the partially protected galactopyranoside (**XV**) gave a mixture of the two possible mono-benzylated and dibenzylated products plus unreacted **XV**. Conversion of **XV** to its O,O'-dibutylstannylidene derivative (**XVI**) followed by treatment with benzyl bromide in DMF gave only the 3-O-benzyl ether (**XVII**) in 66 per cent yield (scheme *xxxviii*; Augé *et al.*, 1976).

Another recent development in organotin reagents has been the use of cyclic distannoxanes (**XVIII**) to prepare macrocyclic tetralactones (scheme *xxxix*; Shanzer *et al.*, 1982; Davies *et al.*, 1984). Although we are concerned here with

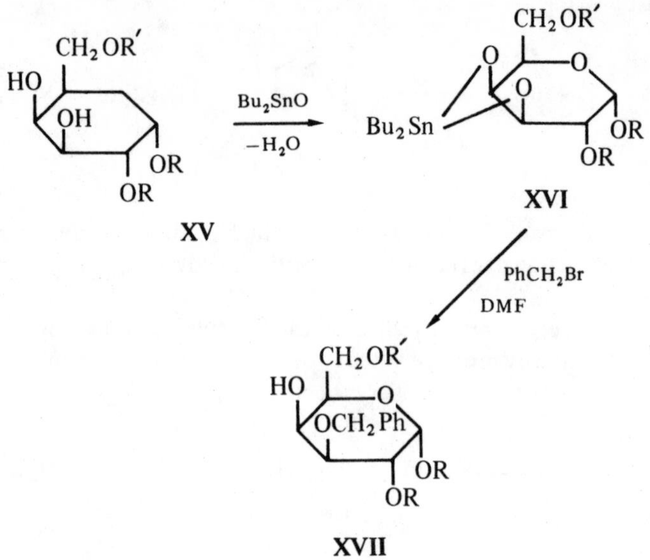

Scheme *xxxviii*

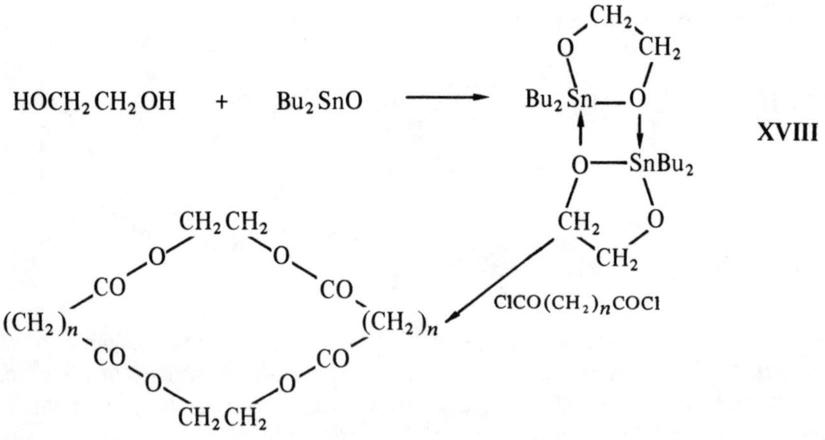

Scheme *xxxix*

organometallic reagents rather than catalysts it is, nevertheless, interesting to note that **XVIII** catalyses the oligomerisation of propionolactone. This reaction was exploited in a recent synthesis of the naturally occurring enterobactin (**XIX**) (scheme *xl*; Shanzer and Libman, 1983).

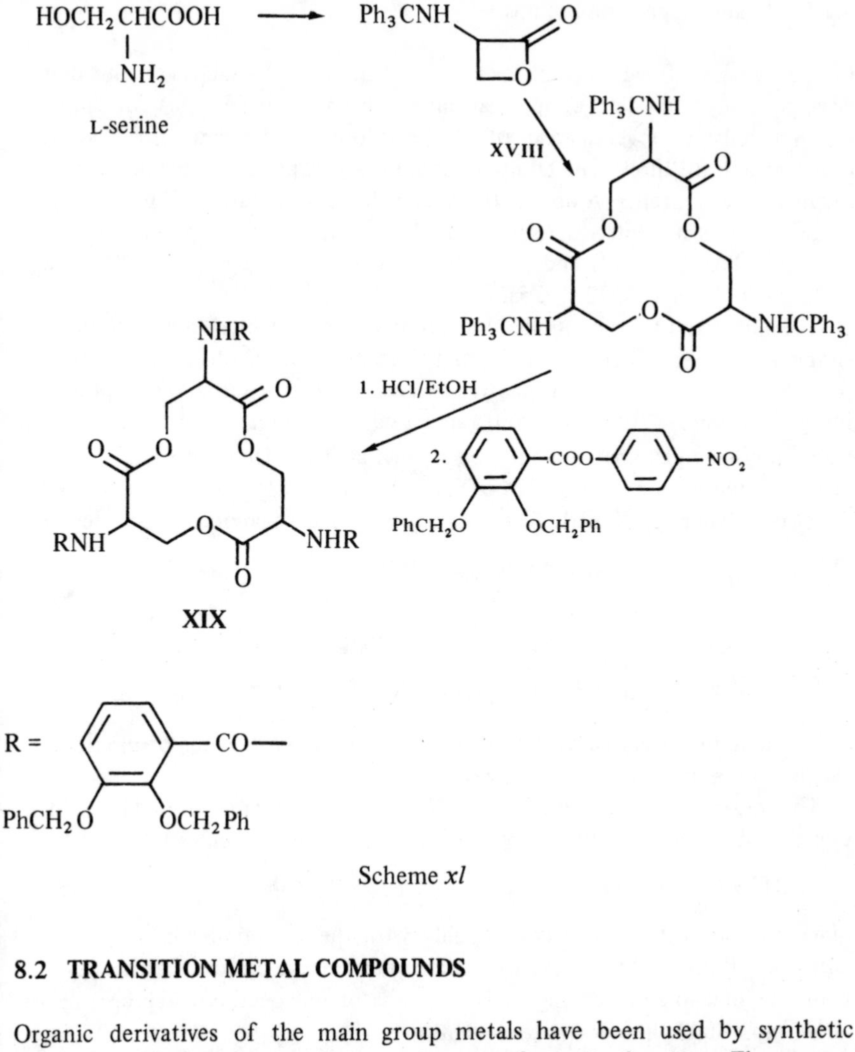

Scheme *xl*

8.2 TRANSITION METAL COMPOUNDS

Organic derivatives of the main group metals have been used by synthetic chemists from the beginning of the modern era of organic chemistry. The organo-transition metal reagents have appeared on the scene much more recently but their effect has been devastating. No particular metal stands out, as does magnesium or lithium among main group elements. Rather, it is the wide range of reactivities and specificities of organotransition metal compounds which simultaneously makes them so valuable in synthesis and so difficult to categorise and discuss. However, the compounds and reactions presented below have been carefully chosen to illustrate the major types of application in synthesis. A more extended treatment of this field is available in the excellent book by Davies (1982) and many valuable shorter reviews are available (for example, Kozikowski and Wetten, 1976).

8.2.1 Organocopper Compounds

Copper, with its filled $3d$ shell, is an atypical transition metal. Its organic derivatives generally contain σ copper–carbon bonds and resemble those of the main group metals. In several areas of synthetic organic chemistry the traditional Grignard and organolithium compounds are being displaced by the more specific organocopper reagents. Amongst the large volume of literature on organocopper reagents is a useful introductory textbook with many recent examples of use in synthesis (Posner, 1980) and a number of excellent review articles (Normant, 1972, 1976; Posner, 1972, 1975).

The most useful compounds are derivatives of copper(I) and are either the simple alkyl (or aryl) species, RCu, or 1:1 complexes, RCu.L. In the latter compounds the ligand may be a neutral species, such as a tertiary phosphine, an inorganic anion X^-, obtained from a metal salt, or an organic anion, R^-, derived from another organometallic compound such as R'Li. Thus there are four types of compound exemplified by **XX–XXIII**. The most widely used reagents are the metal homocuprates (**XXIII**, R = R') but the simple organocopper species (**XX**)

$$RCu \qquad RCu.Bu_3P \qquad MRCuX \qquad MRR'Cu$$

$$\textbf{XX} \qquad\qquad \textbf{XXI} \qquad\qquad \textbf{XXII} \qquad\qquad \textbf{XXIII}$$

$$(M = \text{Li or MgX, etc.; } X = \text{halogen})$$

are valuable in certain cases and their lower reactivity can often be modified by the use of donor solvents or additives.

The organocopper(I) compounds **XX** can be prepared from Grignard or organolithium reagents and a copper(I) salt in ether (equation 8.41).

$$RLi + CuI \longrightarrow R{-}Cu + LiI \qquad\qquad (8.41)$$

Since the product complexes so readily with the organometallic reagent, and with the salt produced, the experimental conditions are critical. Further reaction with the organometallic reagent is the basis of one method of preparing the rather more stable homocuprates (equation 8.42).

$$RLi + RCu \longrightarrow LiR_2Cu \qquad\qquad (8.42)$$

The well-known Ullmann synthesis of biaryls, in which an aryl iodide is heated with copper powder, is now thought to proceed via an arylcopper intermediate (scheme *xli*; Fanta, 1974). The formation of carbon–carbon bonds by reaction between organocopper compounds and organic halides is a reaction of major synthetic utility which can be represented by the general equations (8.43) and (8.44).

$$RCu + R'X \longrightarrow RR' + CuX \qquad\qquad (8.43)$$

$$LiR_2Cu + 2R'X \longrightarrow 2RR' + LiX + CuX \qquad\qquad (8.44)$$

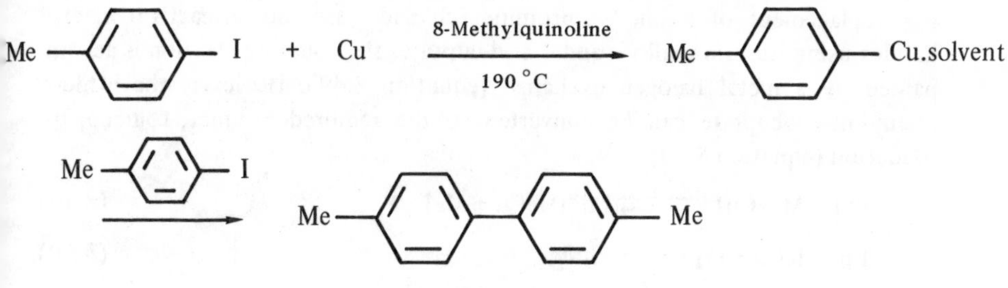

Scheme *xli*

Both the alkylcopper(I) compounds and the dialkylcuprates can be used to obtain ketones from acyl chlorides, even in the presence of other functional groups. In ether the alkylcopper compounds give poor yields unless some donor species is present (equation (8.45); Normant, 1972) but the yields from the dialkylcuprates are acceptable (equation (8.46); Posner and Whitten, 1970).

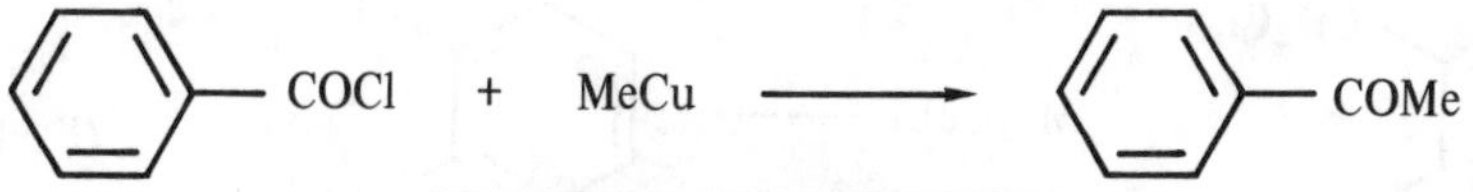

Yield	Solvent
25 per cent	Et_2O
60 per cent	$Et_2O + Bu_3P$

(8.45)

$$tBuCOCl + LiMe_2Cu \xrightarrow{Et_2O} tBuCOMe \qquad (8.46)$$

84 per cent

The order of reactivity for reaction of alkyl halides with dialkylcuprates is primary > secondary > tertiary with iodides giving best yields. As noted above, other functional groups do not usually interfere (equation (8.47); Corey and Posner, 1968).

$$I(CH_2)_{10}CONMePh + LiBu_2Cu \longrightarrow CH_3(CH_2)_{13}CONMePh \qquad (8.47)$$

82 per cent

A striking feature of organocopper reagents is their ability to replace vinyl and aryl halogens, which are notoriously unreactive to nucleophilic attack. As illustrated in reaction (8.48), vinylic halides react with retention of configuration (Corey and Posner, 1967).

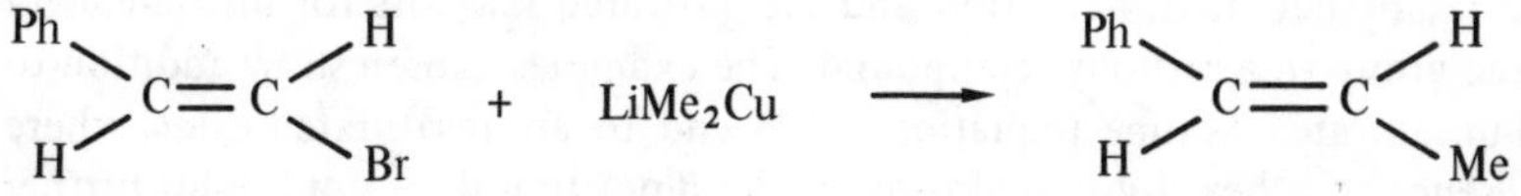

(8.48)

81 per cent

For replacement of iodine or bromine (chlorides are too unreactive) in aryl halides using lithium dialkyl- and diaryl-cuprates the coupling reaction is accompanied by a metal halogen exchange (equation 8.49). However, the lithium phenylmethylcuprate can be converted to the required product, toluene, by oxidation (equation 8.50).

$$PhI + Me_2CuLi \rightleftharpoons LiPhMeCu + MeI \qquad (8.49)$$

$$LiPhMeCu + O_2 \longrightarrow PhMe \qquad (8.50)$$

Generally, high yields of the coupled products can be obtained from aryl iodides and excess of di-n-alkyl- or diaryl-cuprates followed by oxidation of the resulting mixture with nitrobenzene or oxygen. However, this procedure is unsuccessful when di-s-alkyl- and di-t-alkyl-cuprates are used (Whitesides et $al.$, 1969).

Allyl and benzyl chlorides and bromides react readily with organocopper reagents as exemplified in reaction (8.51) (Posner and Brunelle, 1972).

$$(8.51)$$

77 percent

Finally, we note that p-toluenesulphonyloxy (TsO) groups can also be replaced by alkyl with primary substrates giving best yields. The higher reactivity of tosyloxy compared with bromine has been exploited in a two-carbon homologisation procedure (scheme $xlii$; Normant, 1976).

$$RBr \longrightarrow RLi \xrightarrow{CuI} LiR_2Cu \xrightarrow{BrCH_2CH_2OTs} RCH_2CH_2Br$$

Scheme $xlii$

The reactions of α,β-unsaturated carbonyl compounds with organic derivatives of lithium and magnesium have already been discussed (p. 158). Generally, the lithium compounds give 1,2-addition and, although Grignard reagents often give mainly the product from 1,4-addition, the 1,2-mode may also occur, as well as other side reactions. The homocuprates, however, give high yields of products derived from specific 1,4-addition and are favoured reagents for introducing a β-organic group in a carbonyl compound. The examples chosen show addition to an α,β-unsaturated ketone (equation 8.52) and to an unsaturated ester, where the presence of a hexyl group already in the β-position does not inhibit further alkylation (equation (8.53); Posner, 1972).

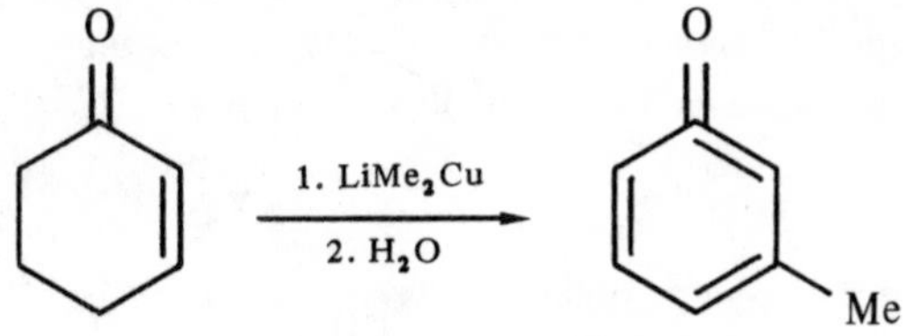

$$(8.52)$$

97 per cent

$$C_6H_{13}CH{=}CHCOOEt \xrightarrow[\text{2. H}_2\text{O}]{\text{1. Me}_2\text{CuLi}} C_6H_{13}CH(Me)CH_2COOEt \qquad (8.53)$$

86 per cent

Reaction between homocuprates and alkynes gives vinylcopper compounds which are useful in synthesis (scheme *xliii*; Normant and Alexakis, 1981). Although the addition is always *cis*, the regioselectivity is often less than complete, leading to mixtures of isomers in the product.

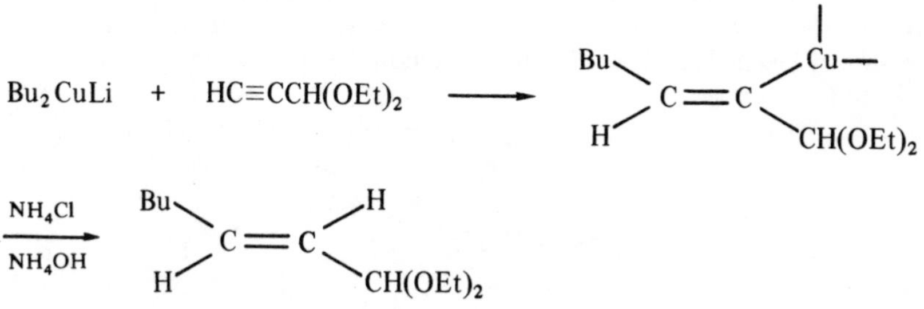

Scheme *xliii*

8.2.2 Organotitanium Compounds

It is difficult to replace the carbonyl oxygen of a carboxylic ester by a methylene group using the Wittig reaction. The reason is that the ester acylates the Wittig reagent (equation (8.54); Le Corre, 1974).

$$Ph_3P{=}CH_2 + RCOOEt \longrightarrow Ph_3P{=}CHCOR + EtOH \qquad (8.54)$$

A more reactive source of methylene (**XXIV**) can be obtained by reaction between trimethylaluminium and di(η^5-cyclopentadienyl)titanium dichloride (Tebbe *et al.*, 1978). This reagent can be used to convert a wide range of esters to vinyl ethers (scheme *xliv*; Pine *et al.*, 1980).

Compound **XXIV** is unusual for a transition metal compound in that it functions as a carbanion source. Other important uses of organotitanium compounds in which the organic groups are σ-bonded to the metal and the reagent

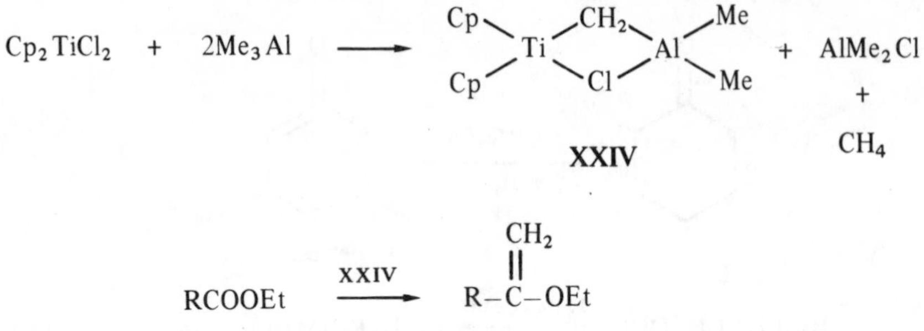

$$Cp_2TiCl_2 \; + \; 2Me_3Al \; \longrightarrow \; \underset{\textbf{XXIV}}{\begin{array}{c} Cp \\ \diagdown \\ Cp \diagup \end{array} Ti \begin{array}{c} CH_2 \\ \diagup \quad \diagdown \\ \diagdown \quad \diagup \\ Cl \end{array} Al \begin{array}{c} \diagup Me \\ \diagdown Me \end{array}} \; + \; AlMe_2Cl \; + \; CH_4$$

$$\underset{RCOOEt}{} \; \xrightarrow{\textbf{XXIV}} \; \underset{R-C-OEt}{\overset{\overset{\displaystyle CH_2}{\parallel}}{}}$$

Scheme *xliv*

functions simply as a source of carbanions, are also atypical. The advantage of an alkyltitanium compound, $RTiX_3$, compared with the corresponding lithium or magnesium analogues, is that by suitable choice of the X groups a range of reactivities and specificities are possible (Reetz, 1982). Thus, methyltitanium triisopropoxide, **XXV**, which is readily prepared by the procedure shown in scheme *xlv*, reacted exclusively with the aldehyde when it was treated with a mixture of hexanal and heptan-2-one (equation 8.55) at low temperatures.

$$TiCl_4 \; + \; 3Ti(OCHMe_2)_4 \; \longrightarrow \; 4ClTi(OCHMe_2)_3$$

$$\xleftarrow{\quad MeLi \quad}$$

$$MeTi(OCHMe_2)_3$$

XXV

Scheme *xlv*

$$Me(CH_2)_5CHO \; + \; Me(CH_2)_4COMe$$

$$\downarrow \begin{array}{l} 1.\ \textbf{XXV},\ -78°C\ to\ -10°C \\ 2.\ H_2O \end{array} \qquad (8.55)$$

$$Me(CH_2)_5CH(OH)Me \; + \; Me(CH_2)_4COMe$$

Compound **XXV** does, however, react smoothly with ketones at room temperature. A number of reactive groups, which would be attacked by lithium or magnesium compounds, are unaffected by titanium reagents such as compound **XXV** (equation 8.56).

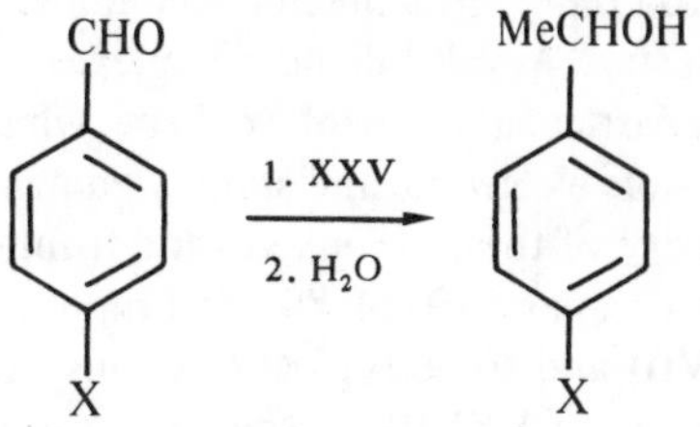

X	Yield (per cent)
NO_2	95
CN	85
COOH	79

(8.56)

8.2.3 Organoiron Compounds

In reactions which are much more typical of transition metal chemistry, alkenes become activated to nucleophilic attack when part of a cationic iron complex. Thus, when the complex cyclopentene cation, **XXVI**, is treated with the diethyl malonate anion, addition occurs and new carbon–carbon and σ carbon–iron bonds are formed to give **XXVII**. Removal of hydride ion from **XXVII** followed by demetallation with iodide regenerates cyclopentene, now carrying a malonate residue in the allylic position (scheme *xlvi*; Lennon *et al.*, 1977).

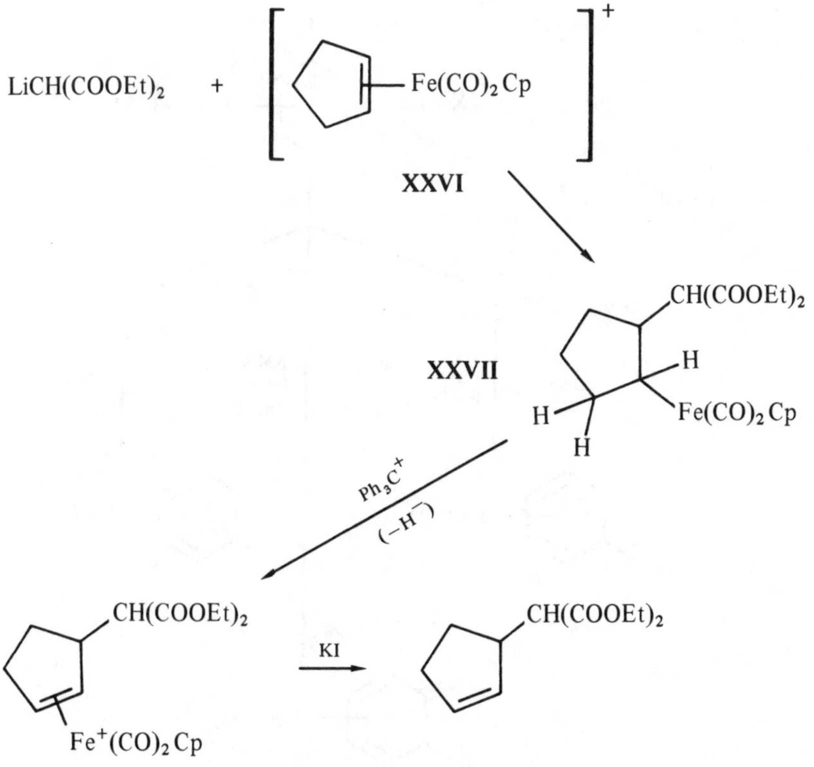

Scheme *xlvi*

Reactions involving diene complexes of iron represent a much larger and very productive area of synthetic organic chemistry. Acyclic diene complexes are known, but it is the adducts of cyclic dienes, particularly, cyclohexadiene, which are currently being exploited in synthesis. Simple electronic and steric arguments are rarely successful in predicting the reactivity of these complexes and frontier orbital considerations offer a better rationale (Pearson, 1981). Preparation of the cyclohexadienyliron tricarbonyl cation **XXVIII** and some typical reactions with nucleophiles are shown in scheme *xlvii*. Attack on **XXVIII** is regio- and stereo-specific with the nucleophile entering the 5-position at the side remote from the metal. Similar reactions of substituted cyclohexadienyliron tricarbonyl com-

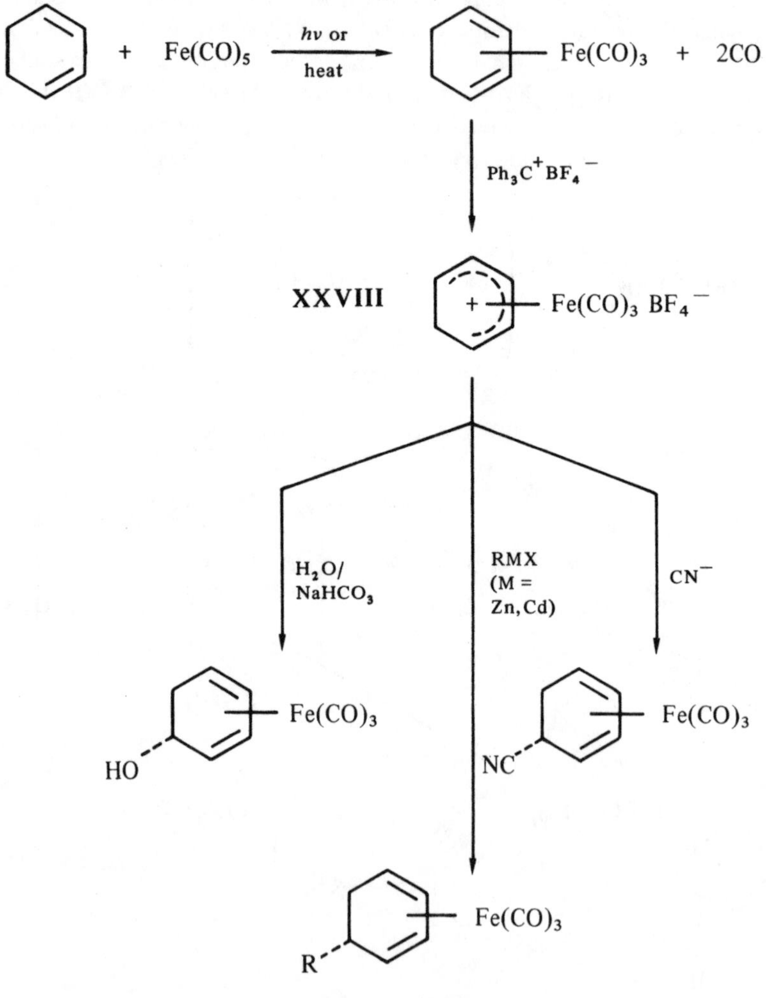

Scheme *xlvii*

plexes are being used for the synthesis of natural products of considerable structural and stereochemical complexity (Pearson, 1981).

The use of complexation with iron to protect one diene system, while reaction occurs at a second diene grouping in the same molecule, is illustrated by the reaction sequence starting from cyclooctatetraene (scheme *xlviii*). The unprotected diene grouping is first protonated and this is followed by rearrangement to a bicyclic system giving the bicyclo[5,1,0]octadienyliron tricarbonyl cation **XXIX**. Nucleophilic addition of hydroxide ion to **XXIX** followed by oxidation of the resulting alcohol, gives, eventually, homotropone (**XXX**), which is of interest in studies of aromaticity (Holmes and Pettit, 1963). The final step in the sequence shown in scheme *xlviii* is liberation of the organic product from the

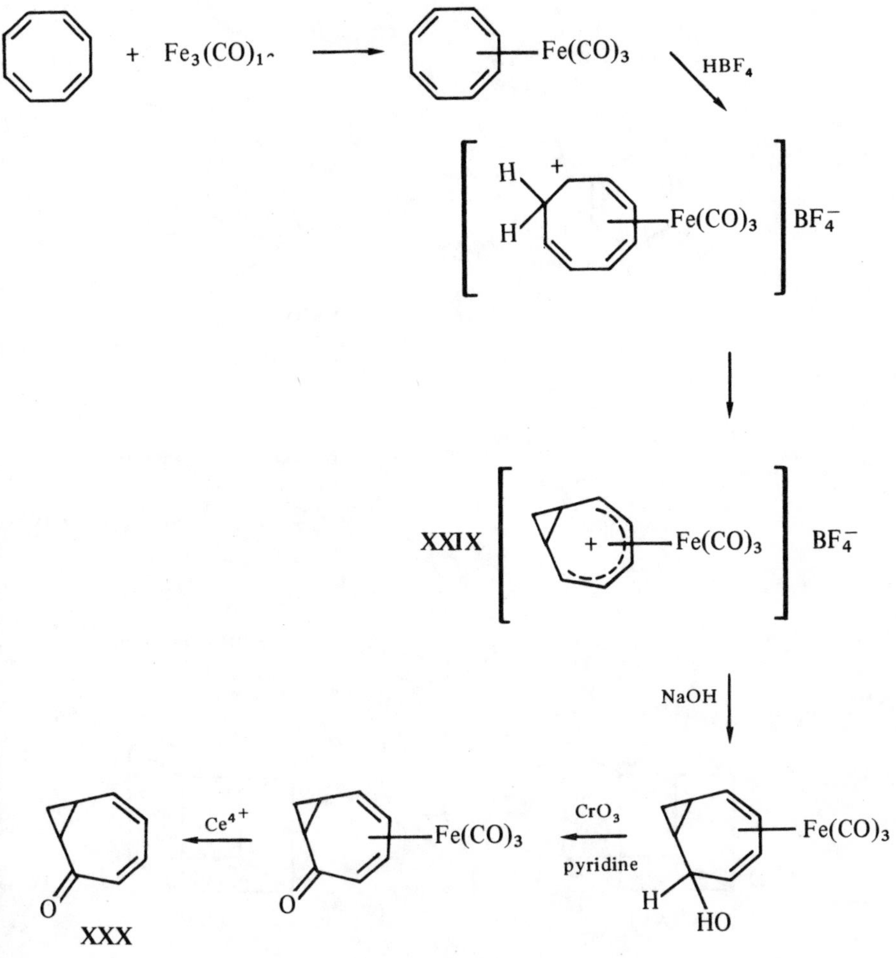

Scheme *xlviii*

metal by treatment with cerium(IV) ions. Decomplexation of dienes is commonly an oxidative process, but the use of more powerful donors is also possible. Cyclobutadiene **(XXXI)** would be expected to be unstable because of severe angle strain and we note that, since it does not obey Hückel's rule (that is it does not have $(4n + 2)$ π-electrons), it will not be aromatic. It is, in fact, unknown in the free state, but forms stable complexes with transition metals. An example is the

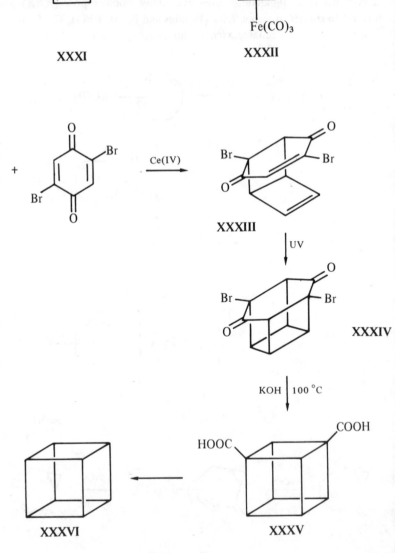

XXXI

XXXII

Fe(CO)$_3$

XXXIII

Ce(IV)

UV

XXXIV

KOH | 100 °C

XXXVI

XXXV

Scheme *xlix*

iron tricarbonyl complex (XXXII) which, on decomplexation with cerium(IV) ion, liberates cyclobutadiene as a transient species which can react with suitable substrates. Thus, decomposition of XXXII in the presence of 2,4-dibromobenzo-quinone gives the Diels–Alder adduct (XXXIII). Irradiation of XXXIII with ultra-violet light causes further cyclisation to XXXIV which, on treatment with potassium hydroxide, gives cubane-1,3-dicarboxylic acid (XXXV) (scheme *xlix*; Barborak *et al.*, 1966). Compound XXXV was converted to its di-*t*-butyl perester which, on thermal decomposition, gave cubane (XXXVI).

Another striking example of the use of complexation to stabilise a labile un-saturated species is the preparation of the iron tetracarbonyl adduct of vinyl alcohol. The free enol form of acetaldehyde is, of course, unknown, but its trimethylsilyl derivative can be isolated (compare p. 176) and converted to an iron complex. Hydrolysis of the complex frees the enolic hydroxyl group which reacts normally with reagents such as methyl isocyanate and the final product, vinyl *N*-methyl carbamate can be liberated from the metal, in this case by treat-ment with the more powerful donor, triphenylphosphine (scheme *l*; Thyret, 1972).

$$Me_3SiOCH=CH_2 \ + \ Fe_2(CO)_9 \longrightarrow \underset{\underset{OSiMe_3}{\overset{\displaystyle CH}{|}}}{\overset{\displaystyle CH_2}{||}}\!\!-\!Fe(CO)_4 \ \xrightarrow{H_2O} \ \underset{\underset{OH}{\overset{\displaystyle CH}{|}}}{\overset{\displaystyle CH_2}{||}}\!\!-\!Fe(CO)_4$$

MeNCO

$$\underset{\underset{OCONHMe}{\overset{\displaystyle CH}{|}}}{\overset{\displaystyle CH_2}{||}}\!\!-\!Fe(CO)_4 \ \xrightarrow{PPh_3} \ CH_2=CHOCONHMe$$

Scheme *l*

Although we are generally excluding simple metal carbonyls from our defini-tion of organometallic compounds, the use of Collman's reagent, $Na_2Fe(CO)_4$, to make aldehydes is included here. This is partly because of its general import-ance, but also because *bona fide* organometallic compounds are essential inter-mediates (Collman *et al.*, 1978). The procedure for conversion of an alkyl halide, RX, to an aldehyde is shown in scheme *li*. The reactions can be modified to yield unsymmetrical ketones, carboxylic acids or esters and the reagent will tolerate the presence of reactive groups which would be attacked by organo-lithium or -magnesium compounds (Collman, 1975).

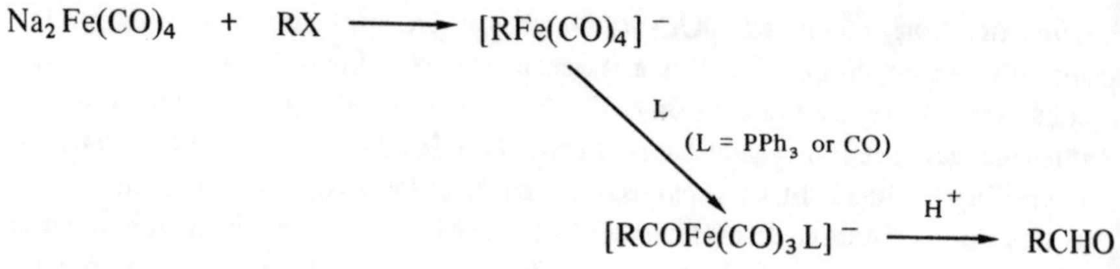

$$Na_2Fe(CO)_4 \quad + \quad RX \quad \longrightarrow \quad [RFe(CO)_4]^-$$

L
(L = PPh$_3$ or CO)

$$[RCOFe(CO)_3L]^- \quad \xrightarrow{H^+} \quad RCHO$$

Scheme *li*

8.2.4 Organocobalt Compounds

As noted elsewhere (p. 221) carbonylation reactions are catalysed by cobalt carbonyls but it is sometimes preferable to use stoicheiometric amounts of organocobalt compounds. Thus, when acyl cobalt tetracarbonyl complexes were treated with conjugated dienes, the acyl group was transferred to the diene to form the π-allyl complex **XXXVII**. The dienone could be recovered by treatment with a suitable base in an atmosphere of carbon monoxide (scheme *lii*; Heck, 1963).

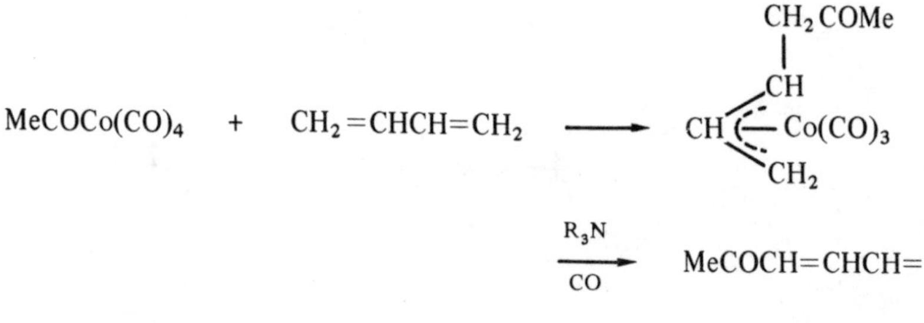

$$MeCOCo(CO)_4 \quad + \quad CH_2{=}CHCH{=}CH_2 \quad \longrightarrow$$

CH$_2$COMe
|
CH
CH $\xleftarrow{}$ Co(CO)$_3$ **XXXVII**
CH$_2$

$$\xrightarrow[CO]{R_3N} \quad MeCOCH{=}CHCH{=}CH_2$$

Scheme *lii*

There are many instances in which the triple bond in a polyfunctional alkyne was protected by complex formation while reactions were carried out elsewhere in the molecule. Thus, specific acid-catalysed hydration of the double bond in 2-methylhept-1-en-3-yn-6-ol (**XXXVIII**) can occur, providing the triple bond is first masked (scheme *liii*; Nicholas and Pettit, 1971). The hydrated product, **XXXIX**, can be recovered by oxidative degradation of the complex using iron(III) nitrate nonahydrate.

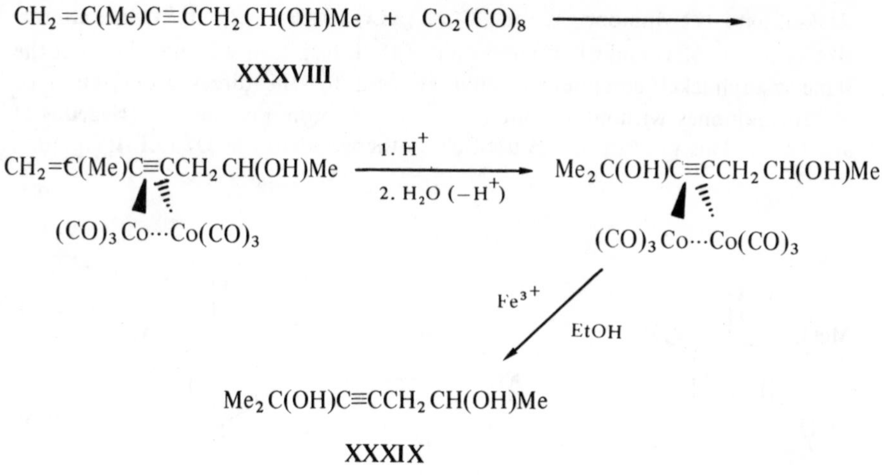

$CH_2=C(Me)C\equiv CCH_2CH(OH)Me$ + $Co_2(CO)_8$ ⟶

XXXVIII

$CH_2=C(Me)C\equiv CCH_2CH(OH)Me$ $\xrightarrow[\text{2. }H_2O\,(-H^+)]{\text{1. }H^+}$ $Me_2C(OH)C\equiv CCH_2CH(OH)Me$

$(CO)_3Co\cdots Co(CO)_3$ $(CO)_3Co\cdots Co(CO)_3$

Fe^{3+}
EtOH

$Me_2C(OH)C\equiv CCH_2CH(OH)Me$

XXXIX

Scheme *liii*

8.2.5 Organonickel Compounds

Halogen atoms in a variety of organic compounds can be replaced by allyl groups using the dimeric π-allylnickel(II) bromides. In the example shown in scheme *liv* an aromatic bromo compound reacts with the (η^3-dimethylallyl)nickel bromide **(XL)** to yield an allyl-aromatic derivative which, by deacetylation and oxi-

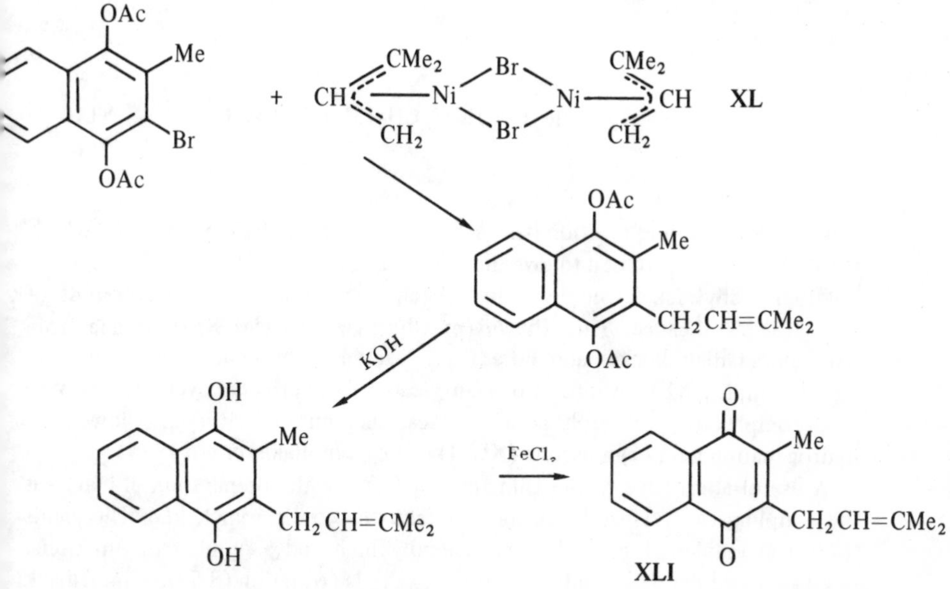

Scheme *liv*

dation, gave the quinone vitamin $K_{2(5)}$ (XLI) (Sato *et al.*, 1973). An alternative synthesis of vitamin K is given on p. 145. It had been shown earlier that the same π-allylnickel compounds could be used for the direct alkenylation of certain quinones without the intermediacy of a bromo compound (Hegedus *et al.*, 1972). This procedure was used to synthesise coenzyme Q_1 (XLII; equation 8.57).

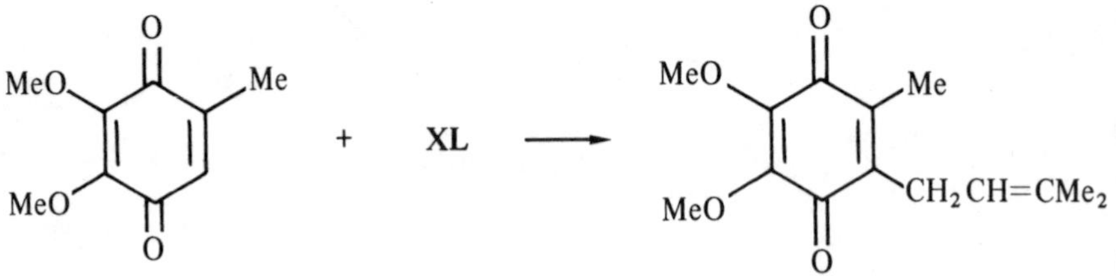

$$
\text{(MeO-quinone with Me)} \quad + \quad \textbf{XL} \quad \longrightarrow \quad \text{(MeO-quinone with Me and } CH_2CH{=}CMe_2)
$$

XLII   (8.57)

Yet another application of this reagent is its reaction with 4-bromo-3-methylbut-2-enyl acetate to give geranyl acetate (XLIII; equation (8.58); Sato *et al.*, 1972).

$$BrCH_2C(Me){=}CHCH_2OAc \quad + \quad \textbf{XL}$$

(8.58)

$$Me_2C{=}CHCH_2CH_2(Me){=}CHCH_2OAc \qquad \textbf{XLIII}$$

This type of coupling reaction has been used to make allylaminobenzenes which were subsequently cyclised to give indoles (p. 206).

Other π-allylnickel complexes have been used to synthesise compounds of biological significance. Thus the $bis(\eta^3\text{-allyl})$nickel species XLIV (made from bis(cyclooctadiene)nickel and butadiene) reacted with allene to give the C_{15}-nickel complex XLV. Addition of t-butyl isocyanide effected cyclisation as well as decomplexation; hydrolysis of the resulting imino-derivative followed by hydrogenation gave ($\pm$)-muscone (XLVI) (scheme *lv*; Baker *et al.*, 1974).

A useful alternative to the Ullmann reaction for the preparation of biaryls is the coupling of aryl halides using bis(cycloocta-1,5-diene)nickel(0). The yields from aryl iodides or bromides are generally high and a number of functional groups can be tolerated as indicated in the example (equation (8.59); Semmelhack *et al.*, 1971).

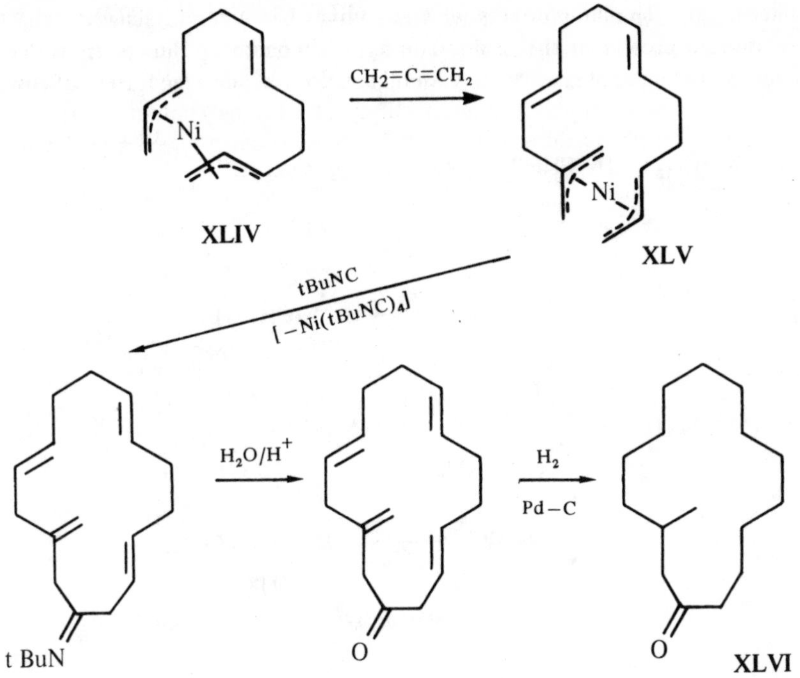

XLIV

CH₂=C=CH₂

XLV

tBuNC
[−Ni(tBuNC)₄]

H₂O/H⁺

H₂
Pd−C

t BuN

O

O   **XLVI**

Scheme *lv*

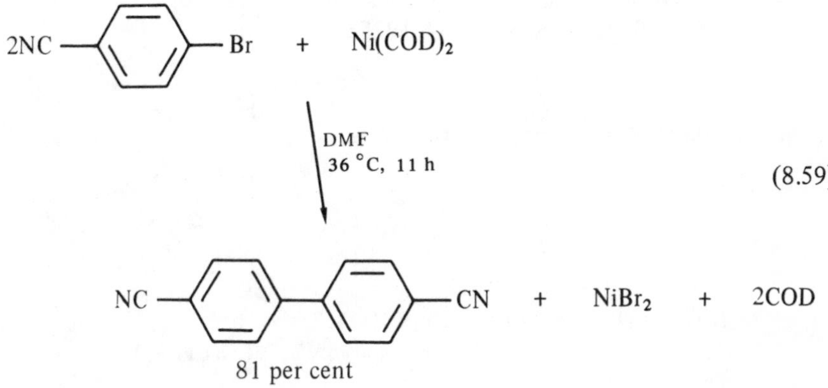

$$2NC \longrightarrow Br \quad + \quad Ni(COD)_2$$

DMF
36 °C, 11 h

(8.59)

$$NC \longrightarrow \longrightarrow CN \quad + \quad NiBr_2 \quad + \quad 2COD$$

81 per cent

8.2.6   Organozirconium Compounds

The zirconium–hydrogen bond of dicyclopentadienylzirconium chloride hydride
(XLVII) is labile and readily adds to alkenes and alkynes in hydrozirconation
reactions. Internal alkenes are effectively isomerised so that these, and terminal

alkenes, give similar products with zirconium becoming attached to the least substituted carbon in the chain. The hydrozirconated products are stable and undergo a number of synthetically useful reactions (scheme *lvi*; Hart and Schwartz,

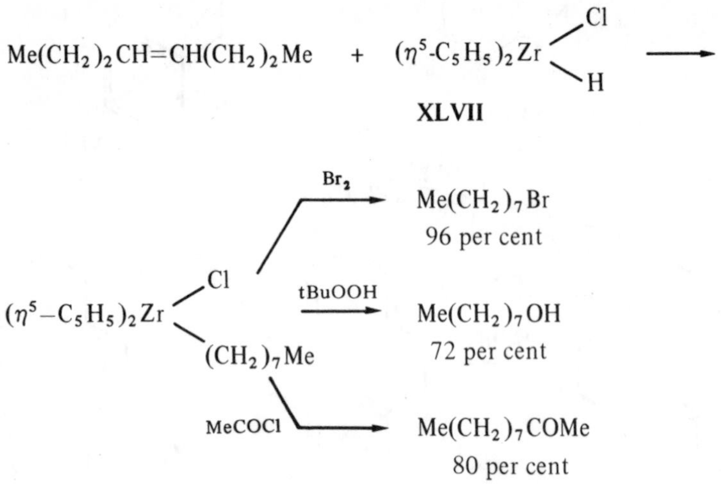

$$Me(CH_2)_2CH=CH(CH_2)_2Me \quad + \quad (\eta^5\text{-}C_5H_5)_2Zr\langle^{Cl}_{H} \longrightarrow$$

XLVII

Scheme *lvi*

1974; Schwartz and Labinger, 1976). The organozirconium compounds show similarities to the organoboranes which are also reactive intermediates in alkene transformations. Like the boranes the hydrozirconation products are readily carbonylated. The acylzirconium products thus obtained can give a variety of products resulting in overall conversion of alkene into a range of carbonyl derivatives (scheme *lvii*; Bertelo and Schwartz, 1975).

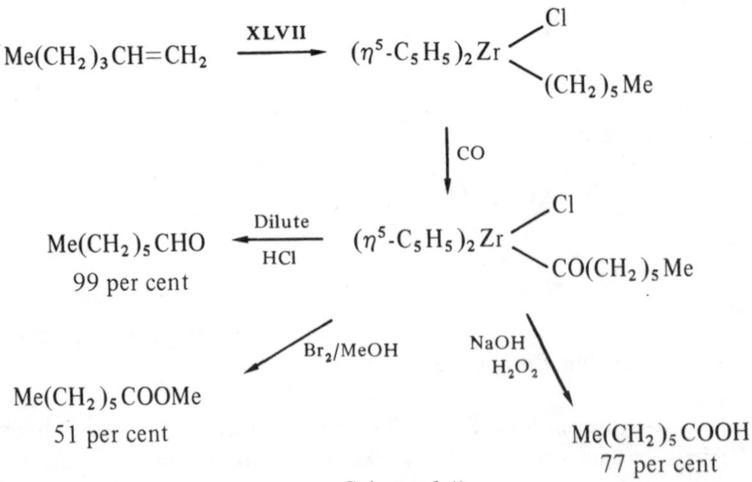

Scheme *lvii*

Hydrozirconation of non-terminal alkynes with **XLVII** occurs by a *cis* mechanism and gives predominantly the product with the metal atom attached to the carbon carrying the smallest alkyl group. There is a high degree of regioselectivity providing the initial product is allowed to equilibrate at room temperature in the presence of catalytic amounts of **XLVII**. This procedure gives a regioselectivity which is generally higher than that achieved using hindered boranes. The metal can be replaced with a halogen atom, as illustrated in scheme *lviii* (Hart *et al.*, 1975).

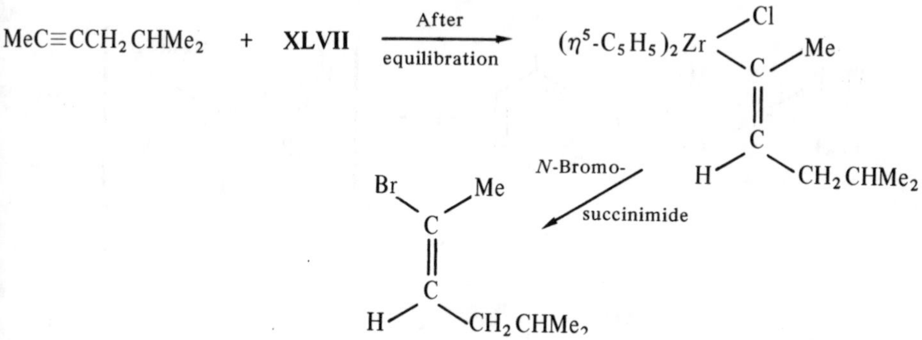
MeC≡CCH₂CHMe₂ + XLVII
After equilibration
(η⁵-C₅H₅)₂Zr
Cl
Me
C
C
H
CH₂CHMe₂
N-Bromo-succinimide
Br
Me
C
C
H
CH₂CHMe₂

Scheme *lviii*

More recently, the nickel-catalysed 1,4-addition of vinylzirconium compounds to α,β-unsaturated carbonyl compounds, illustrated in scheme *lix*, has been found to have advantages over the corresponding reactions with vinylcopper compounds (Schwartz *et al.*, 1980).

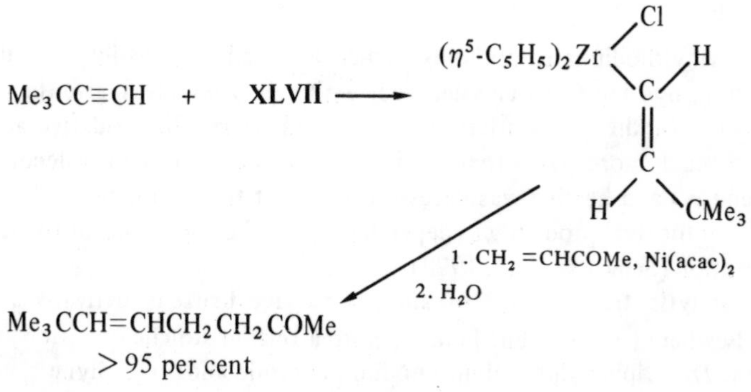
Me₃CC≡CH + XLVII
(η⁵-C₅H₅)₂Zr
Cl
H
C
C
H
CMe₃
1. CH₂=CHCOMe, Ni(acac)₂
2. H₂O
Me₃CCH=CHCH₂CH₂COMe
>95 per cent

Scheme *lix*

8.2.7 Organorhodium Compounds

Methyl*tris*(triphenylphosphine)rhodium converts iodo- and bromo-arenes into their methyl derivatives by an oxidative addition, reductive elimination mechanism (scheme *lx*; Semmelhack and Ryono, 1973; compare scheme *ix*, p. 92).

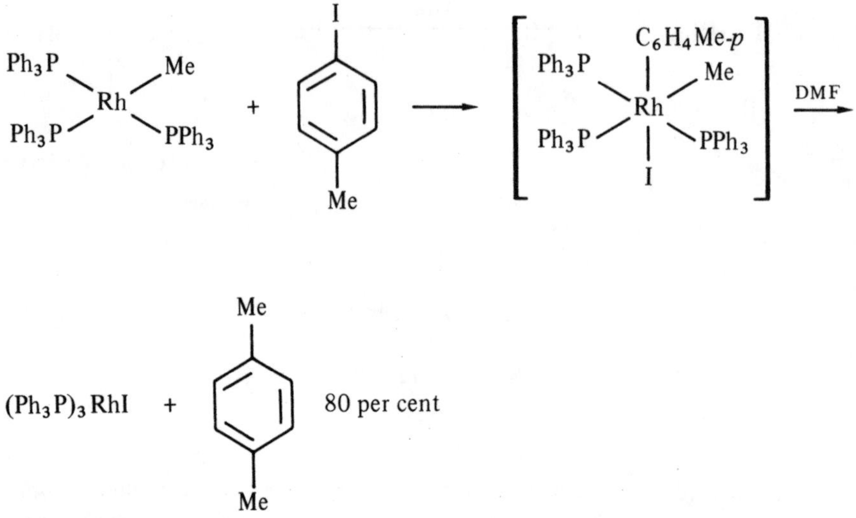

Scheme *lx*

Similarly, vinylrhodium compounds, which are readily accessible by addition of a rhodium hydride to an alkyne, react with methyl iodide to give the methylated alkene. In the preparation shown in scheme *lxi* the oxidative addition product is much more stable than in the previous example (which decomposed spontaneously) and heating was necessary to effect the elimination. The stereochemistry of the final product was dependent upon the conditions of the thermal decomposition (Schwartz *et al.*, 1972).

The catalytic trimerisation of alkynes to give benzene derivatives is discussed elsewhere (p. 234), but here we note a related stoicheiometric synthesis of arenes. *Tris*(triphenylphosphine)rhodium chloride adds to a diyne species to give a metallacycle from which rhodium can be displaced by treatment with an alkyne to form a new aromatic ring (scheme *lxii*; Müller, 1974).

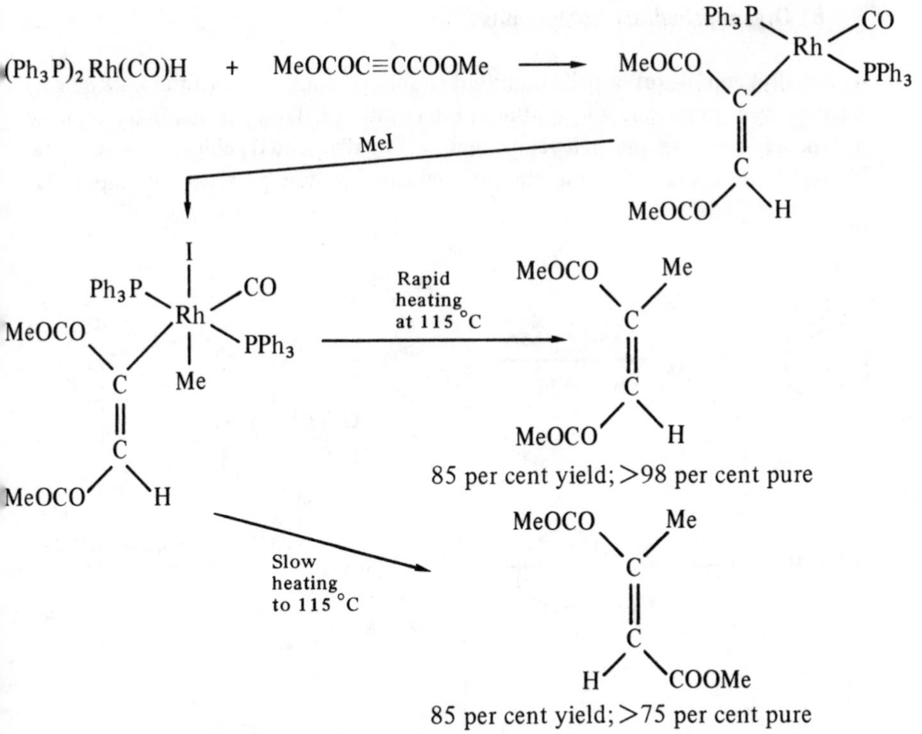

$(Ph_3P)_2 Rh(CO)H$ + $MeOCOC \equiv CCOOMe$ $\longrightarrow$

Scheme *lxi*

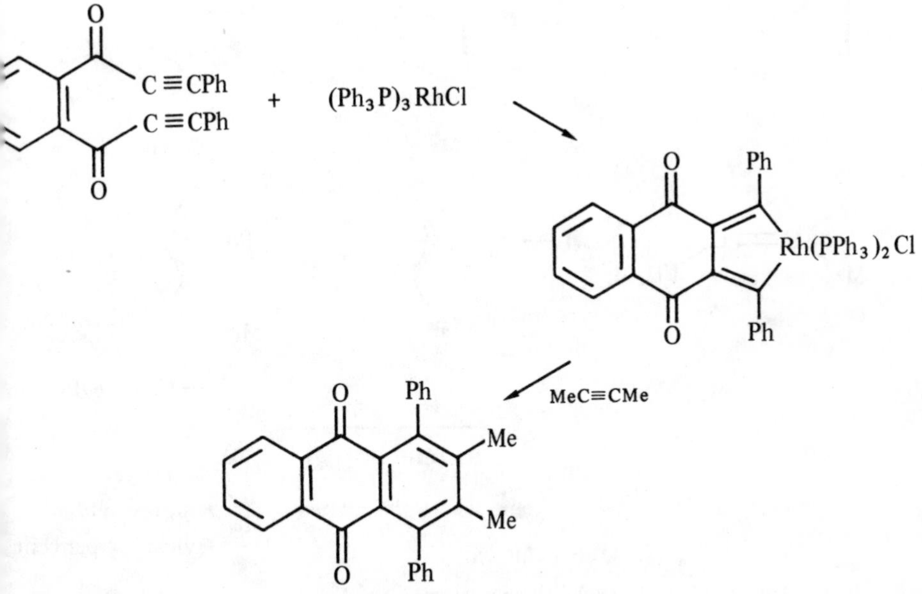

Scheme *lxii*

8.2.8 Organopalladium Compounds

The π-allyl complexes of palladium (which show resemblances to those of nickel) have proved to be valuable synthetic intermediates. They are made by treating an alkene, such as 1-methylcyclohexene, with palladium(II) chloride in an acetic acid–sodium acetate–sodium chloride mixture in the presence of copper(II) chloride (equation 8.60).

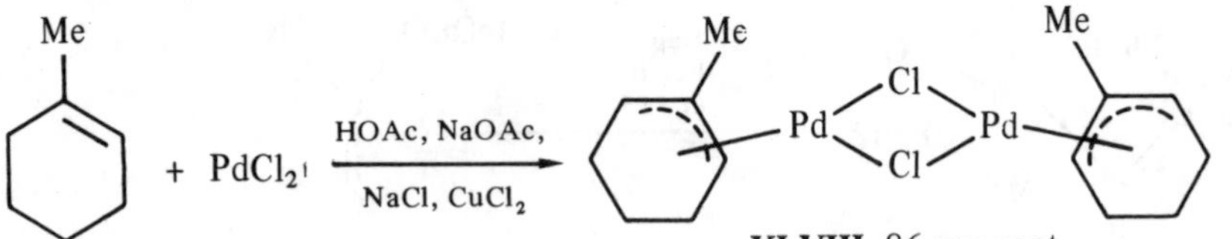
Me
+ PdCl2
HOAc, NaOAc,
NaCl, CuCl2
Me
Pd
Cl
Cl
Pd
Me

XLVIII, 86 per cent (8.60)

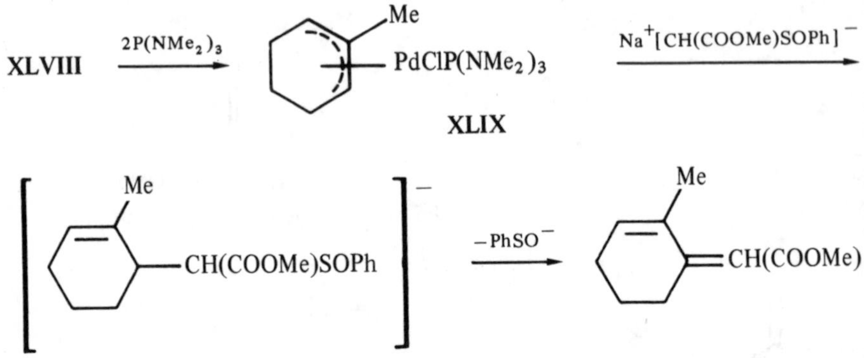
XLVIII
2P(NMe2)3
Me
PdClP(NMe2)3
XLIX
Na+[CH(COOMe)SOPh]−
Me
CH(COOMe)SOPh
−PhSO−
Me
CH(COOMe)

Scheme *lxiii*

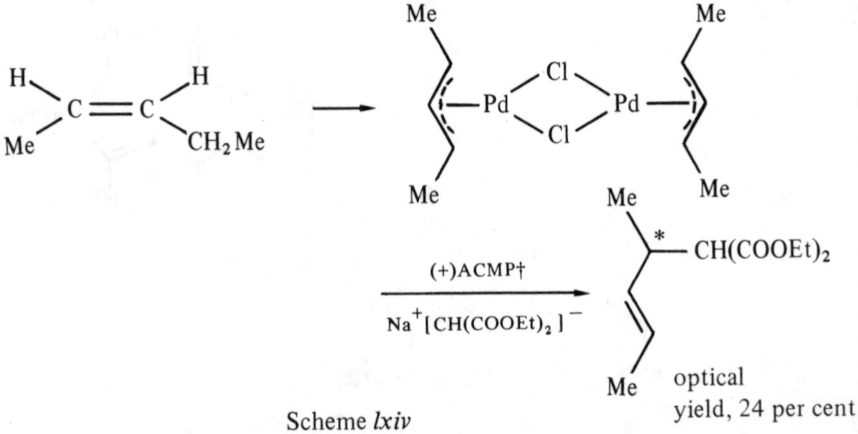
H
Me
C
C
H
CH2Me
Me
Pd
Cl
Cl
Pd
Me
Me
Me
(+)ACMP†
Na+[CH(COOEt)2]−
Me
*
CH(COOEt)2
Me
optical
yield, 24 per cent

Scheme *lxiv*

†ACMP = *o*-anisylcylohexylmethylphosphine.

Adducts such as **XLVIII** are usually treated with a trivalent phosphorus compound to convert them to a monomeric form, such as **XLIX**, before reaction with nucleophiles (scheme *lxiii*; Trost *et al.*, 1974). When a chiral phosphine was used as ligand, attack on the palladium complex of pent-2-ene by the diethyl malonate anion gave an optically active product by asymmetric induction (scheme *lxiv*; Trost and Dietsche, 1973).

Another illustration of the preparative utility of the alkylation of these π-allylpalladium complexes is provided by the synthesis of the dimethyl ester of a pheremone of the Monarch butterfly, **L** (scheme *lxv*; Trost and Weber, 1975).

As well as these π-allyl complexes, alkenes can also be activated to nucleophilic attack by conversion to η^2-alkene palladium complexes. In this case a tertiary amine is used to generate the monomeric form and the carbanion attacks the

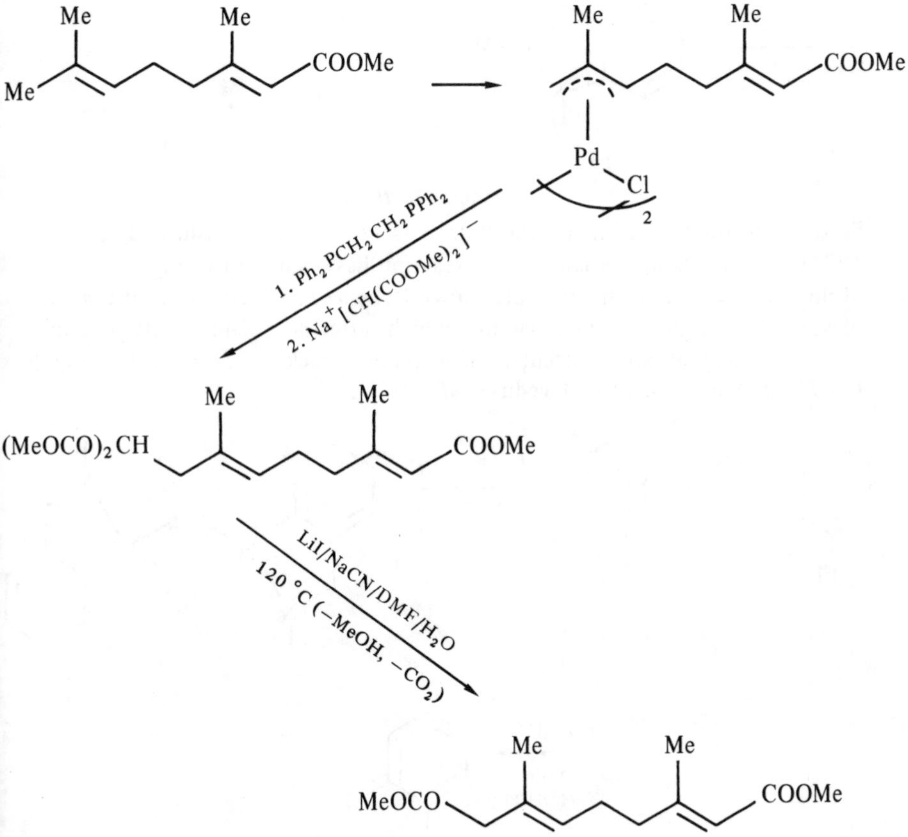

Scheme *lxv*

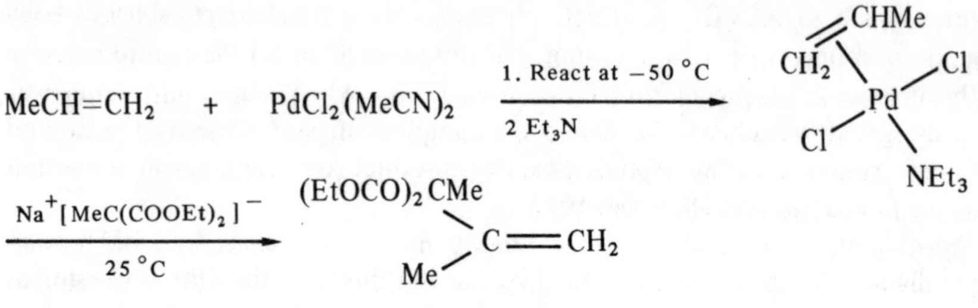

$$MeCH{=}CH_2 \quad + \quad PdCl_2(MeCN)_2 \xrightarrow[\text{2 Et}_3\text{N}]{\text{1. React at } -50\,^{\circ}\text{C}}$$

(product: Pd complex with CHMe, CH₂, Cl, Cl, NEt₃)

$$\xrightarrow[25\,^{\circ}\text{C}]{\text{Na}^+[\text{MeC(COOEt)}_2]^-} \quad \underset{\text{Me}}{\overset{(EtOCO)_2CMe}{>}}C{=}CH_2$$

Scheme *lxvi*

(aromatic ring with Br and NH₂) + (Ni–Br–Ni dimer) $\xrightarrow[\text{50}\,^{\circ}\text{C}]{\text{DMF}}$ (o-allylaniline)

$$\xrightarrow[\text{2. Et}_3\text{N}]{\text{1. PdCl}_2(\text{MeCN})_2}$$ (indole ring with Me substituent, N–H)

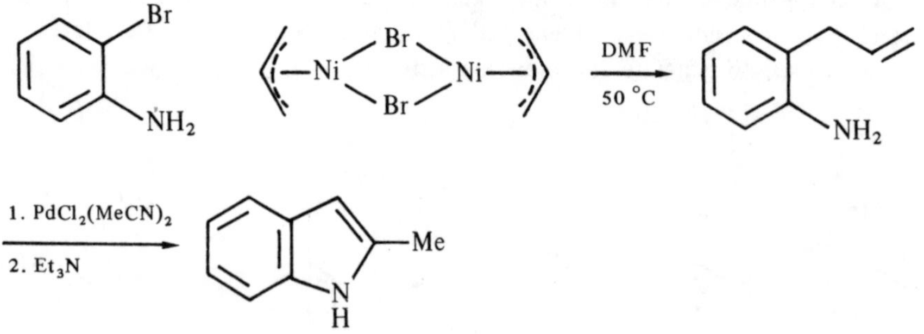

LI

Scheme *lxvii*

most substituted position of the alkene (scheme *lxvi*; Hayashi and Hegedus, 1977). An interesting variant of this reaction has been used to prepare heterocyclic compounds. In the example shown (scheme *lxvii*), arylation of a π-allyl-nickel complex gives an o-allylaniline which, after conversion to its palladium adduct, undergoes intramolecular nucleophilic attack at the activated double bond to give an indole, **LI** (Hegedus *et al.*, 1978).

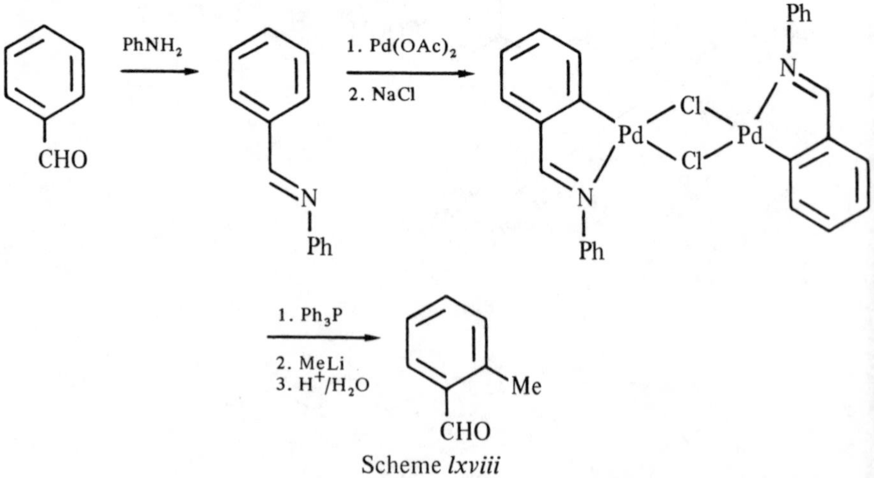

(benzaldehyde, CHO) $\xrightarrow{\text{PhNH}_2}$ (imine, =N–Ph) $\xrightarrow[\text{2. NaCl}]{\text{1. Pd(OAc)}_2}$ (dimeric Pd complex with Cl bridges, N–Ph)

$$\xrightarrow[\substack{\text{2. MeLi} \\ \text{3. H}^+/\text{H}_2\text{O}}]{\text{1. Ph}_3\text{P}}$$ (aromatic ring with Me and CHO)

Scheme *lxviii*

Quite different types of organopalladium compounds have been used to alkylate aromatic aldehydes in the *ortho* position. The aldehyde is first treated with aniline to convert it to the Schiff's base, then complexation of this with palladium results in formation of a σ-bonded compound by cyclometallation (p. 27). This is converted to a monomeric form by treatment with triphenylphosphine which on successive reaction with methyllithium to replace palladium by methyl, and aqueous acid to liberate the aldehyde group, gives *o*-methylbenzaldehyde (scheme *lxviii*; Murahashi *et al.*, 1974).

The cyclotrimerisation of alkynes by palladium compounds is on the borderline between catalytic and stoicheiometric applications. Several intermediates have been isolated and a plausible mechanism for the overall transformation is shown in scheme *lxix* (Maitlis, 1980).

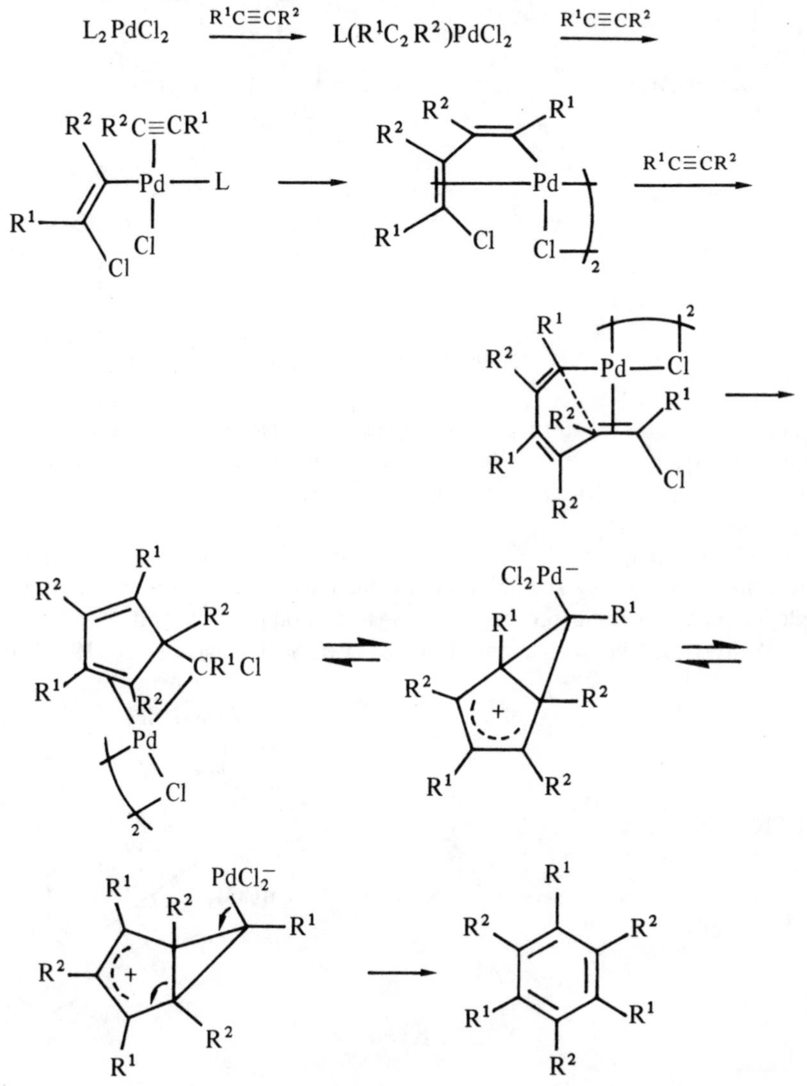

Scheme *lxix*

8.2.9   Organochromium Compounds

When arenes complex with the chromium tricarbonyl group, the metal withdraws electrons from the ring. This produces a number of effects which are of interest to the synthetic organic chemist, particularly that of making substituents more susceptible to nucleophilic attack (Semmelhack, 1976). The effect is similar to that achieved by placing a nitro group in the ring but removal of that group is difficult, whereas chromium can be eliminated by oxidation with Ce(IV) or iodine (scheme *lxx*, compare p. 112). A further advantage of π-bonded chromium

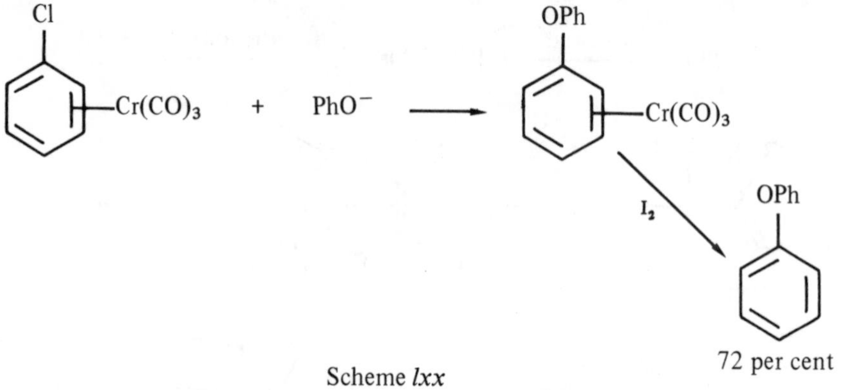

Scheme *lxx*

72 per cent

moieties over σ-bonded nitro groups is that, by replacing one or more of the carbonyl groups by other less effective π-acceptors, the degree of electron withdrawal can be controlled and, indeed, turned into overall electron donation if needed (Davies, 1982).

Another remarkable reaction is direct alkylation of the ring in the η^6-benzene-chromium tricarbonyl by stabilised anions, such as that derived from isobutyro-nitrile, in a nucleophilic process. The organic product is liberated from the complex by treatment with iodine (scheme *lxxi*; Semmelhack *et al.*, 1979). If

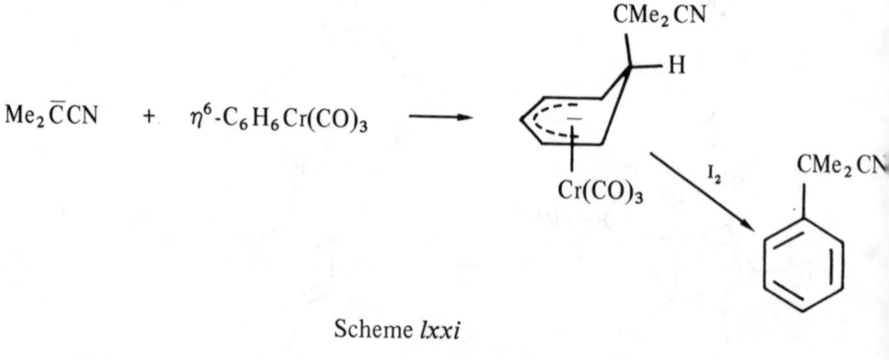

Scheme *lxxi*

the corresponding chlorobenzene derivative is used and reaction with the anion allowed to proceed for only 15 min at $0\,^\circ C$ before quenching with iodine, the major product is that in which alkylation has occurred *meta* to chlorine. In this case, displacement of either hydride or chloride ion can occur and there is also a 19 per cent yield of 2-methyl-2-phenylpropiononitrile (equation (8.61); Semmelhack, 1976).

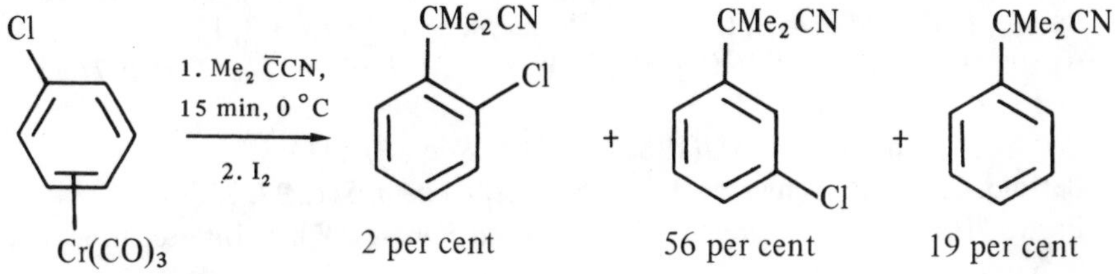

$$\text{(8.61)}$$

Since a benzene ring coordinates to a metal on one side only, there is a loss of symmetry. Advantage of this fact is taken in a stereoselective synthesis of a bridged biphenyl via the mono(chromiumtricarbonyl) complex (scheme *lxxii*; Eyer *et al.*. 1981). Conversion of the anhydride to the imide **LII** occurred with a high degree of stereoselectivity so that, for a given chiral amine, essentially one diastereoisomeric form of **LII** was obtained. Decomplexation of **LII** was accomplished photochemically under mild conditions.

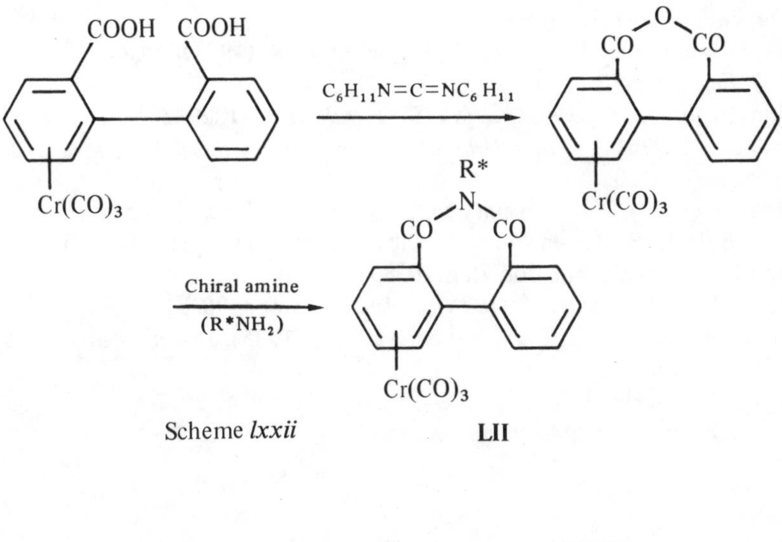

Scheme *lxxii*   **LII**

REFERENCES

Ashby, E. C. (1980). *Pure appl. Chem.* **52**, 545

Ashby, E. C., and Wiesemann, T. L. (1978). *J. Am. chem. Soc.* **100**, 3101

Augé, C., David, S., and Veyrières, A. (1976). *J. chem. Soc. chem. Commun.* 375

Baker, R., Cookson, R. C., and Vinson, J. R. (1974). *J. chem. Soc. chem. Commun.* 515

Barborak, J. C., Watts, L., and Pettit, R. (1966). *J. Am. chem. Soc.* **88**, 1328

Barnes, J. M., and Magos, L. (1968). *Organomet. Chem. Revs* **3**, 137

Barton, D. H. R., and McCombie, S. W. (1975). *J. chem. Soc. Perkin Trans. I* 1574

Bell, R. P., and Smith, P. W. (1966). *J. chem. Soc. (B)* 241

Bertolo, C. A., and Schwartz, J. (1975). *J. Am. chem. Soc.* **97**, 228

Brown, H. C. (1975). *Organic Syntheses via Boranes.* Wiley–Interscience, New York

Collman, J. P. (1975). *Acc. chem. Res.* **8**, 342

Collman, J. P., Finke, R. G., Cawse, J. N. and Brauman, J. I. (1978). *J. Am. chem. Soc.* **100**, 4766

Colvin, E. W. (1981). *Silicon in Organic Synthesis.* Butterworth, London

Corey, E. J., and Posner, G. H. (1967). *J. Am. chem. Soc.* **89**, 3911

Corey, E. J., and Posner, G. H. (1968). *J. Am. chem. Soc.* **90**, 5615

Courtois, G., and Miginiac, L. (1974). *J. organomet. Chem.* **69**, 1

Davies, A. G., Price, A. J., Daws, H. M., and Hursthouse, M. B. (1984). *J. organomet. Chem.* **270**, C1

Davies, S. G. (1982). *Organotransition Metal Chemistry: Applications to Organic Synthesis.* Pergamon Press, Oxford

Dunoguès, J. P. (1982). *Chemtech.* **12**, 373

Eisch, J. J., and Hordis, C. K. (1971). *J. Am. chem. Soc.* **93**, 2974

Eyer, M., Schlögl, K., and Schölm, R. (1981). *Tetrahedron* **37**, 4239

Fanta, P. E. (1974). *Synthesis* 9

Félix, G., Dunoguès, J. P., Pisciotti, F., and Calas, R. (1977). *Angew. Chem. int. Edn Engl.* **16**, 488

Felkin, H., and Swierczewski, G. (1975). *Tetrahedron* **31**, 2735

Fleming, I. (1976). *Frontier Orbitals and Organic Chemical Reactions.* Wiley, New York, p. 80

Fleming, I. (1979). Organosilicon Chemistry in *Comprehensive Organic Chemistry*, Vol. 3 (D. H. R. Barton and W. D. Ollis, eds). Pergamon Press, Oxford

Fleming, I. (1981). *Chem. Soc. Rev.* **10**, 83

Hart, D. W., and Schwartz, J. (1974). *J. Am. chem. Soc.* **96**, 8115

Hart, D. W., Blackburn, T. F., and Schwartz, J. (1975). *J. Am. chem. Soc.* **97**, 679

Hayashi, T., and Hegedus, L. S. (1977). *J. Am. chem. Soc.* **99**, 7093

Heck, R. F. (1963). *J. Am. chem. Soc.* **85**, 3383

Hegedus, L. S., Waterman, E. L., and Catlin, J. (1972). *J. Am. chem. Soc.* **94**, 7155

Hegedus, L. S., Allen, G. F., Bozell, J. J., and Waterman, E. L. (1978). *J. Am. chem. Soc.* **100**, 5800

Hill, E. A. (1975). *J. organomet. Chem.* **91**, 123

Hill, E. A. (1977). *Adv. organomet. Chem.* **16**, 131

Holmes, J. D., and Pettit, R. (1963). *J. Am. chem. Soc.* **85**, 2531

Honeycutt, J. B., and Riddle, J. M. (1960). *J. Am. chem. Soc.* **82**, 3051

Hudrlik, P. F. (1976). *J. organomet. Chem. Library* **1**. 127

Jung, M. E., and Lyster, M. A. (1977). *J. org. Chem.* **42**, 3761

Jung, M. E., and McCombs, C. A. (1976). *Tetrahedron Lett.* 2935

Jung, M. E., Andrus, W. A., and Ornstein, P. L. (1977). *Tetrahedron Lett.* 4175

Kharasch, M. S., and Reinmuth, O. (1954). *Grignard Reactions of Nonmetallic Substances*. Prentice-Hall, New York

Kitching, W. (1972). *Organomet. React.* **3**, 319

Koenig, K. E., and Weber, W. P. (1973). *J. Am. chem. Soc.* **95**, 3416

Kondo, K., Negishi, A., and Tunemoto, D. (1974). *Angew. Chem. int. Edn Engl.* **13**, 407

Kozikowski, A. P., and Wetter, H. F. (1976). *Synthesis* 561

Kuivila, H. G. (1968). *Acc. chem. Res.* **1**, 299

Kumada, M. (1980). *Pure appl. Chem.* **52**, 669

Larock, R. C. (1976). *J. organomet. Chem. Library* **1**, 257

Le Corre, M. (1974). *Bull. Soc. chim. Fr.* 2005

Lehmkuhl, H. (1981). *Bull. Soc. chim. Fr. II*, 87

Lennon, P., Rosan, A. M., and Rosenblum, M. (1977). *J. Am. chem. Soc.* **99**, 8426

Magnus, P. (1980). *Aldrichim. Acta* **13**, 43

Maitlis, P. M. (1980). *J. organomet. Chem.* **200**, 161

Mole, T., and Surtees, J. R. (1964). *Aust. J. Chem.* **17**, 1229

Müller, E. (1974). *Synthesis* 761

Murahashi, S. I., Tanba, Y., Yamamura, M., and Moritani, I. (1974). *Tetrahedron Lett.* 3749

Musker, W. K., and Larson, G. L. (1968). *Tetrahedron Lett.* 3481

Negishi, E.-I. (1976). *J. organomet. Chem. Library* **1**, 112

Negishi, E.-I. (1980). *Organometallics in Organic Synthesis*, Vol. 1. Wiley, New York

Nicholas, K. M., and Pettit, R. (1971). *Tetrahedron Lett.* 3475

Normant, J. F. (1972). *Synthesis* 63

Normant, J. F. (1976). *J. organomet. Chem. Library* **1**, 219

Normant, J. F., and Alexakis, A. (1981). *Synthesis* 841

Pearson, A. J. (1981). *Transition Met. Chem.* **6**, 67

Pillot, J.-P., Dunoguès, J. P., and Calas, R. (1976). *Tetrahedron Lett.* 1871

Pine, S. H., Zahler, R., Evans, D. A., and Grubbs, R. H. (1980). *J. Am. chem. Soc.* **102**, 3270

Posner, G. H. (1972). *Org. React.* **19**, 1

Posner, G. H. (1975). *Org. React.* **22**, 253

Posner, G. H. (1980). *An Introduction to Synthesis using Organocopper Reagents.* Wiley-Interscience, New York

Posner, G. H., and Brunelle, D. J. (1972). *Tetrahedron Lett.* 293

Posner, G. H., and Whitten, C. E. (1970). *Tetrahedron Lett.* 4647

Reetz, M. T. (1982). *Top. curr. Chem.* **106**, 1

Sato, K., Inoue, S., Ota, S., and Fujita, Y. (1972). *J. org. Chem.* **37**, 462

Sato, K., Inoue, S., and Saito, K. (1973). *J. chem. Soc. Perkin Trans. I* 2289

Schwartz, J., and Labinger, J. A. (1976). *Angew. Chem. int. Edn Engl.* **15**, 333

Schwartz, J., Hart, D. W., and Holden, J. L. (1972). *J. Am. chem. Soc.* **94**, 9269

Schwartz, J., Loots, M. J., and Kosugi, H. (1980). *J. Am. chem. Soc.* **102**, 1333

Seebach, D., and Corey, E. J. (1975). *J. org. Chem.* **40**, 231

Seebach, D., and Geiss, K. H. (1976). *J. organomet. Chem. Library* **1**, 1

Semmelhack, M. F. (1976). *J. organomet. Chem. Library* **1**, 361

Semmelhack, M. F., and Ryono, L. (1973). *Tetrahedron Lett.* 2967

Semmelhack, M. F., Helquist, P. M., and Jones, L. D. (1971). *J. Am. chem. Soc.* **93**, 5908

Semmelhack, M. F., Hall, H. T., Farina, R., Yoshifuji, M., Clark, G., Bargar, T., Hirotsu, K., and Clardy, J. (1979). *J. Am. chem. Soc.* **101**, 3535

Seyferth, D. (1972). *Acc. chem. Res.* **5**, 65

Seyferth, D., and Damrauer, R. (1966). *J. org. Chem.* **31**, 1660

Shanzer, A., and Libman, J. (1983). *J. chem. Soc. chem. Commun.* 846

Shanzer, A., Schwartz, E., and Libman, J. (1982). *Rev. Silicon, Germanium, Tin Lead Compounds* **6**, 149

Stork, G., and Hudrlik, P. F. (1968). *J. Am. chem. Soc.* **90**, 4462

Stork, G., Nussim, M., and August, B. (1966). *Tetrahedron Suppl.* **8**, 105

Tebbe, F. N., Parshall, G. W., and Reddy, G. S. (1978). *J. Am. chem. Soc.* **100**, 3611

Thyret, H. (1972). *Angew. Chem. int. Edn Engl.* **11**, 520

Trost, B. M., and Dietsche, T. J. (1973). *J. Am. chem. Soc.* **95**, 8200

Trost, B. M., and Weber, L. (1975). *J. org. Chem.* **40**, 3617

Trost, B. M., Conway, W. P., Strege, P. E., and Dietsche, T. J. (1974). *J. Am. chem. Soc.* **96**, 7165

Weber, W. P. (1982). *Silicon Reagents for Organic Synthesis: Reactivity and Structure.* Vol. 14 Springer, Berlin

Whitesides, G. M., Fischer, W. F., San Filippo, J., Bashe, R. W., and House, H. O. (1969). *J. Am. chem. Soc.* **91**, 4871

Young, W. G., and Roberts, J. D. (1944). *J. Am. chem. Soc.* **66**, 1444

Zweifel, G., and Brown, H. C. (1963). *Org. React.* **13**, 1

GENERAL READING

Alper, H. (1976, 1978). *Transition Metal Organometallics in Organic Synthesis*, Vols I and II. Academic Press, New York

Colquhoun, H. M., Holton, J., Thompson, D. J., and Twigg, M. V. (1984). *New Pathways for Organic Synthesis*. Plenum Press, New York

Davies, S. G. (1982). *Organotransition Metal Chemistry: Applications to Organic Synthesis*. Pergamon Press, Oxford

Kozikowski, A. P., and Wetter, H. F. (1976). Transition metals in organic synthesis. *Synthesis* 561

Negishi, E.-I. (1980). *Organometallics in Organic Synthesis*, Vol. I. Wiley, New York

Scheffold, R. (1983). *Modern Synthetic Methods*, Vol. 3. Wiley, Chichester

Seyferth, D. (1976). New Applications of Organometallic Reagents in Organic Synthesis, *J. organomet. Chem. Library* **1**. Elsevier, Amsterdam

9

CATALYTIC APPLICATIONS OF ORGANOMETALLIC COMPOUNDS

The International Union of Pure and Applied Chemistry defines catalysis as the phenomenon in which a relatively small amount of a foreign material, called a catalyst, augments the rate of a chemical reaction without itself being consumed (Burwell, 1976). In practice this means that a catalyst reduces the activation

Free energy

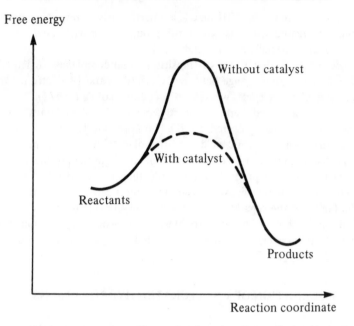

Figure 9.1   Variation of free energy during a reaction.

energy of a reaction which must itself be energetically favoured. This is shown in figure 9.1.

Organometallic compounds (or at least organometallic intermediates) are important in many catalytic reactions involving organic molecules. Catalytic activity occurs widely among the transition metals, but there are also a few instances where main group compounds act as catalysts. In many cases a series of steps can be drawn up in which transition metal complexes undergo a sequence of oxidative addition, migratory insertion and reductive elimination reactions. During this sequence the organic reactants are converted into the product and the transition metal complex is regenerated so that the sequence forms a closed loop called a catalytic cycle. We should, perhaps, point out at this stage that the 'catalyst' which is added to the reaction may not be the 'active catalyst' but a precursor which is converted into the active catalyst under the reaction conditions. For this reason it is often difficult (if not impossible) to recover the catalyst in exactly the form it was added to the reaction. A word of caution about 'intermediates' which are isolated from catalytic reactions is also appropriate. Such compounds may be the products of side reactions and their relevance to the catalytic cycle must be considered very critically. True catalysts bring about transformations precisely because they are very reactive and therefore are not readily isolated (Halpern, 1981).

Complexes of the Group VIII metals are particularly noteworthy for catalysing a range of organic reactions; within this group, metals with the d^8 electronic configuration are especially rich in catalytic chemistry.

The electronic configurations of transition metal complexes participating in catalytic reactions often change between 16 electrons and 18 electrons. We shall see this in several of the cycles we describe in detail (Tolman, 1972).

The systems described here are mostly based on soluble transition metal complexes, and the reactions are said to be homogeneously catalysed (as indeed is the esterification of acetic acid by ethanol catalysed by sulphuric acid). Homogeneous catalysts can be difficult to separate from the products at the end of the reaction. If the products are volatile they can be distilled out of the reaction mixture, but the recovery of solid products can involve difficult separations. To facilitate the separation, efforts have been made to attach the catalyst to organic polymers or inorganic oxides. Such systems are variously called heterogenised homogeneous catalysts, supported catalysts, or anchored catalysts (Hartley and Vezey, 1977).

9.1 HYDROGENATION OF UNSATURATED HYDROCARBONS

The hydrogenation of unsaturated organic compounds is an important industrial reaction. Most large-scale hydrogenations are carried out using heterogeneous catalysts such as Raney nickel, but homogeneous systems are used in the pharmaceutical industry. The subject has proved to be very popular with organometallic chemists. Most of the work has been concerned with the hydrogenation of alkenes, but alkynes, dienes and arenes have also been studied (James, 1979).

9.1.1 Hydrogenation using $(Ph_3P)_3RhCl$

$(Ph_3P)_3RhCl$, which is generally known as Wilkinson's catalyst, acts as a homogeneous catalyst for the hydrogenation of alkenes and alkynes. It does not catalyse the reduction of other organic functional groups, although it is a reactive compound and does react with many organic and inorganic reagents. $(Ph_3P)_3RhCl$ is only slightly dissociated (about 5 per cent) into the 14-electron species $(Ph_3P)_2RhCl$ and PPh_3 in pure solvents of low coordinating power and the absence of all reagents (especially O_2 and H_2). In a catalytic mixture the small initial concentration of $(Ph_3P)_2RhCl$ undergoes a series of oxidative addition and reductive elimination reactions and the equilibrium is displaced (Jardine, 1981).

The catalytic cycle shown in scheme i is very similar to the original mechanism proposed by Wilkinson in 1966 (Halpern, 1981). There are four coordination compounds in the catalytic cycle, II–V. The 14-electron species II is formed by dissociation of a phosphine ligand from I. II has been written with a square in one position to denote a vacant coordination site. This is because stable Rh(I) complexes are generally four-coordinate and are therefore 16-electron species.

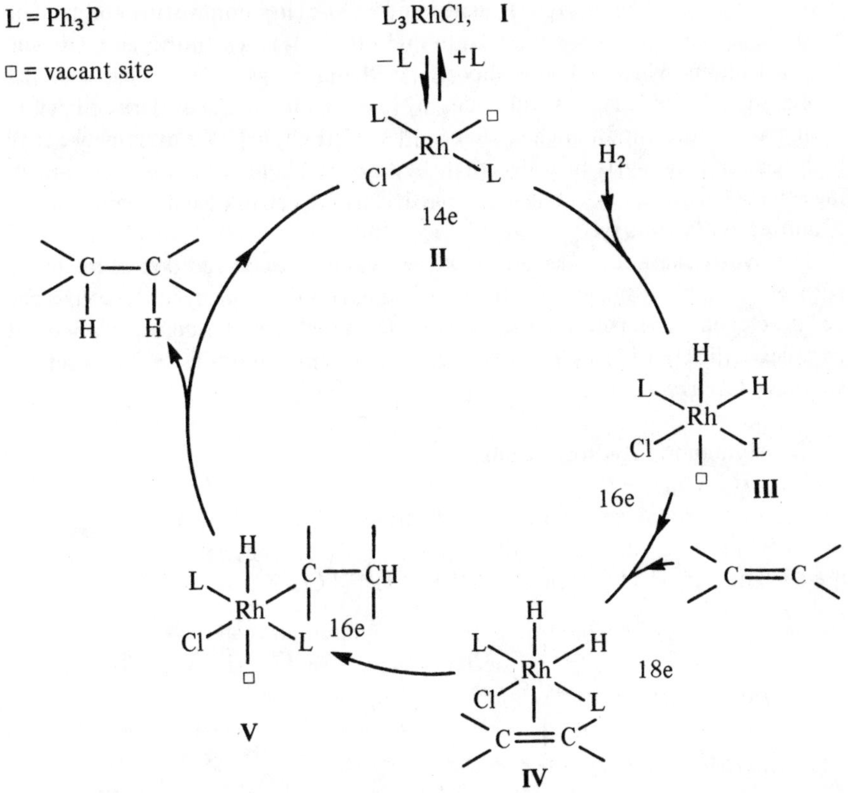

L = Ph$_3$P

☐ = vacant site

L$_3$RhCl,   **I**

−L ↓↑ +L

Scheme *i*

The 14-electron species, **II**, reacts with hydrogen to give **III** by oxidative addition. **III** is still coordinatively unsaturated, and so readily accepts π-electrons from the substrate alkene to give **IV**. **IV** represents a very significant stage in the overall reaction as the substrate alkene, and hydrogen, are now bonded to the same metal atom. The next step is a migratory insertion reaction to give **V**, which then undergoes reductive elimination to release the substrate as the alkane, and regenerate **II**. Although we have used a square to denote an unoccupied coordination site, it is possible that it is occupied by a solvent molecule. It is also possible that **V** might be better described as a trigonal bipyramidal complex rather than the square pyramid as shown, but these are points which are not easily settled by experiment as the complexes only occur transiently. This catalytic cycle has also been studied theoretically (Dedieu, 1981).

9.1.2 Hydrogenation using other Rhodium and Iridium Catalysts

Wilkinson's catalyst, (Ph$_3$P)$_3$RhCl, carries no formal charge. Rhodium complexes which are cationic have also been used as catalysts. The catalyst precursor in this case is a diene complex such as [norbornadiene Rh(PPh$_3$)$_2$]$^+$. In the hydrogenation solution the diene is hydrogenated and so loses its coordinating ability. The catalytic species produced is $[H_2(solv)_2 Rh(PPh_3)_2]^+$ (Schrock and Osborn, 1976). Iridium, which is below rhodium in the periodic table and has a similar coordination chemistry, also provides hydrogenation catalysts. The complexes analogous to the rhodium case, that is $[diene\, Ir(PPh_3)_2]^+$ are extremely active hydrogenation catalysts provided that they are used in polar, but non-coordinating solvents. The iridium catalysts are particularly effective with hindered alkenes (Crabtree et al., 1982).

The hydrogenation of alkynes to alkenes usually gives cis addition of hydrogen with either a homogeneous or heterogeneous (Lindlar) catalyst. Although not yet developed to preparative use, it has been found that the dinuclear rhodium complex $[(iPrO)_3 P]_4 Rh_2(\mu H)_2$ brings about trans addition of hydrogen to alkynes (Burch et al., 1982).

9.1.3 Asymmetric Hydrogenation

The cationic complexes of rhodium described above, $[diene\, Rh(PPh_3)_2]^+$, are the basis of some asymmetric hydrogenation catalysts. The most selective systems use chelating diphosphines such as DIOP (VI) and CHIRAPHOS (VII).

VI, DIOP   VII, CHIRAPHOS

Using the cyclooctadiene complex $[CODRh(CHIRAPHOS)]^+$ an optical purity of 99 per cent was obtained in the hydrogenation of the aminocinnamic acid derivative, VIII, to N-acetylphenylalanine, IX (equation (9.1); Fryzuk and Bosnich, 1977). Similar systems have been used for the manufacture of L-DOPA, a drug used in the treatment of Parkinson's disease (James, 1979).

9.1.4 Arene Hydrogenation

A relatively small number of transition metal complexes catalyse the hydrogenation of arenes. One remarkable discovery was that η^3-allylCo$[P(OCH_3)_3]_3$ catalyses the cis-addition of hydrogen to all six carbon atoms of benzene. Thus using deuterium the all-cis-hexadeuterocyclohexane, X, is formed from benzene (equation 9.2).

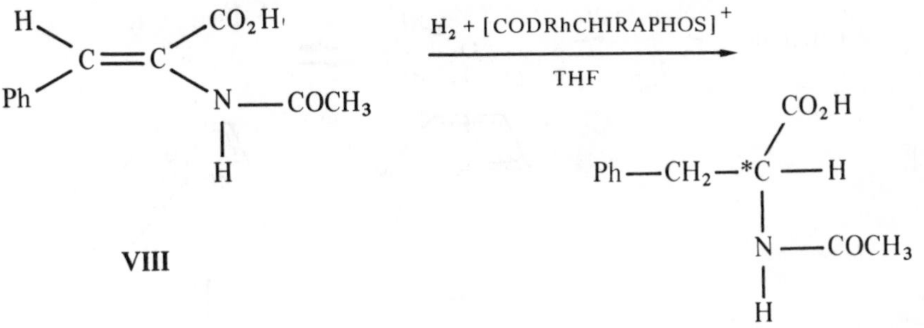

$$
\underset{\text{VIII}}{
\begin{array}{c}
\text{H} \\
\diagdown \\
\text{Ph}
\end{array}
\text{C}=\text{C}
\begin{array}{c}
\diagup \text{CO}_2\text{H} \\
\diagdown \\
\text{N}-\text{COCH}_3 \\
| \\
\text{H}
\end{array}
}
\quad
\xrightarrow[\text{THF}]{\text{H}_2 + [\text{CODRhCHIRAPHOS}]^+}
\quad
\underset{\substack{\textbf{IX}, \text{99 per cent optical purity}}}{
\begin{array}{c}
\text{CO}_2\text{H} \\
\diagup \\
\text{Ph}-\text{CH}_2-{}^*\text{C}-\text{H} \\
| \\
\text{N}-\text{COCH}_3 \\
| \\
\text{H}
\end{array}
}
$$

(9.1)

Benzene + 3D$_2$ $\xrightarrow{\eta^3\text{-}C_3H_5Co[P(OCH_3)_3]_3}$

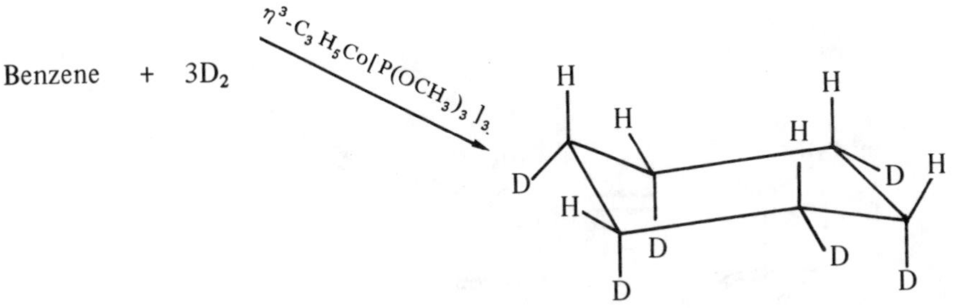

X

(9.2)

The stereospecificity is due to the arene remaining coordinated as a π-complex during the addition of all six deuterium atoms. The allyl ligand alternates between η^3 and η^1 bonding modes, and the cobalt atom undergoes a series of oxidative addition and reductive elimination reactions (see scheme *ii*, p. 220; Muetterties and Bleeke, 1979).

9.2 REACTIONS INVOLVING CARBON MONOXIDE

In chapter 6 we discussed the migratory insertion of carbon monoxide into a metal–carbon bond. This important carbon–carbon bond-forming step occurs in a large number of catalytic processes which are widely used in the chemical industry. The general term 'carbonylation' is often used to describe these reactions.

9.2.1 Hydroformylation

The reaction of an alkene with synthesis gas (CO + H$_2$) to give an aldehyde is called hydroformylation. It is also known as the 'OXO reaction' (equation 9.3).

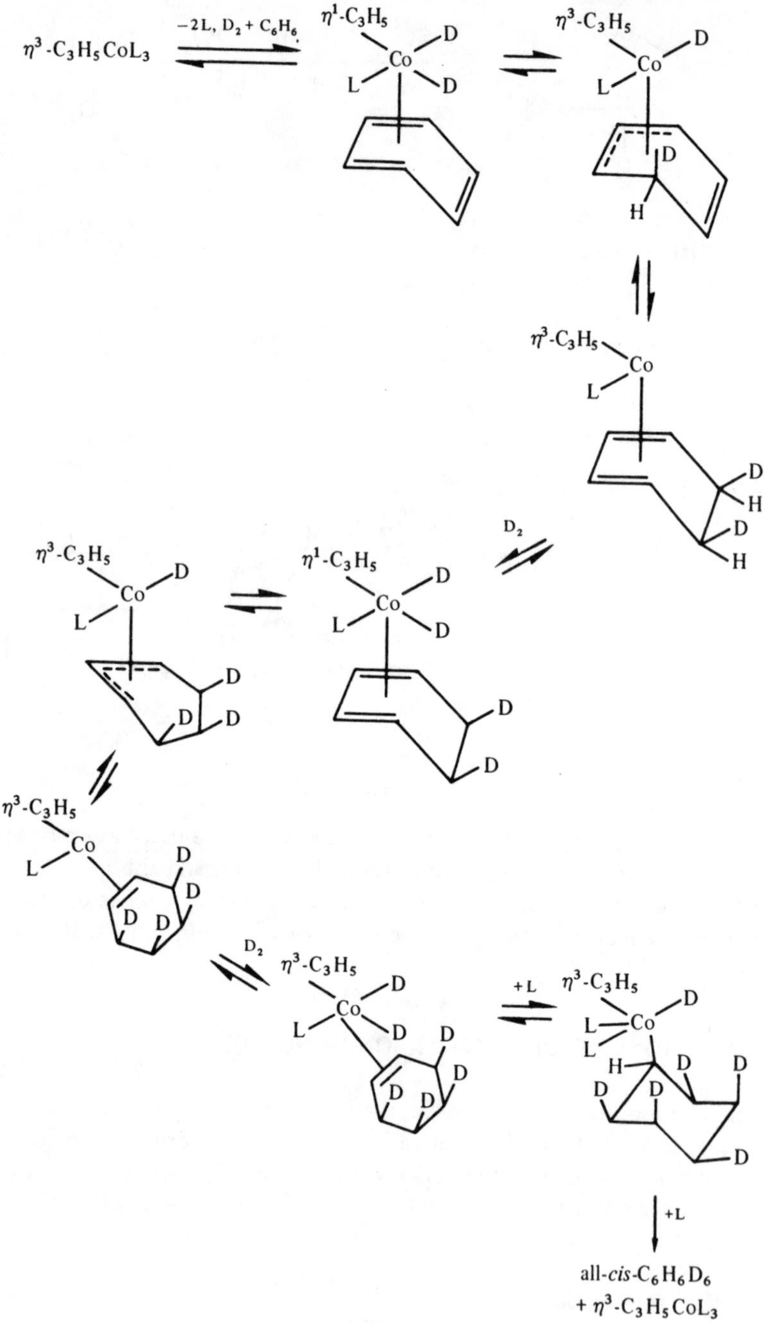

Scheme *ii* (L = P(OMe)₃)

$$RCH=CH_2 \;+\; CO \;+\; H_2 \;\xrightarrow{\text{Catalyst}}\; RCH_2CH_2CHO \;+\; \underset{\overset{|}{CHO}}{RCH} - CH_3$$

$$(9.3)$$

Compounds of several transition metals catalyse hydroformylation to some extent, but the main interest lies in catalysis by cobalt or rhodium compounds (Pruett, 1979). Industrially, the aldehyde products are usually hydrogenated to alcohols either directly or after aldol condensation.

The reaction conditions required for hydroformylation with cobalt carbonyl as catalyst are quite severe, typically 200-300 atm at 130-170 °C, which leads to a high plant cost for the industrial process.

The mechanism of hydroformylation catalysed by cobalt carbonyl is shown in scheme *iii*. The cobalt remains in the +1 oxidation state throughout the cycle,

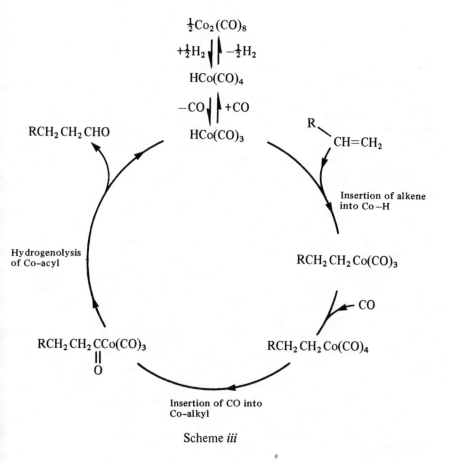

Scheme *iii*

and the sequence involves insertion of the alkene into a Co–H bond followed by insertion of CO in the Co-alkyl bond and finally hydrogenolysis of the Co-acyl bond to give the aldehyde product and regenerate the HCo(CO)$_3$. One difficulty with the hydroformylation reaction is that the product is usually a mixture of isomers because the insertion of the alkene into the Co–H bond can occur with two orientations (equation 9.4).

Markownikoff

$$R\text{—}CH\text{—}CH_3$$
$$\mid$$
$$Co(CO)_3$$
$$\textbf{XI}$$

$$\underset{CH=CH_2}{R}\quad +\quad HCo(CO)_3$$

$$R\text{—}CH_2\text{—}CH_2\text{—}Co(CO)_3$$
$$\textbf{XII}$$
anti-Markownikoff

(9.4)

The two possibilities are usually described as Markownikoff, **XI**, and anti-Markownikoff, **XII**, additions. The direction of addition is important in many commercial hydroformylation reactions, and for most applications the anti-Markownikoff orientation leading to the straight chain aldehyde is preferred. A measure of the desirability of a catalyst is the normal : iso ratio of the product. If a tertiary phosphine is added to the cobalt catalyst, a higher normal : iso ratio can be achieved. This is the basis of the Shell process for hydroformylation and is usually carried out at a lower pressure, but slightly higher temperature than the unmodified process (c. 50–100 atm, at 175 °C). A disadvantage of the Shell process is that some of the alkene is hydrogenated to alkane.

A more marked improvement in the normal : iso ratio can be achieved under yet milder conditions using a rhodium phosphine complex such as HRh(CO)(PPh$_3$)$_3$ as catalyst. A normal : iso ratio of 15 is obtained at 12.5 atm and 125 °C when triphenylphosphine is used both as ligand and solvent. The rhodium-catalysed reaction has been used widely in synthesis (Siegel and Himmele, 1980). A mechanism for rhodium-catalysed hydroformylation is given in scheme *iv*. It is interesting to compare this with the cobalt-catalysed process in scheme *iii*. Both mechanisms involve two insertion steps, but in the rhodium case the metal undergoes a series of oxidative additions and reductive eliminations, while in the cobalt case the oxidation state is +1 throughout the cycle. Not all aspects of the mechanism of hydroformylation are understood. For example, it is not certain that two phosphine ligands remain coordinated to rhodium throughout the cycle. Finally, it should be noted that a study of the iridium system in which the individual steps occur more slowly has provided general support for the mechanisms given above (Whyman, 1975).

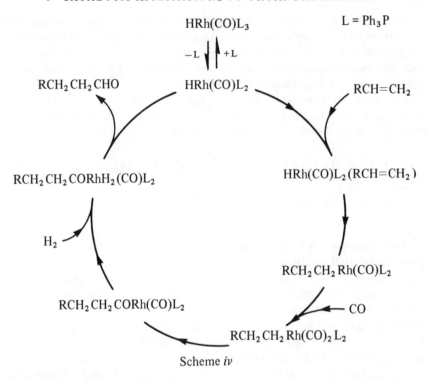

Scheme *iv*

9.2.2 Carboxylation Reactions

The reaction of unsaturated organic compounds with CO and an active hydrogen compound such as H_2O, ROH or RNH_2 in the presence of a catalyst to give a carboxylic acid derivative with addition of hydrogen to the double bond is known as hydrocarboxylation (or the Reppe reaction; equation 9.5).

$$(9.5)$$

Both alkenes and alkynes can be used as substrates, the latter giving rise to α,β-unsaturated carboxylic acid derivatives. Nickel carbonyl is often used as a catalyst for hydrocarboxylation, although cobalt and palladium compounds are also frequently encountered. Nickel is often introduced as the halide, which generates nickel carbonyl and hydrogen halide *in situ*.

The most important examples of hydrocarboxylation are the synthesis of acrylic acid from acetylene (equation 9.6) and the related hydroesterification of propyne in methanol to produce methyl methacrylate (equation 9.7).

$$H-C{\equiv}C-H + CO + H_2O \xrightarrow{\text{Ni catalyst}} CH_2{=}CH-CO_2H \qquad (9.6)$$

$$Me-C{\equiv}C-H + MeOH + CO \xrightarrow{\text{Ni catalyst}} \overset{Me}{\underset{CO_2Me}{\diagdown\diagup}}C{=}CH_2 \qquad (9.7)$$

The mechanism of hydrocarboxylation is less well established than that of other carbonylations. Hydrocarboxylation of unsymmetrical alkynes gives the same mixture of isomers as the nickel carbonyl-catalysed conversion of the corresponding cyclopropenones to acrylic acids, and is therefore assumed to involve common intermediates (scheme *v*).

This suggests that hydrocarboxylation occurs by the catalytic cycle shown in scheme *vi* (Ayrey *et al.*, 1970/71). An alternative mechanism involves the addition of $HNi(CO)_2X$ to the alkyne as the first step (Heck, 1963).

Organic halides can also be carboxylated using transition metal compounds as catalysts. Some examples are shown in equations (9.8)-(9.10) (Cassar *et al.*, 1973; Heck, 1977).

$$CH_2{=}CHBr + CO + MeOH \xrightarrow{Ni(CO)_4} CH_2{=}CH-CO_2Me \qquad (9.8)$$
$$\text{75 per cent yield}$$

$$+ \quad CO \quad + \quad BuOH \xrightarrow[\text{2 atm/60 °C}]{(Ph_3P)_2PdBr_2/Bu_3N} \qquad (9.9)$$

$$\text{89 per cent yield}$$

$$C_8H_{17}I + CO + MeOH \xrightarrow[\text{1 atm/50 °C}]{NaCo(CO)_4/Cy_2EtN} C_8H_{17}CO_2Me \qquad (9.10)$$
$$\text{56 per cent yield}$$

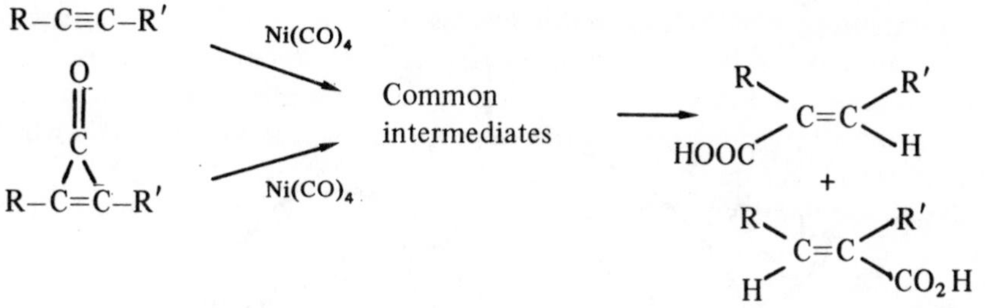
R–C≡C–R'
Ni(CO)4
Common intermediates
R–C=C–R'
Ni(CO)4
HOOC
C=C
R
R'
H
+
CO2H

Scheme *v*

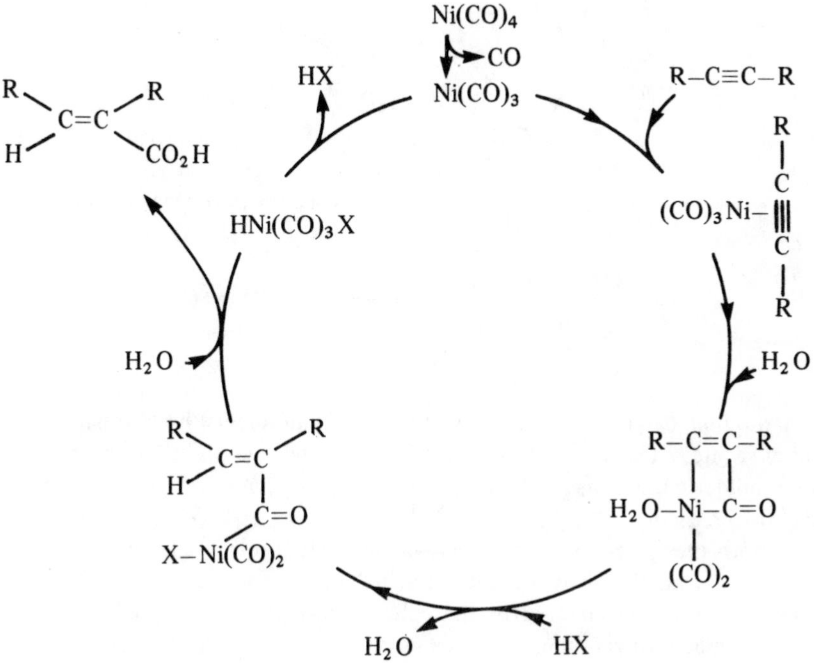
Ni(CO)4
CO
Ni(CO)3
HX
R–C≡C–R
C=C
CO2H
HNi(CO)3X
(CO)3Ni–
H2O
R–C=C–R
H2O–Ni–C=O
(CO)2
C=O
X–Ni(CO)2
H2O
HX

Scheme *vi*

9.2.3 Carbonylation of Methanol

The development of catalysts for the carbonylation of methanol to produce acetic acid has followed a similar historical pattern to that of the hydroformylation of alkenes. The early work involved cobalt catalysts under vigorous conditions and was developed by BASF in Germany (equation (9.11); von Kutepow *et al.*, 1965).

$$CH_3OH + CO \xrightarrow[680\ \text{atm}/250\,^\circ\text{C}]{Co(OAc)_2/CoI_2} CH_3CO_2H \qquad (9.11)$$

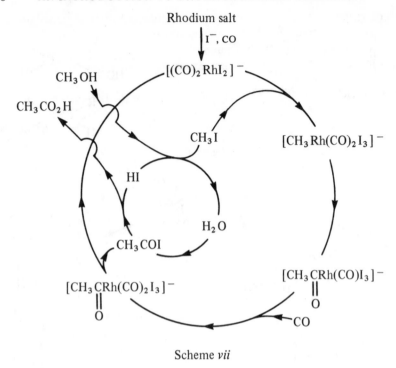

Scheme *vii*

A rhodium-catalysed process was subsequently developed by Monsanto in the United States. The catalyst is $[(CO)_2RhI_2]^-$ and the reaction occurs under much milder conditions ($\sim$ 30 atm/180 °C). The mechanism is shown in scheme *vii* (Forster, 1979).

In both these reactions iodine plays an important role in enabling the methyl group of the methanol to become attached to the transition metal. The chief difference in the mechanism of the cobalt- and rhodium-catalysed reactions is that in the case of rhodium there is a series of oxidative additions and reductive eliminations, whereas cobalt remains in the +1 oxidation state throughout (compare schemes *iii* and *iv*).

9.3   REACTIONS OF UNSATURATED HYDROCARBONS INVOLVING C–C COUPLING

The insertion of alkenes into metal–carbon bonds was described in chapter 6. In this section we shall deal firstly with insertion reactions of aluminium trialkyls (section 9.3.1), but a short historical anecdote shows how the chemistry for sections 9.3.2 and 9.3.3 was discovered.

In 1952 while investigating the growth reaction of aluminium alkyls with ethene, Ziegler and Holzkamp found, on one occasion, that they obtained only butene. This was eventually traced to a nickel impurity. They then experimented with aluminium alkyls in conjunction with other transition metal compounds and found that Group IVa metals, and especially titanium compounds gave long chain polymers (Ziegler, 1968). These systems are the dimerisation catalysts described in section 9.3.2 and the polymerisation catalysts described in section 9.3.3.

9.3.1 Chain Growth of Aluminium Trialkyls and Alkene Oligomerisation

Aluminium trialkyls react with ethene under pressure (temperature 90–120 °C, pressure not less than 100 atm) to give higher homologous alkyls (equation 9.12).

$$Al\begin{array}{c} R \\ R \\ R \end{array} \quad + \quad (x + y + z)CH_2{=}CH_2 \quad \longrightarrow \quad Al\begin{array}{c} (CH_2-CH_2)_x-R \\ (CH_2-CH_2)_y-R \\ (CH_2-CH_2)_z-R \end{array} \quad (9.12)$$

The chain length of the products follows a Poisson distribution, as shown in figure 9.2. If $(x + y + z) = 3n$, n is the most probable number of ethene units per chain.

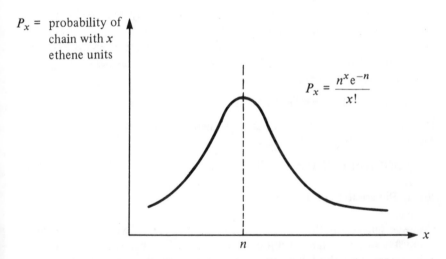

$$P_x = \text{probability of chain with } x \text{ ethene units}$$

$$P_x = \frac{n^x e^{-n}}{x!}$$

Figure 9.2 Product distribution in a chain growth reaction.

There is considerable variation in the reactivity of trialkylaluminium compounds towards ethene. The chain lengths in the products are generally below C_{20}, so that these are oligomerisation rather than polymerisation reactions. Tri(t-butyl)aluminium is particularly reactive. Only trialkylaluminiums react in this way and dialkylaluminium halides do not undergo insertion (Lehmkuhl and Ziegler, 1970, p. 184; Mole and Jeffrey, 1972, p. 130).

The chain growth reactions have been developed into commercial processes for the manufacture of long chain alkenes and alcohols. Alkenes are obtained

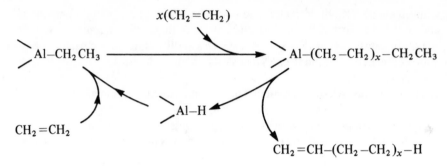

Scheme *viii*

from the long chain trialkylaluminiums by heating. Under these conditions the alkylaluminium undergoes a β-elimination to give, initially, a hydride which then reacts with ethene to generate the alkylaluminium. The factor controlling the chain length of the alkene product is the ratio of the rate of growth to the rate of elimination. The sequence can be arranged to run catalytically (scheme *viii*; Weissermel and Arpe, 1978).

Another process which is also operated commercially but does not involve recirculation of the aluminium is the Alfol process for the production of long chain alcohols. The process uses the growth reaction to produce long chain aluminium alkyls as before and then subjects these to oxidation to give the alkoxide (equation 9.13).

$$R_3 Al + \tfrac{3}{2} O_2 \longrightarrow (RO)_3 Al \qquad (9.13)$$

Hydrolysis of the alkoxide generates the alcohol and alumina (equation (9.14); Lehmkuhl and Ziegler, 1970, p. 207).

$$2(RO)_3 Al + 3H_2 O \longrightarrow 6ROH + Al_2 O_3 \qquad (9.14)$$

9.3.2 Dimerisation of Alkenes

We mentioned above that Ziegler and Holzkamp discovered that nickel-aluminium catalysts cause the dimerisation of ethene. There have been numerous studies of this system since the initial discovery, many of them concerned with propene dimerisation (Bogdanovič, 1979). Unfortunately the catalysts give mixtures of products but some extremely active catalysts have been found. The catalyst made from η^3-allyl Ni(PCy$_3$)Cl and EtAlCl$_2$ has a turnover number of 230 per second at $-55\,^\circ$C (Bogdanovič *et al.*, 1980).

It is also possible to dimerise propene using tripropylaluminium in the absence of any transition metal. The catalytic activity is lower than that of the nickel-aluminium systems, but the product consists of one compound, namely 2-methylpent-1-ene (equation (9.15); Lehmkuhl and Ziegler, 1970, p. 195; Mole and Jeffrey, 1972, p. 143).

$$CH_2=CH-CH_3 \xrightarrow[\text{heat to } 180\,^{\circ}C/200\ atm]{(Pr)_3Al} CH_2=\underset{\underset{CH_3}{|}}{C}-CH_2CH_2CH_3 \qquad (9.15)$$

The dimerisation proceeds by an insertion reaction followed by β-elimination (scheme ix).

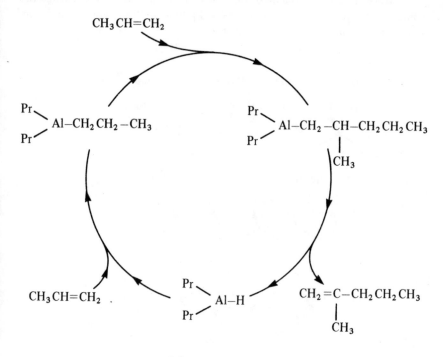

Scheme ix

Another dimerisation system, this time based on a transition metal, is the solution formed from rhodium trichloride in ethanol. The system catalyses the dimerisation of ethene to butene. At high concentrations of rhodium chloride the rhodium(I) complex of ethene $[(C_2H_4)_2RhCl]_2$ precipitates out from solution and the catalytic cycle is thought to involve complexes of this type (Cramer, 1965).

9.3.3 Polymerisation of Alkenes

We mentioned at the beginning of this section Ziegler's discovery of titanium-aluminium catalysts for alkene polymerisation. The catalysts used by Ziegler, such as $TiCl_4/Et_3Al$, allow ethene to be polymerised without the need for the enormous pressures (1000 atm) which are needed in the free radical-initiated ICI process. The Ziegler systems operate below 30 atm.

A second discovery by Natta and his group in Italy, that propene can be poly-merised using titanium–aluminium systems and the resulting polymer fractionated by solvent extraction into three components, provided complementary know-ledge of great importance. Ziegler and Natta were awarded the Nobel Prize for Chemistry in 1963 (Pino, 1980).

The three fractions separated by Natta differ in their stereochemistries, specifically in the orientation of the methyl groups with respect to the carbon chain. The components are described as follows:

(a)   Isotactic (regular orientation of CH_3 groups)

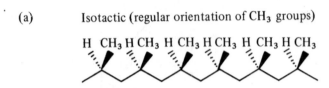

(b)   Syndiotactic (alternating orientation)

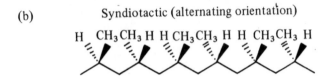

(c)   Atactic (random orientation)

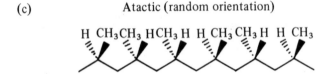

Isotactic polypropylene is insoluble in heptane and is the commercially desir-able product. Catalysts based on α-$TiCl_3$, in which the chlorine atoms are hex-agonal close packed, produce a high degree of isotacticity. One gram of catalyst mixture α-$TiCl_3$ + Et_2AlCl will give up to 1 kg of polypropylene with 95 per cent isotacticity. The addition of Lewis bases (so-called 'third components') to the catalyst tends to improve the stereospecificity, but lowers the activity of the catalyst.

The mechanism of polymerisation is thought to involve (a) alkylation of the titanium by the aluminium alkyl; (b) a series of insertion reactions into the titanium–alkyl bond. The fact that the isotacticity of the product depends on the transition metal compound employed is caused by the local geometry at the active site. The system is not fully homogeneous, although carried out in the liquid phase, so that the crystal lattice of the titanium trichloride will influence the geometry at the active site (scheme x; Cossee, 1964).

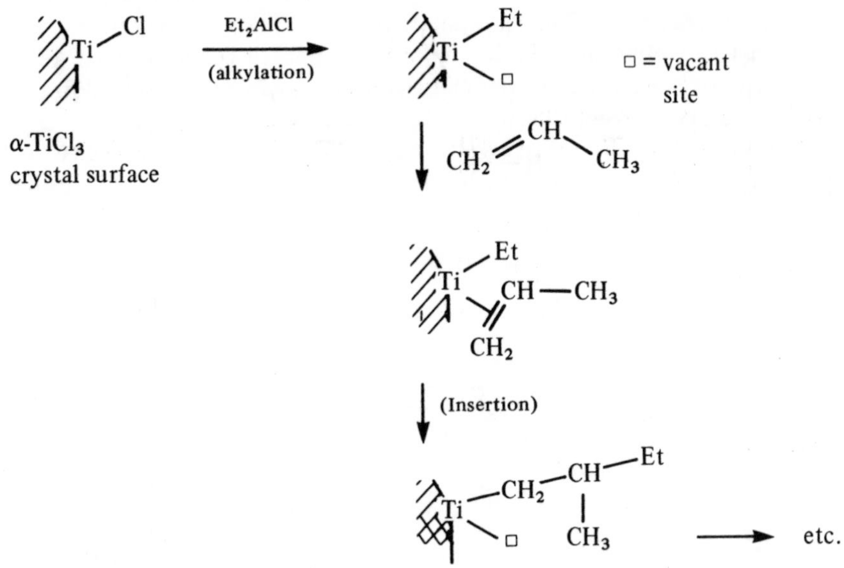

α-TiCl₃
crystal surface

Et₂AlCl
(alkylation)

□ = vacant
site

(Insertion)

etc.

Scheme *x*

There have been many developments in the area of alkene polymerisation since the early systems described above were discovered. The main difficulty in the search for new catalysts is combining high activity with good stereochemical control. Gas-phase systems using so-called 'high mileage' catalysts provide the most successful solution (Candlin, 1981). The polymerisation of ethene requires no stereochemical control and a number of catalysts have been developed. Zirconium compounds have featured in two intensively studied systems: (a) tetrabenzylzirconium supported on silica (Ballard, 1973); (b) Cp_2ZrMe_2 with methylaluminoxane (Sinn *et al.*, 1980). The latter catalyst is extremely active; the rate of polymerisation of ethene corresponds to 10 000 insertions per second at each zirconium centre.

The polymerisation of alkenes is generally accepted to involve the insertion of the alkene into a metal–carbon bond. There is, however, another mechanism which has been gaining ground in recent years (Ivin *et al.*, 1978). The alternative mechanism involves the formation of a carbene and coupling with the alkene to give a metallacycle which then undergoes ring opening (scheme *xi*). The mechanism is similar in some ways to that of olefin metathesis (p. 150) and has been supported by work on tantalum alkylidene complexes (Turner *et al.*, 1983).

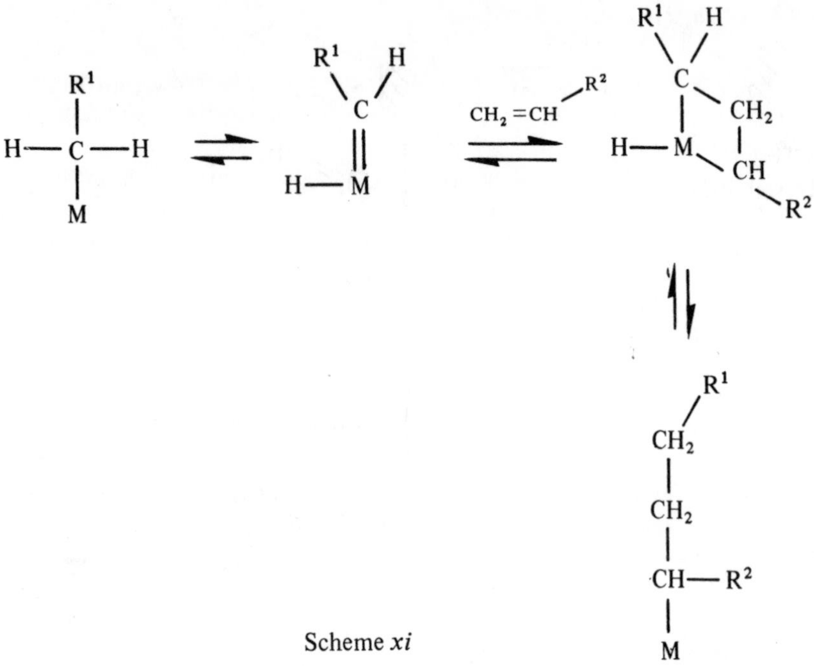

$$
H-\underset{\underset{M}{|}}{\overset{\overset{R^1}{|}}{C}}-H \;\rightleftharpoons\; \underset{\underset{M}{\overset{\|}{\vphantom{|}}}\!-\!H}{\overset{R^1\quad H}{\diagdown C \diagup}} \;\xrightarrow{CH_2=CH\diagup^{R^2}}\; H-\underset{}{M}\Big\langle{\overset{\overset{R^1\quad H}{\diagdown C \diagup}}{\quad}}{\underset{CH\diagdown R^2}{CH_2}}
$$

$$
\updownarrow
$$

$$
\underset{M}{\overset{CH_2\diagup^{R^1}}{\underset{\underset{\cdot CH-R^2}{CH_2}}{|}}}
$$

Scheme *xi*

9.3.4 Polymerisation of Butadiene

Organolithium reagents can be used in the presence of tertiary amines to poly-merise ethene, but a more important use is in diene polymerisation (Halasa *et al.*, 1980). A special feature of lithium alkyl initiated anionic polymerisation, and the related reaction initiated by free radical anions derived from sodium metal, is that the polymerisation does not have a termination step. The chain reaction proceeds until all the monomer is consumed, but when more monomer is added chain growth restarts. The polymers which have consumed all available monomer but not undergone termination are called 'living polymers'. If a second monomer is added to the living polymer a block copolymer is formed (scheme *xii*; Szwarc, 1983).

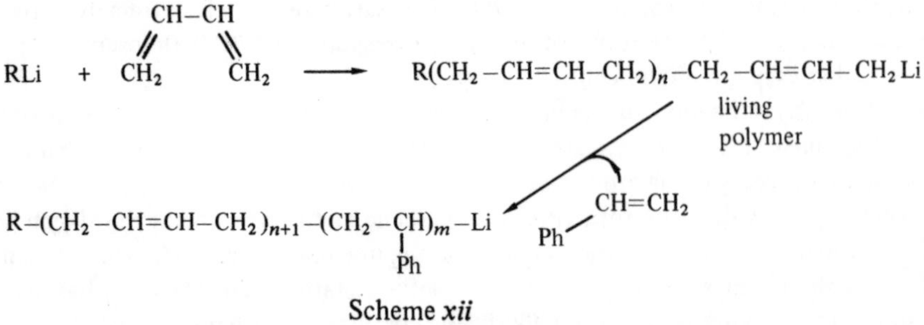

$$
RLi \;+\; \overset{CH-CH}{\underset{CH_2\quad\quad CH_2}{\diagup\diagup\quad\diagdown\diagdown}} \;\longrightarrow\; R(CH_2-CH=CH-CH_2)_n-CH_2-CH=CH-CH_2Li
$$

living
polymer

$$
R-(CH_2-CH=CH-CH_2)_{n+1}-(CH_2-\underset{Ph}{\overset{|}{CH}})_m-Li \quad\xleftarrow{\underset{Ph}{\overset{CH=CH_2}{\diagup}}}
$$

Scheme *xii*

9.3.5 Oligomerisation of Butadiene

The work of Wilke and his group in Mülheim on the chemistry of butadiene in conjunction with nickel complexes has revealed a fascinating area of chemistry with several catalytic reactions (Wilke, 1963). A number of nickel starting materials can be used to initiate catalytic reactions with butadiene. What they have in common is that in solution they produce 'naked nickel', that is nickel which is free from strongly coordinating ligands. *Bis*(cyclooctadiene)nickel, $Ni(COD)_2$, is a convenient source of naked nickel and with butadiene it can bring about dimerisation or trimerisation depending on the presence or absence of additional ligands. The products are cyclic.

In the absence of additional ligands $Ni(COD)_2$, **XIII**, causes the cyclotrimerisation of butadiene to give all-*trans*-cyclododeca-1,5,9-triene, **XVII**, as major product, as shown in scheme *xiii*. The key intermediate is the η^3, η^1-diallyl compound **XIV**; in the absence of stronger donors this 14-electron species takes on an additional butadiene molecule to give **XV**, which now has three butadiene molecules bonded to the nickel atom. Subsequent coupling and ring closure leads to the product. It should be pointed out that a titanium–aluminium Ziegler system will also bring about the cyclotrimerisation of butadiene to give the *cis,-trans,trans* isomer as the major product.

In the presence of additional phosphorus ligands, L, dimers rather than trimers are formed because the ligand occupies a coordination site to give **XVIII**, which cannot take on a third molecule of butadiene. The formation of dimers is shown

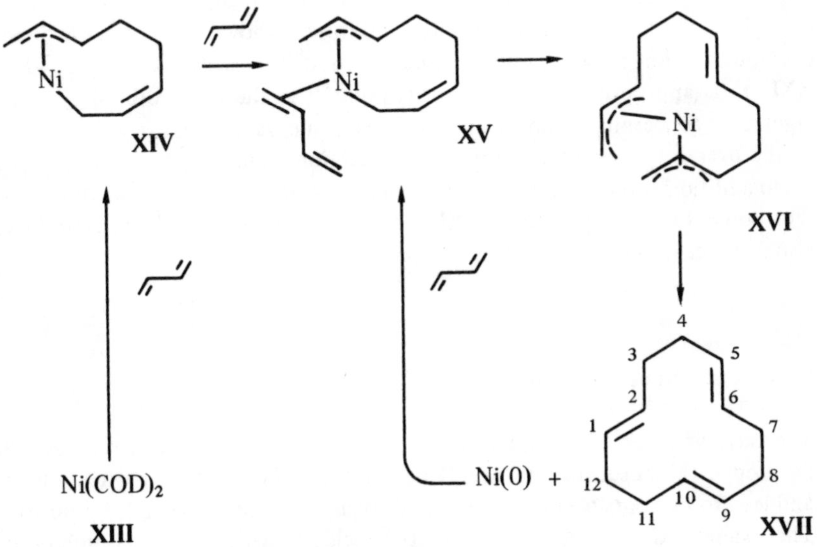

Scheme *xiii*

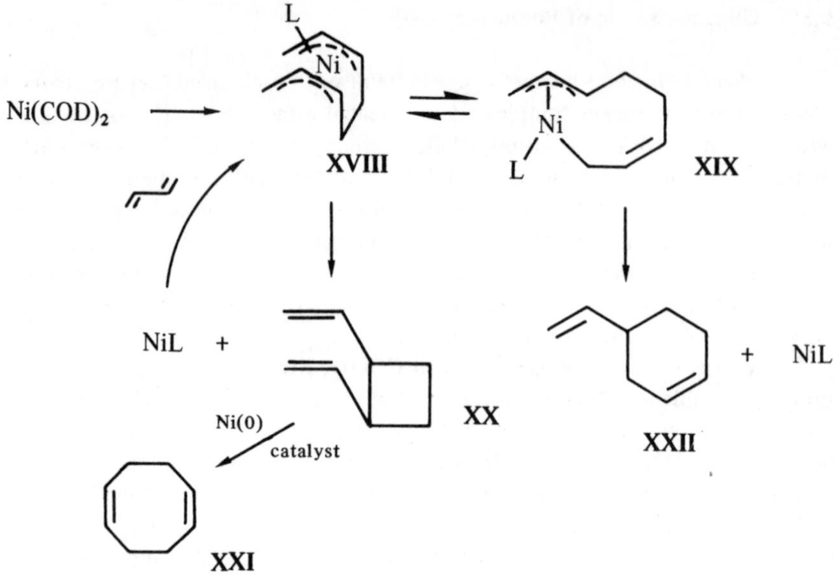

Scheme *xiv*

in scheme *xiv*. Many compounds have been used as the ligand L. Tri(*o*-phenyl-phenyl)phosphite is effective in producing relatively large amounts of *cis*-1,2-divinylcyclobutane, **XX**. If the reaction mixture is allowed to stand, the divinyl-cyclobutane undergoes catalytic isomerisation to *cis,cis*-cycloocta-1,5-diene, **XXI**. It is appropriate to state at this point that the usual orbital symmetry arguments concerning cyclo-addition reactions in organic chemistry do not apply to the overall process of these multistep reactions. The appearance of a [2 + 2] cyclo-addition product, **XX**, does not mean that the rules have broken down. In fact, some 4-vinylcyclohexene, **XXII**, the [4 + 2] symmetry-allowed product, is also produced, as shown in scheme *xiv* (Buchholtz *et al.*, 1972).

9.3.6 Cyclotrimerisation of Alkynes

The early work on the cyclotrimerisation of alkynes was carried out in Germany by Reppe. Monosubstituted acetylenes undergo the reaction particularly well and lead to two possible isomers, with substituents in the 1,2,4 or 1,3,5 positions. The example quoted in equation (9.16) yields roughly equal amounts of each isomer (Reppe *et al.*, 1969).

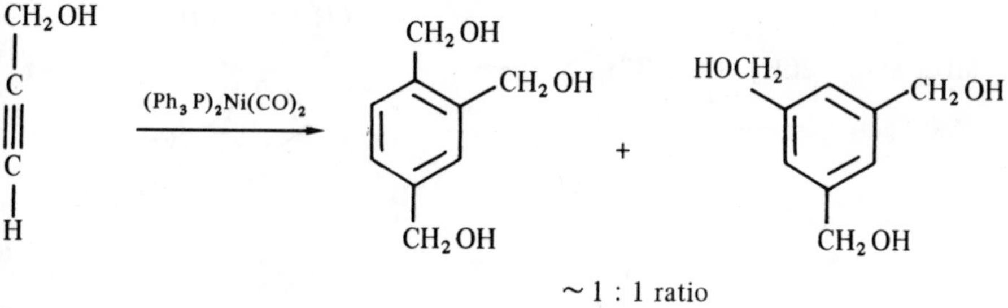

$$
\begin{array}{c}
CH_2OH \\
| \\
C \\
||| \\
C \\
| \\
H
\end{array}
\xrightarrow{(Ph_3P)_2Ni(CO)_2}
\text{(benzene with CH}_2\text{OH at 1,2,4)}
+
\text{(benzene with CH}_2\text{OH at 1,3,5)}
$$

$\sim 1 : 1$ ratio

(9.16)

Disubstituted alkynes can also be employed in the reaction if carbonyl-free nickel(0) compounds such as $Ni[P(OR)_3]_2$ or $Ni(COD)_2$ are used (Reppe et al., 1969; Muetterties et al., 1978).

The reaction is not always straightforward, even with monosubstituted alkynes, and but-1-yne with $(Ph_3P)_2Ni(CO)_2$ produces acyclic products (equation (9.17); Meriwether et al., 1961).

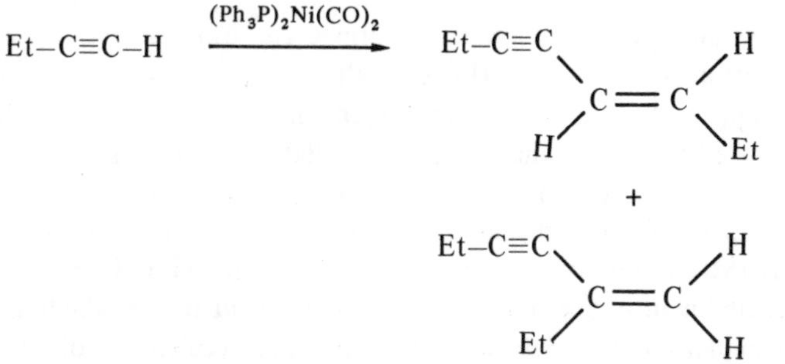

$$
Et-C\equiv C-H \xrightarrow{(Ph_3P)_2Ni(CO)_2}
\begin{array}{c}
Et-C\equiv C \\
\diagdown \\
H \diagup C = C \diagup H \\
\diagdown Et
\end{array}
$$

$$
+
\begin{array}{c}
Et-C\equiv C \\
\diagdown \\
Et \diagup C = C \diagup H \\
\diagdown H
\end{array}
$$

(9.17)

Compounds of nickel in the +2 oxidation state may also be used for the cyclisation of alkynes. Nickel(II) cyanide catalyses the cyclisation of acetylene itself to cyclooctatetraene (equation 9.18) but the phosphine complex $(Ph_3P)_2Ni(CN)_2$ leads to benzene (equation (9.19); Schrauzer et al., 1964).

$$
H-C\equiv C-H \xrightarrow{Ni(CN)_2} \text{(cyclooctatetraene)}
$$

(9.18)

$$
H-C\equiv C-H \xrightarrow{(Ph_3P)_2Ni(CN)_2} \text{(benzene)}
$$

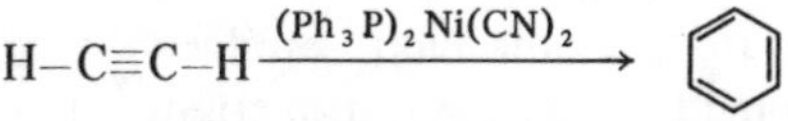

(9.19)

Cyclooctyne undergoes cyclotrimerisation when heated under reflux with nickel bromide in THF (Wittig and Fritze, 1968).

Many compounds other than those of nickel catalyse the cyclotrimerisation of alkynes (Hoogzand and Hübel, 1968; Yer'eva, 1974).

The mechanism of alkyne cyclotrimerisation is unlikely to be the same for all catalytic systems. In 1969 Whitesides and Ehmann carried out the cyclotrimerisation of $CD_3-C\equiv C-CH_3$ using a variety of catalysts. An analysis of the isomeric

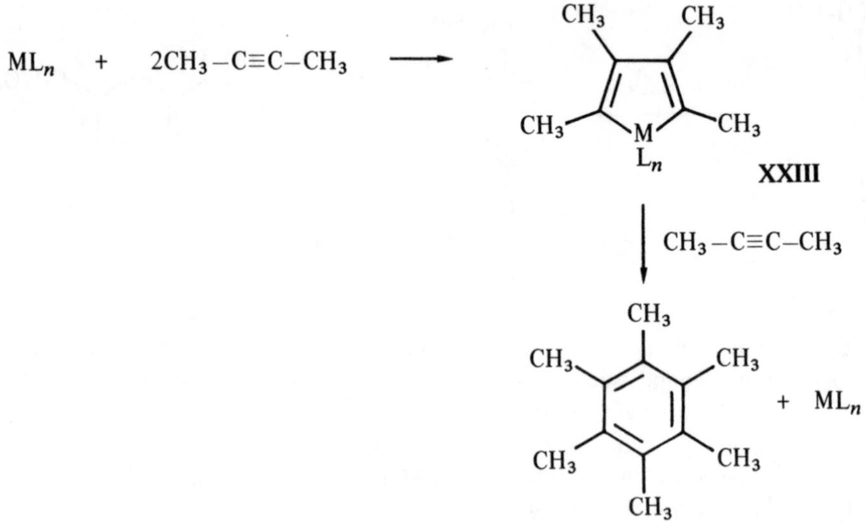

$$ML_n \quad + \quad 2CH_3-C\equiv C-CH_3 \quad \longrightarrow$$

XXIII

$$CH_3-C\equiv C-CH_3$$

$$+ \quad ML_n$$

Scheme *xv*

arenes produced allowed cyclobutadiene intermediates to be ruled out for all the first row transition metal systems. However, the aluminium trichloride catalysed reaction did appear to involve cyclobutadiene intermediates. Whitesides and Ehmann suggested that a metallacyclic intermediate, **XXIII**, was involved in the cyclisation of but-2-yne (scheme *xv*; Whitesides and Ehmann, 1969). There is considerable support for this mechanism from other work, but it should be said that chain growth followed by cyclisation (a mechanism not considered by Whitesides and Ehmann) would produce the same result in the labelling studies as the metallacyclic path which they favoured. The mechanism of the cyclo-octatetraene synthesis using nickel dicyanide has also been shown not to involve cyclobutadiene intermediates (Colborn and Vollhardt, 1981). The cyclotrimeri-sation of alkynes by palladium complexes is discussed on p. 207.

The synthetic applications of alkyne cyclotrimerisation have been explored in recent years. Cyclopentadienylcobalt dicarbonyl has been the chosen catalyst for much of this synthetic work (Funk and Vollhardt, 1980).

In scheme *xv* we have shown the metallacycle reacting with a third alkyne molecule to produce an arene. An important modification of this reaction is the incorporation of an organic cyanide into the molecule to produce the pyridine skeleton. Cobalt catalysts are most suited for the co-cyclotrimerisation (equation (9.20); Bönnemann, 1978).

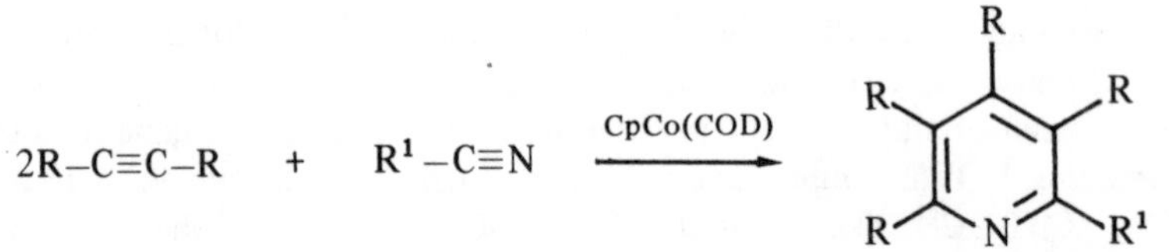

$$2R-C\equiv C-R \quad + \quad R^1-C\equiv N \quad \xrightarrow{\;CpCo(COD)\;} \quad$$

(9.20)

9.3.7 Isomerisation of Alkenes

A large variety of transition metal compounds bring about double bond migration in hydrocarbons. The reaction is seldom used preparatively as a mixture of products is generally formed. Exceptions to this are the preparation of olefin complexes preceded by ligand isomerisation and the isomerisation of allylic alcohols to carbonyl compounds.

Isomerisation often occurs under the conditions of hydroformylation (for example using HCo(CO)$_4$) and this case has been thoroughly studied. The mechanism is thought to involve the intermediacy of an alkyl cobalt carbonyl, **XXIV** (Taylor and Orchin, 1971; scheme *xvi*).

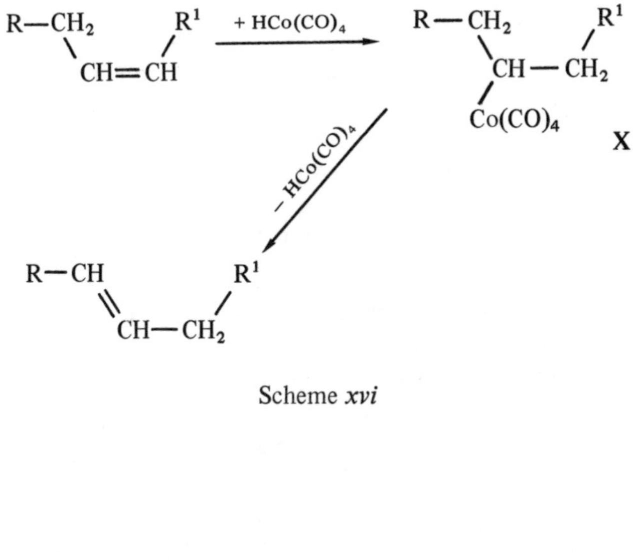

$$
\begin{array}{ccc}
R-CH_2 \quad\ \ R^1 & \xrightarrow{\ +\ HCo(CO)_4\ } & R-CH_2 \quad\ \ R^1 \\
\diagdown\ \diagup & & \diagdown\ \diagup \\
CH=CH & & CH-CH_2 \\
& & \diagup \\
& & Co(CO)_4 \qquad \textbf{XXIV}
\end{array}
$$

$$
\begin{array}{c}
\xleftarrow{-\,HCo(CO)_4} \\
R-CH \qquad\qquad R^1 \\
\diagdown\diagdown \qquad\qquad \diagup \\
CH-CH_2
\end{array}
$$

Scheme *xvi*

Iron carbonyls catalyse isomerisation of alkenes. Allyl alcohol isomerises to propionaldehyde on treatment with iron pentacarbonyl. The mechanism is thought to involve a π-allyl complex, **XXV**, formed by hydrogen transfer from the ligand. The initial product is a vinylic alcohol which spontaneously isomerises to an aldehyde (Hendrix *et al.*, 1968; scheme *xvii*). This reaction has been applied to a range of unsaturated alcohols (Damico and Logan, 1967).

The iron tricarbonyl group has a preference for coordinating to conjugated dienes. However, it is not necessary to use a conjugate diene as the starting material to prepare a complex, as isomerisation precedes complexation (Arnet and Pettit, 1961; equation 9.21).

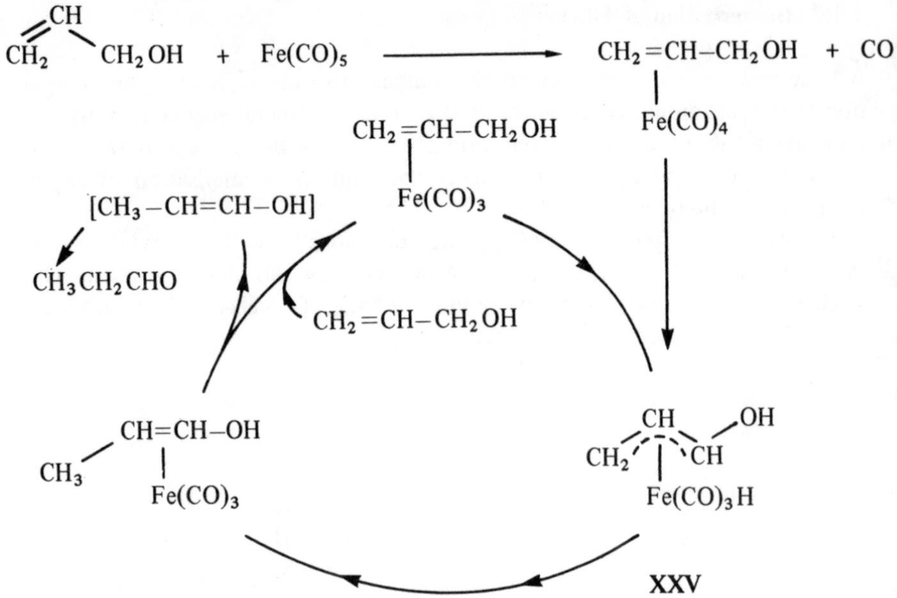

$$CH_2{=}CH{-}CH_2OH \quad + \quad Fe(CO)_5 \longrightarrow CH_2{=}CH{-}CH_2OH \quad + \quad CO$$

$$\begin{array}{c} CH_2{=}CH{-}CH_2OH \\ | \\ Fe(CO)_3 \end{array} \qquad \begin{array}{c} | \\ Fe(CO)_4 \end{array}$$

$$[CH_3{-}CH{=}CH{-}OH]$$

$$CH_3CH_2CHO$$

$$CH_2{=}CH{-}CH_2OH$$

$$\begin{array}{c} CH=CH{-}OH \\ CH_3 \qquad | \\ Fe(CO)_3 \end{array} \qquad \begin{array}{c} CH \qquad OH \\ CH_2 \qquad CH \\ | \\ Fe(CO)_3H \end{array} \quad \mathbf{XXV}$$

Scheme *xvii*

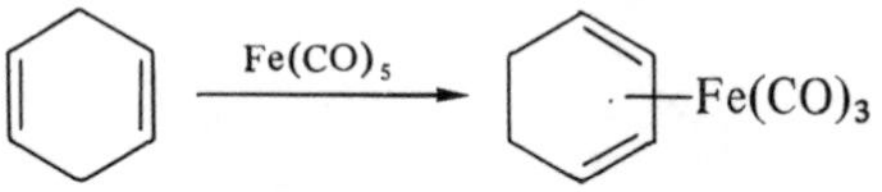

$$\text{(cyclohexadiene)} \xrightarrow{Fe(CO)_5} \text{(diene)}{-}Fe(CO)_3 \qquad (9.21)$$

In contrast to iron, rhodium prefers to coordinate to non-conjugated dienes and the double bonds isomerise out of conjugation to form the complex **XXVI** (equation (9.22); Rhinehardt and Lasky, 1964).

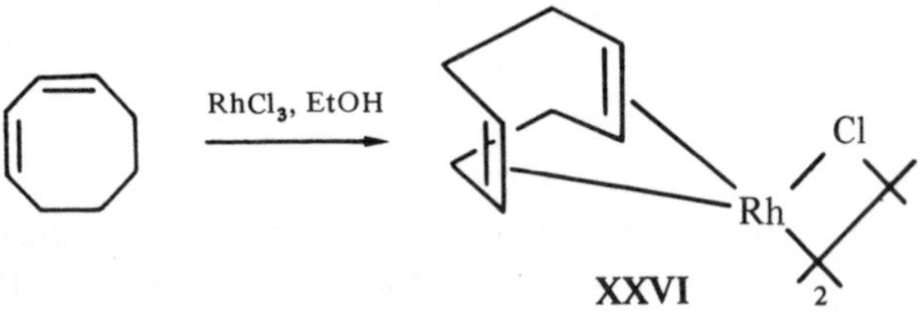

$$\text{(cyclooctadiene)} \xrightarrow{RhCl_3,\ EtOH} \left[\text{(ligand)} {-}Rh \begin{array}{c} Cl \end{array} \right]_2 \qquad (9.22)$$

XXVI

9.4 OXIDATION OF ALKENES

The oxidation of ethene to acetaldehyde using a palladium catalyst is an established industrial process known as the Wacker process (equation 9.23).

$$C_2H_4 + \tfrac{1}{2}O_2 \xrightarrow{Pd\ catalyst} CH_3CHO \qquad (9.23)$$

The development of this process owes much to the ingenuity of the research team at the Consortium für Electrochemische Industrie in Munich. This is because the reaction is not inherently catalytic, but was made so by introducing a second stage to the organometallic reaction. The second stage uses Cu(II) to oxidise palladium metal to Pd(II), and a third stage employs oxygen to reoxidise the Cu(I). The overall process is thus made up of three principal reactions (equations 9.24–9.26).

$$PdCl_4^{2-} + C_2H_4 + H_2O \longrightarrow CH_3CHO + Pd + 2HCl + 2Cl^- \qquad (9.24)$$

$$Pd + 2CuCl_2 + 2Cl^- \longrightarrow PdCl_4^{2-} + 2CuCl \qquad (9.25)$$

$$2CuCl + \tfrac{1}{2}O_2 + 2HCl \longrightarrow 2CuCl_2 + H_2O \qquad (9.26)$$

Summation of these three reactions gives the overall process (equation 9.23). The copper based re-oxidation cycle thus allows the use of palladium which would be prohibitively expensive if used without recycling.

The reaction is generally run in aqueous solution containing chloride ions. At very high chloride concentrations the product is chloroethanol, a point we shall return to later. When run in acetic acid as solvent, vinyl acetate is the product (Henry, 1980).

The organometallic interest in the Wacker process lies in equation (9.24). This simple equation summarises a complicated chain of events. The kinetics of the oxidation step have been investigated and the rate expression is

$$-\frac{d[C_2H_4]}{dt} = \frac{k\,[PdCl_4^{2-}]\,[C_2H_4]}{[Cl^-]^2\,[H_3O^+]}$$

This expression has generally been interpreted to mean that there are three consecutive equilibria (equations 9.27–9.29) which determine the concentration of the intermediate $[C_2H_4PdCl_2OH]^-$, **XXVII**.

$$PdCl_4^{2-} + C_2H_4 \rightleftharpoons C_2H_4PdCl_3^- + Cl^- \qquad (9.27)$$

$$C_2H_4PdCl_3^- + H_2O \rightleftharpoons C_2H_4PdCl_2OH_2 + Cl^- \qquad (9.28)$$

$$C_2H_4PdCl_2OH_2 + H_2O \rightleftharpoons C_2H_4PdCl_2OH^- + H_3O^+ \qquad (9.29)$$

XXVII

The rate determining step is the hydroxypalladation in which **XXVII** rearranges within the coordination sphere to give the hydroxyethyl complex **XXVIIIa** (equation 9.30).

$$C_2H_4PdCl_2OH^- \longrightarrow HO-CH_2-CH_2PdCl_2^- \qquad (9.30)$$

XXVII   **XXVIIIa**

The next step, in which the hydroxyethyl compound **XXVIIIb** decomposes to release acetaldehyde, must account for the fact that when the oxidation is carried out in deuterium oxide as solvent, no deuterium incorporation occurs.

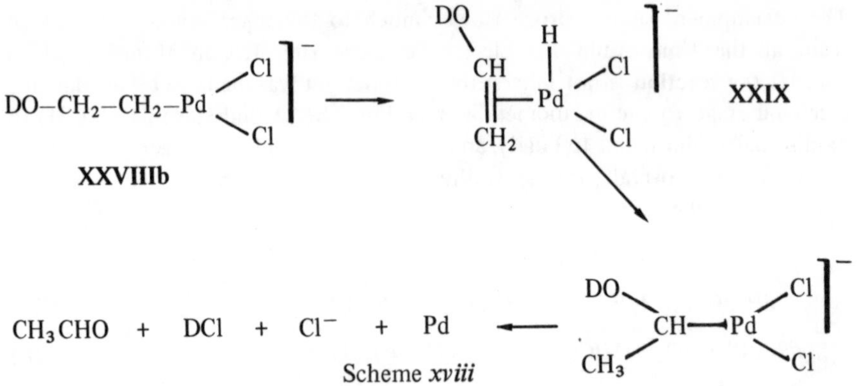

$$\text{DO–CH}_2\text{–CH}_2\text{–Pd} \begin{bmatrix} \diagup Cl \\ \diagdown Cl \end{bmatrix}^- \longrightarrow \begin{bmatrix} \underset{CH_2}{\overset{\overset{\displaystyle DO}{|}}{\underset{\|}{CH}}}\overset{\overset{\displaystyle H}{|}}{-Pd}\begin{matrix} \diagup Cl \\ \diagdown Cl \end{matrix} \end{bmatrix}^- \quad \textbf{XXIX}$$

XXVIIIb

$$\text{CH}_3\text{CHO} + \text{DCl} + \text{Cl}^- + \text{Pd} \longleftarrow \begin{bmatrix} \underset{CH_3}{\overset{DO}{\diagdown}}CH\text{—Pd}\begin{matrix}\diagup Cl \\ \diagdown Cl\end{matrix}\end{bmatrix}^-$$

Scheme *xviii*

This is because a β-elimination reaction gives the palladium hydride, **XXIX**, which then transfers hydrogen to the carbon not bonded to oxygen (scheme *xviii*).

Controversy surrounds the hydroxypalladation step (equation 9.30). The kinetic evidence was interpreted to favour the reaction occurring by rearrangement of $C_2H_4PdCl_2OH$, **XXVII**, within the coordination sphere to give $HOCH_2CH_2PdCl_2^-$, **XXVIIIa**. Some doubt has been cast upon this by experiments with *trans*-1,2-dideuteroethene at high copper and chloride concentrations. Under these conditions the *threo* isomer of dideuterochloroethanol, **XXXI**, is obtained. The formation of this isomer implies that the attack by oxygen occurred from outside the coordination sphere (scheme *xix*; Bäckvall *et al.*, 1979). However, recent results suggest that the chloroethanol is not derived

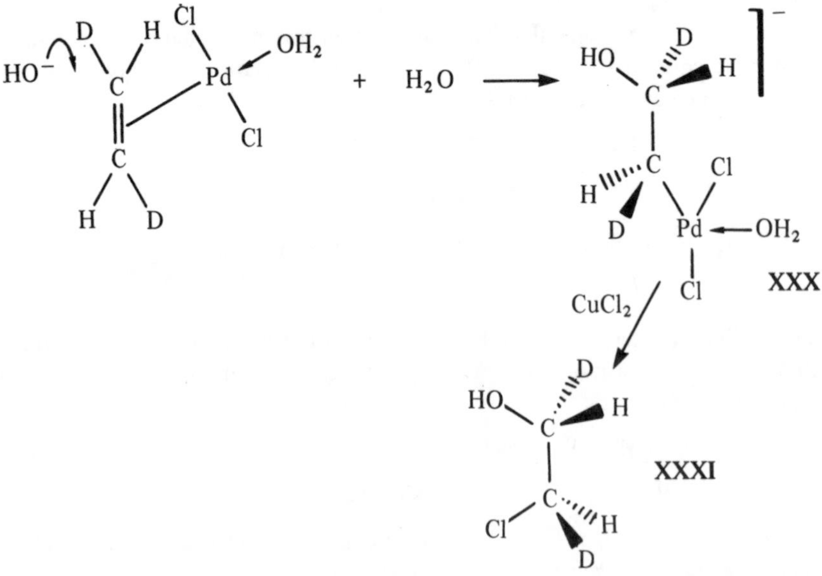

Scheme *xix*

from **XXX** (Gragor and Henry, 1981). Furthermore, if attack from oxygen is from outside the coordination sphere then the rate determining step in the formation of acetaldehyde becomes the loss of chloride from **XXX**, whereas it can be shown that the hydroxypalladation itself is rate determining (Wan *et al.*, 1982).

9.5 HYDROSILATION

Hydrosilation is the addition of $\geq$Si–H to a substrate. The most important case is addition to alkenes, but addition to alkynes, carbonyl compounds and other groups is also known (Harrod and Chalk, 1977).

Platinum is the most frequently used element in hydrosilation catalysts, but several other Group VIII elements are also effective. It is interesting to note that in some platinum catalysts the oxidation state of the platinum alternates between +2 and +4, whereas in others it alternates between 0 and +2. In both cases the fundamental steps are (a) oxidative addition of $\geq$Si–H to the lower oxidation state complex, (b) insertion of the alkene into the Pt–H bond, and (c) reductive elimination leading to formation of the new Si–C bond (scheme *xx*; Green *et al.*, 1977).

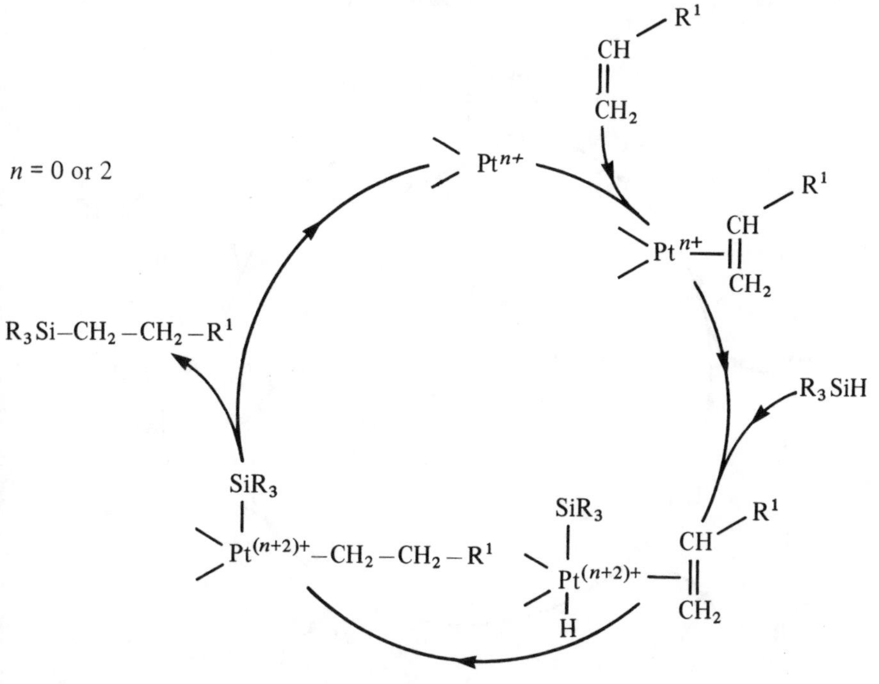

Scheme *xx*

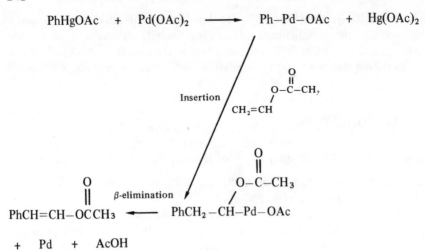

Scheme *xxi*

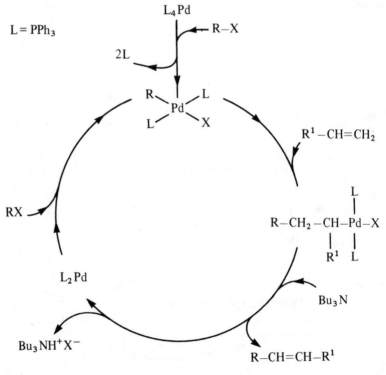

Scheme *xxii*

9.6 SYNTHESIS OF SUBSTITUTED ALKENES USING PALLADIUM COMPLEXES

An area of palladium chemistry has been developed which allows the introduction of substituents into alkenes (Heck, 1979).

The reaction can take a variety of forms depending on which palladium compound is used. The early work involved initial preparation of a Ph–Pd bond using phenylmercuric acetate. The palladium aryl then reacts with alkenes by insertion followed by elimination, thus effecting substitution of the alkene (scheme *xxi*; Heck, 1968). The reaction is not catalytic in this form, but the palladium can be recycled using a Cu(II) oxidant as explained in section 9.4.

The reaction can also be made catalytic by using an alkyl halide as alkylating agent and the Pd(0) complex, Pd(PPh$_3$)$_4$, as catalyst. In this case the palladium can be kept in solution as its phosphine complex, rather than precipitating out as the metal. The alkyl halide then reacts with the complex to complete the catalytic cycle. A tertiary amine is required to remove HX from the reaction (scheme *xxii*). The overall reaction is shown in equation (9.31).

$$R-X + R^1-CH=CH_2 + Bu_3N \xrightarrow{\text{Pd catalyst}} R-CH=CH-R^1 + Bu_3\overset{+}{N}H\, X^- \quad (9.31)$$

9.7 TRANSITION METAL CATALYSIS WITH GRIGNARD REAGENTS

Many transition metal coordination compounds and halides catalyse the coupling of Grignard reagents with organic halides (equation 9.32).

$$RMgX + R^1X \xrightarrow{L_nMX_m} R-R^1 + MgX_2 \quad (9.32)$$

Nickel(II) and Cu(I) compounds have been used most extensively. The mechanisms of these reactions are uncertain. The reactions have been explained by a sequence of oxidative addition and reductive eliminations which may or may not involve paramagnetic intermediates (Kochi, 1978). These and other transition metal-catalysed Grignard reactions are discussed more fully on p. 164.

9.8 MAIN GROUP CATALYSIS

In addition to the lithium-catalysed polymerisation reactions discussed on p. 232, there are several examples of organic reactions which are catalysed by main group organometallic compounds. These reactions often utilise oxygen-containing organic compounds and the mechanisms do not, in general, involve the breaking and re-making of metal–carbon bonds. The role of the organometallic compound is to provide a metal centre with the correct oxidation state and coordination environment for the reaction to occur.

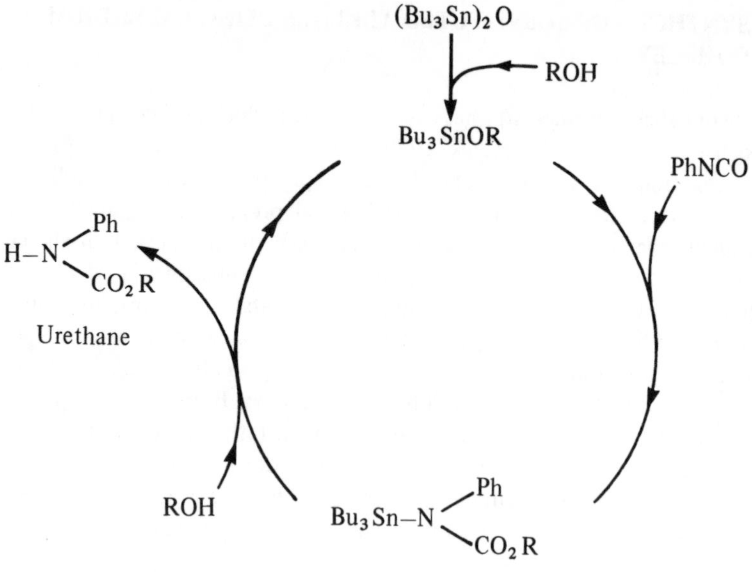

XXXII

Scheme *xxiii*

One example where a mechanism has been proposed is the reaction between isocyanates and alcohols to form urethanes. This reaction is catalysed by trialkyltin-alkoxides which may themselves be generated *in situ* from the alcohol and an organotin oxide (Bloodworth and Davies, 1965; scheme *xxiii*).

The catalysis proceeds via formation of a stannylcarbamate intermediate, **XXXII**, which is more reactive than the isocyanate itself towards the alcohol.

Organotin compounds are used commercially in the production of polyurethanes (Ulrich, 1983). Another use for main group organometallic catalysts is in the esterification of terephthalic acid with ethylene glycol to produce polyesters (equation 9.33). This reaction is catalysed by dialkyltin oxides. Transition metal compounds such as titanium tetraalkoxides are also effective (Fradel and Maréchal, 1982).

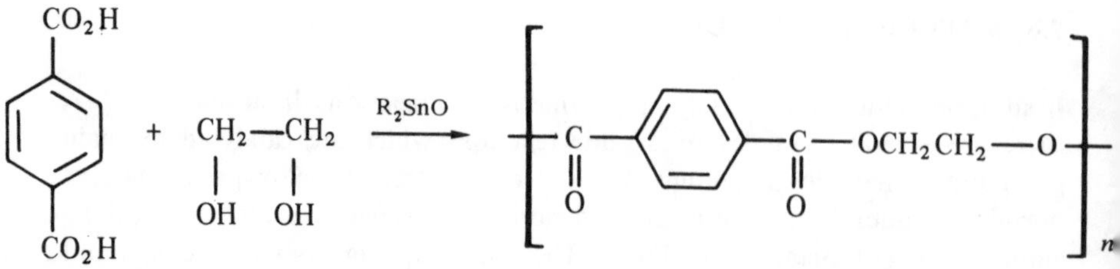

$$(9.33)$$

Epoxides can be polymerised with combinations of organozinc and tin compounds such as $(Et_2 Zn + Ph_3 SnOH)$ or $(Et_2 Zn + Ph_2 SnO)$ (Buys *et al.*, 1983).

REFERENCES

Arnet, J. E., and Pettit, R. (1961). *J. Am. chem. Soc.* **83**, 2954

Ayrey, G., Bird, C. W., Briggs, E. M., and Harmer, A. F. (1970/71). *Organomet. Chem. Synth.* **1**, 187

Bäckvall, J. E., Akermark, B., and Ljunggren, S. O. (1979). *J. Am. chem. Soc.* **101**, 2441

Ballard, D. G. H. (1973). *Adv. Catal.* **23**, 263

Bloodworth, A. J., and Davies, A. G. (1965). *J. chem. Soc.* 5238

Bogdanovič, B. (1979). *Adv. organomet. Chem.* **17**, 105

Bogdanovič, B., Spliethoff, B., and Wilke, G. (1980) *Angew. Chem. int. Edn Engl.* **19**, 622

Bönnemann, H. (1978). *Angew. Chem. int. Edn Engl.* **17**, 505

Buchholz, H., Heimbach, P., Hey, H.-J., Selbeck, H., and Weise, W. (1972). *Coord. Chem. Rev.* **8**, 129

Burch, R. R., Muetterties, E. L., Teller, R. G., and Williams, J. M. (1982). *J. Am. chem. Soc.* **104**, 4257

Burwell, R. L. (1976). *Pure appl. Chem.* **46**, 71

Buys, H. C. W. M., Overmars, H. G. J., and Noltes, J. G. (1983). In *Coordination Polymerisation* (C. C. Price and E. J. Vandenberg, eds). Plenum, New York, p. 75

Candlin, J. P. (1981). In *Catalysis and Chemical Processes* (R. Pearce and W. R. Patterson, eds). Blackie, Edinburgh, p. 219

Cassar, L., Chiusoli, G. P., and Guerrieri, F. (1973). *Synthesis* 509

Colborn, R. E., and Vollhardt, K. P. C. (1981). *J. Am. chem. Soc.* **103**, 6259

Cossee, P. (1964). *J. Catal.* **3**, 80

Crabtree, R. H., Demou, P. C., Eden, D., Mihelcic, J. M., Parnell, C. A., Quirk, J. M., and Morris, G. E. (1982). *J. Am. chem. Soc.* **104**, 6994

Cramer, R. (1965). *J. Am. chem. Soc.* **87**, 4717

Damico, R., and Logan, T. J. (1967). *J. org. Chem.* **32**, 2356

Dedieu, A. (1981). *Inorg. Chem.* **20**, 2803

Forster, D. (1979). *Adv. organomet. Chem.* **17**, 255

Fradel, A., and Maréchal, E. (1982). *Adv. Polymer Sci.* **43**, 51

Fryzuk, M. D., and Bosnich, B. (1977). *J. Am. chem. Soc.* **99**, 6262

Funk, R. L., and Vollhardt, K. P. C. (1980). *Chem. Soc. Rev.* **9**, 41

Gragor, N., and Henry, P. M. (1981). *J. Am. chem. Soc.* **103**, 681

Green, M., Spencer, J. L., Stone, F. G. A., and Tsipis, C. A. (1977). *J. chem. Soc. Dalton Trans.* 1519

Halasa, A. F., Schulz, D. N., Tate, D. P., and Mochel, V. D. (1980). *Adv. organomet. Chem.* **18**, 55

Halpern, J. (1981). *Inorg. chim. Acta* **50**, 11

Harrod, J. F., and Chalk, A. J. (1977). In *Organic Synthesis via Metal Carbonyls*, Vol. 2 (I. Wender and P. Pino, eds). Wiley, New York, p. 673

Hartley, F. R., and Vezey, P. N. (1977). *Adv. organomet. Chem.* **15**, 189

Heck, R. F. (1963). *J. Am. chem. Soc.* **85**, 2013

Heck, R. F. (1968). *J. Am. chem. Soc.* **90**, 5518

Heck, R. F. (1977). *Adv. Catal.* **26**, 323

Heck, R. F. (1979). *Acc. chem. Res.* **12**, 146

Hendrix, W. T., Cowherd, F. G., and von Rosenberg, J. L. (1968). *Chem. Comm*, 97

Henry, P. M. (1980). *Palladium Catalysed Oxidation of Hydrocarbons*. Reidel, Dordrecht

Hoogzand, C., and Hübel, W. (1968). In *Organic Synthesis via Metal Carbonyls*, Vol. 1 (I. Wender and P. Pino, eds). Wiley, New York, p. 343

Ivin, K. J., Rooney, J. J., Stewart, C. D., Green, M. L. H., and Mahtab, R. (1978). *J. chem. Soc. chem. Commun.* 604

James, B. R. (1979). *Adv. organomet. Chem.* **17**, 319

Jardine, F. H. (1981). *Prog. inorg. Chem.* **28**, 63

Kochi, J. K. (1978). *Organometallic Mechanisms and Catalysis*. Academic Press, New York, p. 372

Lehmkuhl, H., and Ziegler, K. (1970). In *Methoden der Organischen Chemie*, Vol. 13/4 (E. Müller, ed.). G. Thieme, Stuttgart

Meriwether, L. S., Colthup, E. C., and Kennerly, G. W. (1961). *J. org. Chem.* **26**, 5163

Mole, T., and Jeffrey, E. A. (1972). *Organoaluminium Compounds*. Elsevier, Amsterdam

Muetterties, E. L., Pretzer, W. R., Thomas, M. G., Beier, B. F., Thorn, D. L., Day, V. W., and Anderson, A. B. (1978). *J. Am. chem. Soc.* **100**, 2090

Muetterties, E. L., and Bleeke, J. (1979). *Acc. chem. Res.* **12**, 324

Pino, P. (1980). *Angew. Chem. int. Edn Engl.* **19**, 857

Pruett, R. L. (1979). *Adv. organomet. Chem.* **17**, 1

Reppe, W., von Kutepow, N., and Magin, A. (1969). *Angew. Chem. int. Edn Engl.* **8**, 727

Rhinehardt, R. E., and Lasky, J. S. (1964). *J. Am. chem. Soc.* **86**, 2516

Schrauzer, G. N., Glockner, P., and Eichler, S. (1964). *Angew. Chem. int. Edn Engl.* **3**, 185

Schrock, R. R., and Osborn, J. A. (1976). *J. Am. chem. Soc.* **98**, 2134

Siegel, H., and Himmele, W. (1980). *Angew. Chem. int. Edn Engl.* **19**, 178

Sinn, H., Kaminsky, W., Vollmer, H.-J., and Woldt, R. (1980). *Angew. Chem. int. Edn Engl.* **19**, 390

Szwarc, M. (1983). *Adv. Polymer Sci.* **49**, 1

Taylor, P., and Orchin, M. (1971). *J. Am. chem. Soc.* **93**, 6504

Tolman, C. A. (1972). *Chem. Soc. Rev.* **1**, 337

Turner, H. W., Schrock, R. R., Fellman, J. D., and Holmes, S. J. (1983). *J. Am. chem. Soc.* **105**, 4942

Ulrich, H. (1983). In *Kirk-Othmer Encyclopedia of Chemical Technology*, Vol. 23. Wiley, New York, p. 576

von Kutepow, N., Himmele, W., and Hohenschutz, H. (1965). *Chem. -Ing.-Tech.* **37**, 383

Wan, W. K., Zaw, K., and Henry, P. M. (1982). *J. molec. Catal.* **16**, 81

Weissermel, K., and Arpe, H.-J. (1978). *Industrial Organic Chemistry*. Verlag Chemie, Weinheim, p. 68

Whitesides, G. M., and Ehmann, W. J. (1969). *J. Am. chem. Soc.* **91**, 3800

Whyman, R. (1975). *J. organomet. Chem.* **94**, 303

Wilke, G. (1963). *Angew. Chem. int. Edn Engl.* **2**, 105

Wittig, G., and Fritze, H. P. (1968). *Annalen* **712**, 79

Yer'eva, L. P. (1974). *Russ. Chem. Rev. (Engl. Transl.)* **43**, 48

Ziegler, K. (1968). *Adv. organomet. Chem.* **6**, 1

GENERAL READING

Bird, C. W. (1967). *Transition Metal Intermediates in Organic Synthesis*. Logos Press, London

Braterman, P. S. (1980). Orbital correlation in the making and breaking of transition metal–carbon bonds. *Top. curr. Chem.* **92**, 150

Falbe, F. (ed.) (1980). *New Syntheses with Carbon Monoxide*. Springer, Berlin

Masters, C. (1981). *Homogeneous Transition-metal Catalysis*. Chapman and Hall, London

Nakamura, A., and Tsutsui, M. (1980). *Principles and Applications of Homogeneous Catalysis*. Wiley, New York

Parshall, G. W. (1980). *Homogeneous Catalysis*. Wiley, New York

Ugo, R. (ed.) (1970-81). *Aspects of Homogeneous Catalysis*, Vols 1-4. Reidel, Dordrecht

Various authors (1968). *Homogeneous Catalysis*, Advances in Chemistry Series no. 70. American Chemical Society, Washington, D.C.

Various authors (1974). *Homogeneous Catalysis*, Advances in Chemistry Series no. 132. American Chemical Society, Washington, D.C.

Various authors (1977-79). *Fundamental Research in Homogeneous Catalysis*, Vols 1-3. Plenum Press, New York

Various authors (1977-82). *Specialist Periodical Reports on Catalysis*, Vols 1-5. Royal Society of Chemistry, London (tend to emphasise heterogeneous catalysis)

Wender, I., and Pino, P. (eds) (1968-77). *Organic Synthesis via Metal Carbonyls*. Vols. 1-2. Wiley, New York

INDEX